Food and Water Security

BALKEMA – Proceedings and Monographs
in Engineering, Water and Earth Sciences

Food and Water Security

Editor

U. Aswathanarayana

*Mahadevan International Centre for
Water Resources Management
Hyderabad, India*

CRC Press
Taylor & Francis Group
Boca Raton London New York

CRC Press is an imprint of the
Taylor & Francis Group, an **informa** business

CRC Press
Taylor & Francis Group
6000 Broken Sound Parkway NW, Suite 300
Boca Raton, FL 33487-2742

First issued in paperback 2019

©2008 by Taylor & Francis Group, LLC
CRC Press is an imprint of Taylor & Francis Group, an Informa business

No claim to original U.S. Government works

ISBN-13: 978-0-415-44018-9 (hbk)
ISBN-13: 978-0-367-38844-7 (pbk)

Typeset by Charon Tec Ltd (A Macmillan company), Chennai, India
Printed and bound in Great Britain by Antony Rowe Ltd
(CPI-Group), Chippenham, Wiltshire

Library of Congress Cataloging-in-Publication Data

Food and water security / edited by U. Aswathanarayana.
    p. cm.
  Includes index.
  ISBN 978-0-415-44018-9 (hardcvover : alk. paper)
  1. Sustainable agriculture–Developing countries. 2. Food supply–Developing countries.
3. Water supply–Developing countries. I. Aswathanarayana, U.
  S494.5.S86F66 2007
  630.9172′4–dc22
                              2007024278

Visit the Taylor & Francis Web site at
http://www.taylorandfrancis.com

and the CRC Press Web site at
http://www.crcpress.com

Cover Illustration:
Indonesian farmer holding two rice plants of same variety and maturity, the one on left grown with SRI practices and the one on right grown with conventional practices. (Picture courtesy of Shuichi Sato, Nippon Koei, Jakarta, Indonesia, quoted by Norman Uphoff, in this volume)

Table of Contents

*Section 3: Governance of Food Security in Different Agroclimatic
and Socioeconomic Settings*

Preface

Food security is a fundamental human right. US AID defines food security as follows: "When all people at all times have both physical and economic access to sufficient food to meet their dietary needs in order to lead a productive and healthy life". While there are indeed undernourished people in the industrialized countries, it can be safely stated that virtually all the undernourished people in the world are in developing countries. Economically-viable, ecologically-sustainable and people-participatory technologies are indeed available to realize hunger-free developing countries. The difficult part is for the developing countries to integrate combinations of technologies with policy interventions, economic instruments, and managerial systems consistent with the agroclimatic, socioeconomic situations and food preferences in a country, or in some cases, parts of a country. The purpose of this volume is to place on the present selection of techno-socio-economic options, from among which the developing countries could choose combinations to suit their particular needs.

What the great Chinese philosopher, Confucius, said 2500 years ago with profound prescience, "Despite the many accomplishments of mankind, we owe our existence to the thin layer of top soil, and the fact that it rains", is as valid as ever. Food security is critically dependent upon optimizing the use of soil and water. Food security and water security are inseparable, as food cannot be grown without water. We must find ways of growing more crop per drop, and getting more meat and milk per drop, and more fish per drop. What constitutes food security changes, as food habits of people change. Food availability is a necessary but not a sufficient condition for food security. It can happen that food security at the aggregate level may not translate into food security at family level – for instance, China is not only self-sufficient in food grains, but is an exporter of food grains, but still, according to UN sources, about 142 million Chinese are food-insecure, though this number has been coming down rapidly.

The chapters in the volume are based on the presentations made in the Panel Discussion on the Biophysical and Socioeconomic dimensions of Food Security in the developing countries, organized by the editor on Jan. 6, 2006 in the ANGR Agricultural University, Hyderabad, as a part of the Event, "Science and the UN Millennium Development Goals", plus some solicited papers. The key challenge of growing more food with less water has both science-based and people-based dimensions. Regrettably, there has not been adequate interaction between the practitioners of the two sets of dimensions. This volume deals with ways and means of management of food and water security in various agroclimatic environments through the integration of research and development, training, people participation, agronomic practices, economic instruments, and administrative policies, etc.

The problems of food security in India, and the attempts that are being made to address them, are relevant to the developing countries generally. India has 100 million farming families, i.e. ten times the number in all the OECD countries, including USA, who incidentally get a support of USD one billion a day.

Because of the advantage of repetitive coverage and capability of synoptic overview, satellite remote sensing has emerged as a powerful and cost effective tool covering all aspects of soil, water and crop management, wasteland reclamation, etc. Advances in data assimilation methodologies have made it possible to customize synoptic remote sensing data to farmer-specific use. Through the instrumentality of the Village Knowledge Centres (*Gyan Chaupals*), Information and Communication Technology (ICT) is playing a crucial multipurpose development role in regard to food and water security, health, sanitation, habitation, etc. in the villages. Young, *practicing* farmers who are trained in the measurement of soil and water quality parameters, agronomic practices (e.g. SRI), use of fertilizers and pesticides, etc. have proved effective in promoting science-based, productive agriculture.

In this book three basic key issues are treated:

Soil: Preparation of soil health cards, to enable the farmer to manage the fertility of the soil in his farm, with particular reference to micro-nutrients, such as zinc, boron and sulphur, which have a marked effect on the productivity of dryland agriculture.

Water: Rainwater harvesting, aquifer recharge, conjunctive use of surface and groundwater, monitoring and management of soil moisture, sprinkler and drip irrigation. Preparation of hydroclimatic calendar for use in crop planning. Reuse of waste water.

Agronomic practices: System of Rice Intensification (SRI), and Aerobic rice which yield more rice with less water, development of salt-tolerant and drought-resistant seed varieties, through the use recombinant DNA technology and methods of transferring salt-tolerant genes (e.g. from mangroves) into important food crops, such as rice.

Market-linked farming systems to improve the income of the farmers. ICT empowers poor, rural women to get self-employed by producing various food-based goods and services. Food fortification (methods of processing foods to improve their nutritional value) whereby (say) the same quantity of cereals could provide nutritious diet to more people.

I thank Dr. Malin Falkenmark, the doyen of water scientists in the world, for her perceptive Foreword. *Gangamai* (Mother Ganga) is the goddess of water. In Sanskrit, Ganga refers not only to River Ganga, but also to all surface and groundwater in India. I have a fanciful notion that the great fervour with which Malin Falkenmark pursued water issues all her life, could be attributed to her being a devotee of *gangamai* in her previous birth!

The care of my wife, Vijayalakshmi, and my physician friends, Drs. Bhaskar Reddy and Surya Prakash, kept me in good health. I am grateful to them.

This broad-spectrum volume would be useful to university students and public officials concerned with water, soil and crop sciences, remote sensing, agrometerology, food processing, food distribution, technology transfer, sociology, economics, etc.

U. Aswathanarayana
Editor
September, 2007

Foreword

The global food security issue is gathering increasing interest, especially in view of the recent IPCC report's projections on future water scarcity outlooks. The challenge is particularly large in irrigation dependent regions of the South, where river basins are in a state of closure, and hunger and poverty still common.

The dilemma is that food demand tends to grow more rapidly than population, driven by socio-economic changes, new food preferences, and altered food composition for nutritional reasons. One particular dilemma is of course that animal protein, which demands one order of magnitude more water than grain to produce, is fundamental for both growing children and breast-feeding mothers.

While food water requirements rapidly increase, the availability of irrigation water is moving in the opposite direction. One reason is that irrigation has added a consumptive use element into the water availability, altering it with time. This has contributed to the fact that, today, one basin after the other is suffering from streamflow depletion, entering a state of river basin closure.

In the mean time, with surface water heavily exploited, farmers have been turning more and more to groundwater, securing more immediate access to irrigation water when and where needed. This approach is sustainable as long as only the amount of groundwater recharged is being withdrawn and consumptively used. But as soon as the aquifer is being overused, water table continuously drops. At present, the scale of water overuse is already dramatic – it has even been suggested that the groundwater withdrawals in S Asia would have to be reduced by as much as 70 percent to move back towards a sustainable situation.

Indian authorities now indicate the need to double food production in ten years only – a major step in terms of agricultural approaches. Adding up all these dramatic challenges – population growth, changing food preferences towards more water-consuming food items, closing river basins, and severe groundwater overdraft – we arrive at a situation where the expression by Will Steffen, earlier Executive Director of the IGBP, is in fact highly pertinent, when he rightly refers to humanity's environmental challenge as the era of Great Acceleration.

What will be needed is an almost completely new mode of management of the whole food chain from production to consumption. In this process, many of the components discussed in this publication will have to be included. First of all, there are in the food chain huge losses of both water and food, that can be minimised by resolute agronomic efforts, reduction of post-harvest losses, and an effective enough food security-oriented governance.

Secondly, even if water is at the core of food production growth, it is fortunately NOT necessarily more irrigation water that is the critical issue. It is rather to find various ways to secure enough soil moisture/green water in the root zone, that is the crucial task. First of all the huge water losses in current agriculture will have to be minimised. Only when deficient, the green water resource may be complemented by additional water from locally harvested rain or runoff, or from blue water withdrawn from rivers or aquifers, to increase the water available to the roots. Wise upgrading of rainfed agriculture will indeed have to be a core component of the new production mode.

This publication is therefore extremely timely with its discussion of a whole set of steps possible to take. On the one hand to reduce the large yield gap, still characterising Indian agriculture; better use of remote sensing, biotechnology and GMO's; new approaches to improved crop varieties including aerobic rice; to soil microorganisms and geochemistry; to rainfed agriculture and green water conservation; and to expanded irrigation based on wastewater reuse, a source that in fact grows with increasing urbanisation. And on the other hand reduction of post harvest losses in the various segments of the food chain, including household waste.

A strategy that considers what benefits can be reached though improvement of both food production and consumption could result in a win-win-win solution where both water, environmental and public health can all benefit.

Malin Falkenmark
Professor
Stockholm International Water Institute (SIWI)
& Stockholm Resilience Centre
Stockholm, April 12th, 2007

List of Figures

List of Tables

List of Plates

Biophysical Dimensions of Food Security

Chapter I

Remote sensing methodologies for ecosystem management

R. Nemani[1], P. Votava[2], A. Michaelis[2], M. White[3], F. Melton[2],
C. Milesi[2], L. Pierce[2], K. Golden[4], H. Hashimoto[2], K. Ichii[5],
L. Johnson[2], M. Jolly[6], R. Myneni[7], C. Tague[8], J. Coughlan[1],
& S. Running[9]

[1] NASA Ames Research Center, Moffett Field, CA; [2] California State University, Monterey Bay, Seaside, CA;
[3] Utah State University, Logan, UT; [4] Google Inc., Mountain View, CA; [5] San Jose State University, San Jose, CA;
[6] U.S. Forest Service, Missoula, MT; [7] Boston University, Boston, MA; [8] San Diego State University, San Diego, CA;
[9] University of Montana, Missoula, MT

1.1 Introduction

The latest generation of NASA Earth Observing System (EOS) satellites has brought a new dimension to monitoring the living part of the Earth system – the biosphere. EOS data can now measure weekly global productivity of plants and ocean chlorophyll and related biophysical factors, such as changes to land cover and to the rate of snowmelt. However, a greater economic benefit would be realized by forecasting biospheric conditions (Clark *et al.*, 2001). Such predictive ability would provide an advanced decision-making tool to be used in the mitigation of natural hazards or in the exploitation of economically advantageous trends. Imagine if it were possible to accurately predict shortfalls or bumper crops, epidemics of vector-borne diseases such as malaria and West Nile virus, or wildfire danger as much as 3 to 6 months in advance. Such a predictive tool would allow improved preparation and logistical efficiencies. Forecasting provides decision-makers with insight into the future status of ecosystems and allows for the evaluation of the *status quo* as well as alternatives or preparatory actions that could be taken in anticipation of future conditions. Whether preparing for the summer fire season or for spring floods, knowledge of the magnitude and direction of future conditions can save time, money, and valuable resources. Space and ground-based observations have significantly improved the ability to monitor natural resources and to identify potential changes, but these observations can describe current conditions only. This information is useful, but many resource managers often need to make decisions months in advance for the coming season. Recent advances in climate forecasting have elicited strong interest in a variety of economic sectors: agriculture (Cane *et al.*, 1994), health (Thomson *et al.*, 2005) and water resources (Wood *et al.*, 2001). The climate forecasting capabilities of coupled ocean-atmosphere global circulation models (GCMs) have steadily improved over the past decade (Zebiak, 2003).

Given observed anomalies in sea-surface temperatures (SSTs) from satellite data, GCMs are now able to forecast general climatic conditions, including temperature and precipitation trends, 6 to 12 months into the future with reasonable accuracy (Goddard *et al.*, 2001; Robertson *et al.*, 2004).

While such climatic forecasts alone are useful, the advances in ecosystem modeling allow specific exploration of the direct impacts of these future climate trends on the ecosystem. One day predictions made in March might accurately forecast whether Montana's July winter wheat harvest will be greater or less than normal, and whether the growing season will be early or late.

One of the key problems in adapting climate forecasts to natural ecosystems is the "memory" that these systems carry from one season to the next. For example, soil moisture levels, plant seed banks, and fire fuel build-up are all affected by cumulative ecosystem processes that occur over many seasons or years. Simulation models are often the best tools to carry forward information about this spatio-temporal memory. The ability of models to describe and to predict ecosystem behavior has advanced dramatically over the last two decades, driven by major improvements in process-level understanding, climate mapping, computing technology, and the availability of a wide range of satellite- and ground-based sensors (Waring and Running, 1998). In this chapter, we summarize the efforts of the Ecological Forecasting Group at NASA Ames Research Center over the past six years to integrate advances in these areas and develop an operational ecological forecasting system.

1.2   Background of ecological forecasting

Ecological Forecasting (EF) predicts the effects of changes in the physical, chemical, and biological environments on ecosystem state and activity (Clark et al., 2001). EF is pursued with a variety of tools and techniques in different communities. For example, for community ecologists EF commonly includes methods for describing or predicting the ecological niche for various species. Much of invasive species forecasting falls in this area, where a set of conditions associated with the presence/absence of a species is derived and then these empirical relations are used to predict the occurrence or potential for occurrence of that species within a landscape. Similarly, bio-geographers use EF to predict species/community compositions in response to changes in long-term climate or geo-chemical conditions. Climate change and carbon cycling research falls in this category of predicting the state and/or functioning of ecosystems over long-lead times of decades to centuries. In contrast, the eco-hydrological community uses EF as a way of extending weather/climate predictions, with lead times ranging from days to months, for use in operational decision-making. Examples include forecasts of frost damage, flood/streamflow, crop yield, and pest/disease outbreaks. Though such forecasts are age-old among practitioners of various trades, there has been much subjectivity in the decision-making process that is hard to quantify and pass on to later generations. Providing an operational forecasting capability brings a new level of complexity to creating, verifying, and distributing information that is worth acting upon.

1.3   Components of ecological forecasting for eco-hydrological applications

Increasing interest in ecological forecasting is evident from several recent applications ranging from streamflows (Wood et al., 2001), crop yields (Can et al., 1994) and human health (Thompson et al., 2006) These attempts tend to focus on specific watersheds or a geographic location with a very specific application; therefore they do not deal with EF as a broad theme associated with certain tools and technologies. Our past heritage in eco-hydrology and NASA's strengths in global observations and technology led us to focus on the development of a general data and modeling system that allows producing operational nowcasts and forecasts relevant for many in the ecohydrological community. Here we briefly review the important components that make our approach to EF possible, extensible, and economically viable.

Ecosystem Models: As in numerical weather prediction, models form the basis for EF. These models range in complexity, computational requirements, and in the representation of the spatio-temporal details of a given process or system. For example, biogeochemical cycling models are often complex and versatile in the sense that the basic ingredients that they simulate (carbon, water, nutrient cycling) form the core information for a variety of biospheric activities of economic value. For example, changes in carbon cycling expressed as net primary production (NPP) can be a key indicator of crop yields, forage production or production of board feet of wood. These models use the Soil-Plant-Atmosphere continuum concept to estimate various water (evaporation, transpiration, stream flows, and soil water), carbon (net photosynthesis, plant growth) and nutrient flux (uptake and mineralization) processes. They are adapted for all major biomes exploiting their unique eco-physiological principles such as drought resistance, cold tolerance, etc. (e.g. BIOME-BGC, Waring and Running, 1998; CASA, Potter *et al.*, 2003). The models are initialized with ground-based soil physical properties and satellitebased vegetation information (type and density of plants).

Following the initialization process, daily weather conditions (max/min temperatures, solar radiation, humidity, and rainfall) are used to drive various ecosystem processes (e.g., soil moisture, transpiration, evaporation, photosynthesis, and snowmelt) that can be translated into drought, crop yields, NPP, and water yield estimates. We currently use a diagnostic (with satellite data input) version of BGC to produce nowcasts and a prognostic (without satellite data inputs) version of BIOME-BGC to produce forecasts of carbon and water related fluxes.

Extensive discussion on types of ecosystem models and their relevant applications can be found in Waring and Running (1998) and Canham *et al.* (1997).

Microclimate mapping from surface weather observations: Access to reliable weather data is a pre-requisite for ecosystem modeling. The availability of weather observations has been a key obstacle in the development of real-time EF systems. Historically, weather data was made available on tapes or CDs months after it was collected and corrected for errors. This time lag precluded real-time simulations, a precursor to developing forecasting capability. Through the World Wide Web, however, there are now thousands of on-line weather stations providing real-time weather data. These real-time data include ground-based observations of max/min/dew temperature and wind speed, satellite-based solar radiation, and spatially continuous rainfall fields produced by weather agencies. Another important advancement for EF is the ability to grid point observations onto the landscape at various spatial resolutions, as observations are rarely sufficient to represent the spatial variability. Models such as PRISM, DAYMET, and SOGS (Daly *et al.*, 1994; Thornton *et al.*, 1997; Jolly *et al.*, 2004) ingest point surface observations, and use topography and other ancillary information to compute spatially continuous meteorological fields (temperature, humidity, solar radiation, and rainfall) that can be directly used in ecosystem modeling.

Weather/Climate Forecasts: There is considerable optimism among the climate community about our ability to forecast climate into the future (Trenberth, 1996). This optimism stems from several recent advancements in climate modeling, such as improvements in GCMs that have allowed realistic reproduction of observed global climate (Roads *et al.*, 1999), adaptation of new forecasting strategies, demonstration of the links between El Niño/Southern Oscillation (ENSO) and short-term climate, and the ability to forecast ENSO 12–18 months in advance. Barnett *et al.* (1994) showed that with the above improvements, GCMs could be used successfully to predict air temperature, precipitation, and solar radiation at extended lead times over many parts of the world. Research as well as operational agencies that currently produce and disseminate climate forecasts includes the NOAA's National Center for Environmental Prediction, Columbia University's International Research Institute, the Scripps Institute of Oceanography's Experimental Climate Prediction Center, and others.

Satellite remote sensing: A number of studies over the past two decades have shown the utility of satellite data for monitoring vegetation (type, density, and production), extent of flood damage, wildland fire detection, and monitoring snow and drought conditions. However, many of the products generated from satellite data have been experimental, and did not have a wide distribution among natural resource managers. Over the past five years, through NASA's EOS program, there have been substantial improvements in the way satellite data is acquired, processed, converted to products, and delivered (Table1.1). For example, weekly maps of leaf area index (LAI, area of leaves per unit ground area) and vegetation indices, key inputs for many ecosystem models, are being generated and distributed from the NASA/MODIS sensor. A number of other key land products such as NPP, fire occurrence, snow cover, and surface temperature are available globally at 1-km resolution every 8 days (Justice *et al.*, 1998; Myneni *et al.*, 2002). Without this near realtime observing capacity, systems such as TOPS would never have materialized.

Integrated modeling: Numerous studies over the past two decades addressed the logical steps for modeling land surface processes over various spatial scales, by integrating ecosystem models with satellite, climate data, and other ancillary information (Waring and Running, 1998). One such attempt that many of us have been part of was the development of the Regional Hydro-Ecological Simulation System [RHESSys, Band *et al.*, 1993; Nemani *et al.*, 1993; Tague and Band, 2004). RHESSys has been used in various studies for estimating soil moisture, stream flows, snow pack, and primary production (Waring and Running, 1998). Much of the work using RHESSys has been retrospective, using past climate and satellite data, mainly to evaluate various issues related to the parameterization of key variables, scaling and determining the suitability of RHESSys outputs for use by resource managers (Waring and Running, 1998). While this type of retrospective analysis is useful for long-term management decisions, only a realtime analysis can provide data necessary for dynamic decision making such as the assessment of fire risk. Our work has focused on the development of the Terrestrial Observation and Prediction System (TOPS) to provide this capability for real-time analysis, which is essential for forecasting ecological conditions desired by decision makers.

1.4 The terrestrial observation and prediction system (TOPS)

TOPS is a data and modeling software system designed to seamlessly integrate data from satellite, aircraft, and ground sensors with weather/climate models and application models to expeditiously produce operational nowcasts and forecasts of ecological conditions (Plate 1.1). TOPS provides reliable data on current and predicted ecosystem conditions through automation of the data retrieval, pre-processing, integration, and modeling steps, allowing TOPS data products to be used in an operational setting for a range of applications.

Implementation of TOPS over a region consists of first developing the parameterization scheme for the area of interest. Parameterization inputs include data on soils, topography, and satellite derived vegetation variables (land cover and LAI). Observational weather data, gridded from point data or downscaled from previously gridded data to the appropriate resolution, are then used to run a land surface model, such as BIOME-BGC [Waring and Running 1998; White *et al.*, 2000]. Finally, weather and climate forecasts are brought into the system as gridded fields and downscaled to the appropriate resolution to drive the land surface model and generate predictions of future ecosystem states.

Given the diversity of data sources, formats, and spatio-temporal resolutions, system automation is critical for the reliable delivery of data products for use in operational decision-making. TOPS has been engineered to automatically ingest various data fields required for model simulations (Figure 1.1). Ingested data go through a number of preprocessing filters in which each parameter is mapped to a list of attributes (e.g., source, resolution, and quality). This results in each data field being self-describing to TOPS component models such that any number of land surface models can be run without extensive manual interfacing. Similarly, the model outputs also pass through a specification interface, facilitating post-processing so that model outputs can be presented as actionable information, as opposed to just another stream of data. TOPS derives its flexibility and automation capability from two key software components: JDAF (*Java Distributed Application Framework*) and the ImageBot planner.

1.4.1 *JDAF and IMAGEbot*

The TOPS software is implemented using a flexible framework that enables fast and easy integration of new models and data streams into an automated system. The core components of this framework are *JDAF* (Java Distributed Application Framework) and *IMAGEbot*. JDAF consists of a large set of data processing and image analysis algorithms that are deployed to pre-process and post-process inputs and outputs of the TOPS ecosystem models. When we want to process new data with our existing models, we reuse the JDAF algorithms to create intermediate datasets that adhere to the model's input specifications so that we can execute our models without having to alter the science implementation. Because the pre-processing itself can be a very complex process, involving for example data acquisition, mosaicking, reprojection, subsetting, scaling etc. We have developed a planner-based agent (IMAGEbot), which automatically generates the sequence of processing steps needed to perform the appropriate data transformations.

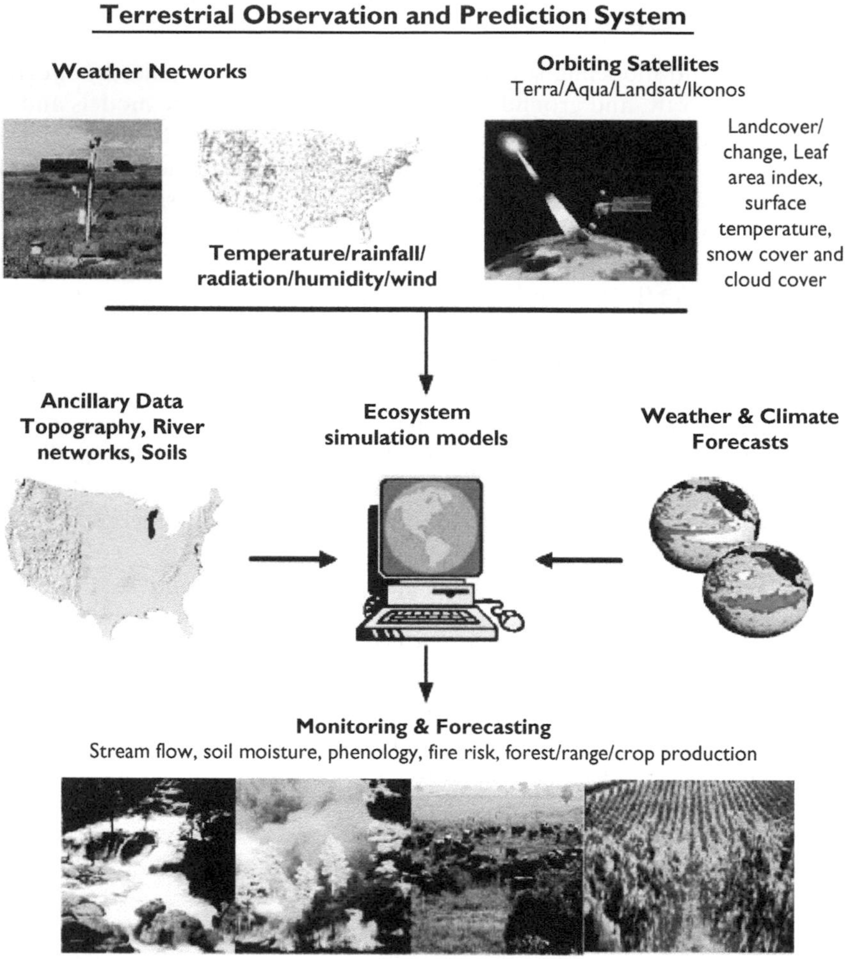

Terrestrial Observation and Prediction System

Weather Networks

Orbiting Satellites
Terra/Aqua/Landsat/Ikonos

Temperature/rainfall/
radiation/humidity/wind

Landcover/
change, Leaf
area index,
surface
temperature,
snow cover and
cloud cover

Ancillary Data
Topography, River
networks, Soils

**Ecosystem
simulation models**

**Weather & Climate
Forecasts**

Monitoring & Forecasting
Stream flow, soil moisture, phenology, fire risk, forest/range/crop production

Plate 1.1 The Terrestrial Observation and Prediction System (TOPS) integrates a wide variety of data sources at various space and time resolutions to produce spatially and temporally consistent input data fields, upon which ecosystem models operate to produce ecological nowcasts and forecasts needed by natural resource managers. (See *Colour plate 1.1*)

In other words, JDAF provides all the processing components of the system and IMAGEbot determines how they fit together to achieve the desired goal, creates a plan, and executes it. This gives a great flexibility to the TOPS software and speeds-up significantly the integration of both datasets and models into new applications. Additionally, JDAF provides interface to the database system and to web services capabilities for seamless access to both data and services provided by TOPS. As currently deployed within TOPS, JDAF and ImageBot perform two dynamic functions critical for the real-time monitoring, modeling and forecasting of ecosystem conditions: gridding of weather observations to create continuous fields of climatic parameters, and acquisition and processing of satellite data for initializing or verifying the models.

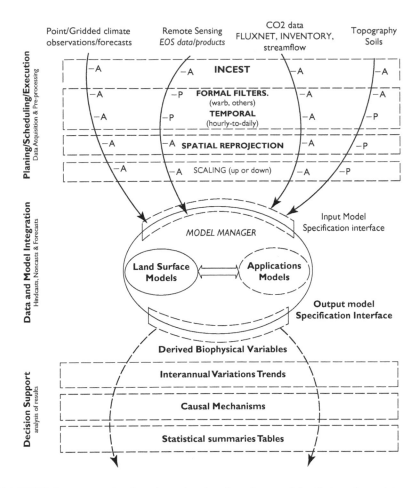

Figure 1.1 TOPS data processing flowchart showing three key modules that perform data acquisition and pre-processing, data and model integration and decision support through analysis of model inputs and outputs. Application front-ends, predictive models, and data mining algorithms are modular and can be easily swapped out as needed. The modular architecture also allows for the concurrent use of multiple ecosystem models to generate forecasts for different parameters.

1.4.2 *Climate gridding*

To produce gridded climate fields, the user specifies a geographic area of interest and the spatial resolution for the gridded fields. The ImageBot planner uses these specifications to create a data processing plan comprised of a series of requirements and corresponding actions. For example, ImageBot will identify the acquisition of topographic data as a requirement, evaluate the possible sources for this data from the data library, identify the required resolution, and create the set of actions required to obtain the data at the appropriate resolution. These actions are then passed to JDAF, which fetches the data from the source, and reformats and reprojects the data to meet the user-specified requirements. Similarly, for meteorological data, ImageBot produces a list of weather networks available for the region, a list of variables available from

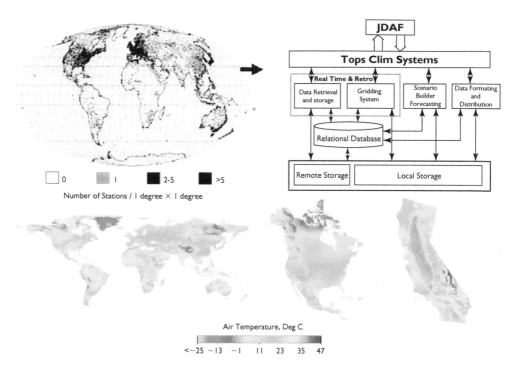

Plate 1.2 Flow diagram of the SOGS. Three main components that comprise the system are: data retrieval and storage, interpolation and output handling. Data retrieval is configured to automatically retrieve the most recent data available and insert those data into the SQL database. Interpolation methods are modular and allow maximum flexibility in implementing new routines as they become available. Outputs are generated on the prediction grid that is determined by the latitude, longitude, elevation and mask layers. Another key feature of the SOGS implementation is scenario generation where long-term normal station data can be perturbed according to climate model forecasts. Weather data from over 6000 stations distributed globally is ingested into TOPS database where it is gridded to a variety of resolutions, globally at 0.5 degree, continental U.S at 8 km and at 1 km over California. (See *Colour plate 1.2*)

each network, and the frequency of observations available from the network. From this information and the user-defined set of constraints, ImageBot again formulates a series of actions specifying which networks and what variables need to be retrieved and input to the database. After receiving these instructions, JDAF fetches the necessary data, checks for consistency against historical averages, fills-in missing values from additional sources, flags missing values, and finally interfaces these observations with the Surface Observation and Gridding System (SOGS, Jolly *et al.*, 2004, Plate 1.2), a component layer within TOPS. SOGS is an operational climate-gridding system, and an improvement upon DAYMET (Thornton *et al.*, 1997), that uses maximum, minimum, and dewpoint temperatures, in addition to rainfall, to create spatially continuous surfaces for air temperatures (e.g. Plate 1.2), vapor pressure deficits, and incident radiation. The cross-validation statistics returned from SOGS allow ImageBot to decide if the user-specified requirements for accuracy have been achieved, or if alternative gridding methods need to be found.

Table 1.1 Data sources, derived products and their usage within TOPS.

Data Source	Products	Usage
Satellite data		
MODIS-TERRA/AQUA (250/500/1000 m)	Surface reflectances	Vegetation monitoring
	Land surface Temperature	Drought/Fire/Snow
	Leaf are index	Vegetation monitoring
	Snow cover	Hydrology
	Vegetation indices	Vegetation monitoring
	Fire	Burnt area/recovery
AVHRR (1000 m)	Vegetation index	Vegetation monitoring
	Land surface temperature	Drought/Fire/snow
AMSR-E (25 km)	Brightness temperature	Hydrology monitoring
	Surface soil moisture	
Thematic Mapper (30 m)	Reflectances, land cover maps	Vegetation monitoring
SSM/I (25 km)	Brightness temperatures	Hydrology monitoring
ASTER (10–20 m)	Reflectances, land surface temperature	Vegetation monitoring
SRTM (30–50 m)	Topography	Drainage, climate mapping
Ground observations		
NWS (hourly)	Weather	Model inputs
SNOTEL (daily)	Weather, snow	Model inputs, validation
RAWS (hourly)	Weather	Model inputs
CIMMIS (hourly)	Weather, ETo	Model inputs
FLUXNET (fortnightly)	Weather, ET, GPP	Inputs, validation
SCAN (daily)	Soil moisture	Validation
USGS-Gauges (daily)	Streamflow	Validation
Model results		
NCEP (daily)	Global weather/climate forecasts	Model inputs
DAO (daily)	Global weather	Model inputs
ECPC (weekly)	Climate forecasts	Model inputs

1.4.3 *Acquiring and processing of satellite data*

TOPS has access to a number of satellite data sets (Table 1.1), produced and processed by either NASA or NOAA. This access involves machine-to-machine, webbased ordering, and FTP pushes for routine data sets such as those from the NOAA Geostationary Operational Environmental Satellites (GOES). Similar to the climate gridding process described in the previous section, ImageBot defines a set of actions pertaining to satellite data and products based on user-defined constraints and requirements. The requirements in this case may include, for example, obtaining LAI and snow cover data with the following constraints: a minimum resolution of 1 km, a weekly time interval, and a specification to obtain the highest possible quality data available. From the data library, ImageBot creates a list of sites that provide LAI. ImageBot sends a command to JDAF to fetch all of the metadata files relating to the LAI product to be evaluated and screened for quality. A list of 'tiles' (the 1200×1200 km area used in MODIS processing) covering the geographic area of interest and meeting the quality criteria is prepared. A request is sent to the archival site (for example, the USGS Eros Data Center). When the order is ready for download, JDAF collects the order and updates the internal database. JDAF then initiates a series of actions on the tile data available locally, including the creation of mosaics, filling-in of missing values, regridding, and reprojection. An example of LAI output from this procedure is shown in Plate 1.3b.

In many cases, data available from the Distributed Active Archive Centers (DAACs) may be 2–8 days old. While this may not pose a significant problem for geophysical fields such as LAI that vary slowly, snow cover can change dramatically in a week. To deal with these situations, TOPS has the ability to ingest and use MODIS data from Direct Broadcast readouts available throughout the United States.

1.4.4 Interfacing new models in TOPS

Interfacing the pre-processed climate and satellite data with models is the next step in producing TOPS nowcasts and forecasts. In order to facilitate the integration of new models, TOPS provides a system for describing new models in terms of their inputs and their outputs. These descriptions include specifications for the format, resolution, variables, and temporal and spatial extent of model parameters. These descriptions are then embedded in the domain descriptions of the model using the DPADL language (Golden, 2003). While this method still lacks robustness and is not fully automated, it enables TOPS to integrate new models into the system faster than the manual integration that would otherwise be required. To improve the automation of this process, we are currently designing an applications programming interface for model integration.

1.4.5 TOPS Nowcasts

On a daily basis, TOPS uses the technologies described above to produce a set of 30 variables including gridded climate, satellite-measured and modeled fields. For the current TOPS implementation for the State of California at 1 km and the continental U.S at 8 km, the data products that are most widely requested are gridded climatic data consisting of daily max/min temperatures, vapor pressure deficits, rainfall, and incident shortwave radiation (Plate 1.3a). Satellite products include daily GOES-based infra-red and visible reflectances and 8-day MODIS-based leaf area index (Plate 1.3b) and other sensor products (Table 1.1). One key difference between TOPS satellite products and products from various DAACs is that TOPS pre-processes the data for immediate use. In addition to the meteorological data, TOPS provides ecosystem nowcasts for California including maps of soil moisture, snow cover and depth, gross and net primary production (e.g., Plate 1.3c), growing season dynamics (leaf on and leaf off), evapotranspiration, and streamflow.

1.4.6 TOPS forecasts

TOPS currently produces two types of ecological forecasts: 1) model-based, where the model is run into the future, and 2) those based on historical associations. Model-based forecasts are either short-term (3–7 days), as in the case of vineyard irrigation, or experimental long-term (3 months) based on outputs from GCMs. In both cases, land surface states are continuously updated using satellite and ground-based observations. Forecasts based on historical associations include those derived from weather-phenology

Gridded Air Temperature, April 27, 2003

(a)

Air Temperature, Deg C

<-25 -13 -1   11   35   47

MODIS Leaf Area Index April 23-30, 2003

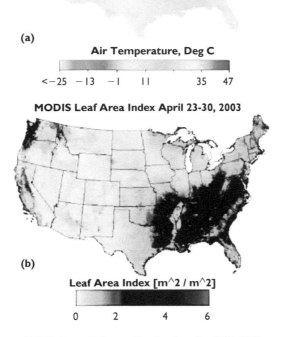

(b)

Leaf Area Index [m^2 / m^2]

0   2   4   6

TOPS Gross Primary Production, April 27, 2003

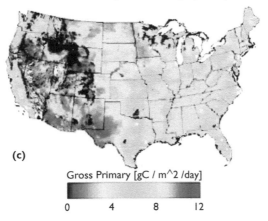

(c)

Gross Primary [gC / m^2 /day]

0   4   8   12

Plate 1.3 Examples of TOPS nowcasts. (a) Patterns of maximum air temperature over the conterminous U.S., produced using over 1400 stations on April, 27, 2004. (b) MODISderived leaf area index after pre-processing through TOPS, and (c) Model estimated gross primary production, the amount of photosynthate accumulated on April 27, 2004 over the conterminous U.S. (See *Colour plate 1.3*)

and weather-fire risk. For both applications, we developed empirical models that provided reasonable predictions of the parameter of interest, i.e., the start and end of
the growing season, and relative fire risk. In each, we used a combination of climate
data and spatially continuous historical satellite data to develop the predictive models
(https://ecocast.arc.nasa.gov).

1.4.7 Verification & validation

TOPS outputs are continuously compared against observed data to assess spatiotem-
poral biases and general model performance. In the case of snow pack dynamics, for
example, we perform a three-way comparison among model-,observation-, and satellit-
ederived fields of snow cover expansion and contraction ($r = 0.91$, Figure 1.2a). Water
and carbon related variables such as evapotranspiration and gross and net primary production (Figure 1.2b and Figure 1.2c) fields are tested against FLUXNET-derived data
at selected locations representing a variety of landcover/climate combinations. Similarly, the SCAN network of soil moisture, USGS streamflow, and SNOTEL provide
valuable data for verifying the hydrology predictions from TOPS.

1.5 TOPS applications

1.5.1 TOPS helping the california wine industry

The impetus for developing TOPS came from NASA's research in Napa Valley,
California, which explored the relationship between climate and wine quality and
the application of remote sensing and modeling in vineyard management. Analysis of
longterm climate records and wine ratings showed that interannual variability in climate has a strong impact on the yearly $30 billion California wine industry. Warmer
SSTs observed from satellite along the California coast were found to help wine quality
by modulating humidity, reducing frost frequency, and lengthening the growing season
(Nemani et al., 2001). Because changes in regional SSTs persist for 6 to 12 months,
predicting vintage quantity and quality from previous winter conditions appears to be
possible (Nemani et al., 2001).

TOPS also helps vintners during the growing season as a real-time vineyard management tool. For example, satellite remote sensing data during the early growing
season helps vineyard managers to locate areas for pruning so that an optimum canopy
density is maintained. Similarly, LAI derived from satellite data is used in ecosystem
process models to compute water use and irrigation requirements to maintain vines
at given water stress levels. Research suggests that vines need to be maintained at
moderate water stress to maximize fruit quality (Johnson et al., 2003). By integrating
leaf area, soils data, daily weather, and weekly weather forecasts, TOPS can estimate
spatially varying water requirements within the vineyard so that managers can adjust
water delivery from irrigation systems (Plate 1.4a). A number of Napa valley vintners
presently participate in our experimental irrigation forecast program, helping us verify
the utility of the forecasts, the packaging and delivery of information, and assess the
economic value of the forecasts. Satellite imagery at the end of the growing season also
helps growers in delineating regions of similar grape maturity and quality so that dif-
ferential harvesting can be employed to optimize wine blending and quality (Johnson
et al., 2003).

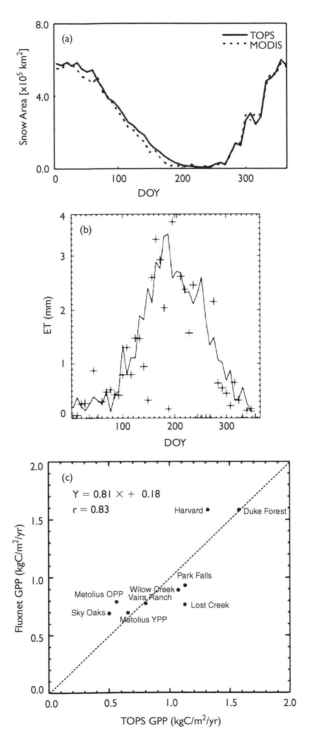

Figure 1.2 Testing TOPS products against satellite and network observations: (a) TOPS snow cover against MODIS-derived snow cover, (b) TOPS Evapotranspiration against FLUXNET observations at Harvard Forest, and (c) TOPS Gross Primary Production shown against FLUXNET observations at a number of sites across the U.S.

Forecasted Irrigation, September 7th, 2005

(a)

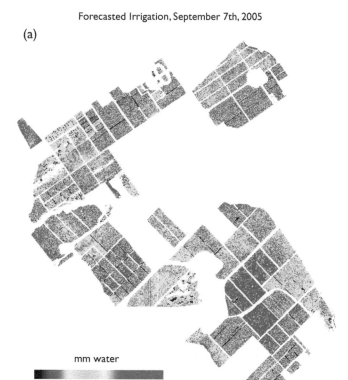

mm water

0   15   30

Near Realtime Biospheric Monitoring, May 2005
(b)   Changes in Vegetation Prodcution and Sea Surface Temperatures

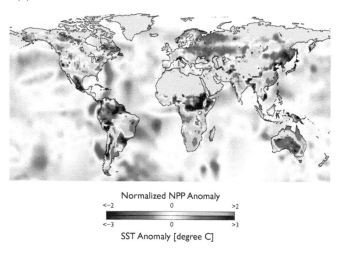

Normalized NPP Anomaly
<−2   0   >2

<−3   0   >3
SST Anomaly [degree C]

Plate 1.4 (a) Application of TOPS over Napa valley vineyards showing the recommended irrigation amounts to keep the vines at a stress level of −12bars for the week of September 7, 2004. (b) A global application of TOPS for monitoring and mapping net primary production anomalies over land and sea surface temperature anomalies over global oceans. NPP and SST anomalies for May 2005 are based on monthly means from 1981–2000. (See *Colour plate 1.4*)

1.5.2 *TOPS monitors global ecosystems*

NPP, the net result of photosynthesis and respiration by plants, forms the basis for life on earth, and provides food, fiber and shelter for humanity. Continuous monitoring of NPP therefore is in our best interest (Running *et al.*, 2004) as the biosphere responds to a variety of pressures from changing climate, atmospheric chemistry, agricultural and land use practices. Spatially continuous assessments of global NPP have been possible only in the past two decades with the availability of NOAA/AVHRR data. One such assessment spanning 1982 to 1999 showed significant increases in global NPP, attributed to a variety of changes in climatic conditions (Nemani *et al.*, 2003). While this is good news, the underlying message of this study is that interannual and decadal changes in NPP can be dramatic and require regular monitoring (Milesi *et al.*, 2005). Using TOPS, we extended our historical analysis of global NPP into global NPP nowcasts (Plate 1.4b). Every 8 days, TOPS brings together the latest MODIS data on land cover, LAI/FPAR, climate data from NCEP, regrids the data to 0.5 degree resolution, estimates NPP, and expresses the output as weekly/monthly anomalies from long-term normals. Because SSTs have a strong association with land surface climate (e.g. ENSO), we also produce maps of global SST anomalies. Animations of these anomalies provide information regarding the location, magnitude, and persistence of anomalies that need further exploration using high resolution data sets. When a persistent anomaly is detected, TOPS can be tasked to perform a higher resolution model run for that region using the best possible data sets. An extended analysis of the anomaly helps us to understand whether the estimated NPP anomaly is related to changes in climate or land use (Hashimoto *et al.*, 2004).

On-going applications of TOPS include nowcasting and forecasting of snow dynamics in the Columbia River Basin (Northwestern U.S.), mapping fire risk across the continental U.S., mapping NPP at 250 m in protected areas such as U.S. National Parks, carbon and water management in urban ecosystems (Milesi *et al.*, 2005) and monitoring and forecasting mosquito abundance and outbreaks of West Nile virus in California. TOPS products are available in WMS format so these data can be accessed and visualized using NASA's WorldWind (http://worldwind.arc.nasa.gov) software system designed explicitly for educational purposes.

1.6 Summary

In the past, ecological forecasting has been largely anecdotal. Its transformation into a rigorous, scientific endeavor is now possible through the observing capacity of operational satellites, the speed and flexibility of the internet, the use of high performance computing for complex modeling of living systems, and mining of large quantities of data in search of relations that could offer potential predictability. Unlike the case of weather and climate forecasting, EF can be as diverse as the number of weather-influenced phenomena. We hope our efforts at EF can provide the necessary guidance for future applications, since the basic infrastructure needed to enable ecological forecasts appears to be similar across different domains. Our experience with EF has been that producing the forecasts may be the easy part, convincing users and conveying the uncertainty associated with the forecasts has been a challenge. Much work is needed along these lines to realize the full potential of ecological forecasting.

Acronyms

AMSR-E	Advanced Microwave Scanning Radiometer-E Observing System
ASTER	Advanced Spaceborne Thermal Emission and Reflect Radiometer
AVHRR	Advanced Very High Resolution Radiometer
BIOME-BGC	Biome-biogeochemistry
CIMIS	California Irrigation Management Information System
DAAC	Distributed Active Archive Center
DAO	Data Assimilation Office
ECPC	Experimental Climate Prediction Center
EF	Ecological Forecasting
ENSO	El Niño–Southern Oscillation
ET	Evapotranspiration
ETo	Reference Evapotranspiration
FLUXNET	Network of eddy covariance towers
FPAR	Fraction of Photosynthetic Active Radiation
GCM	General Circulation Model
GOES	Geostationary Operational Environmental Satellites
GPP	Gross Primary Production
ImageBot	heuristic-search constraint-based planner
JDAF	Java-based Distributed Application Framework (executing pains and for interfacing with DAACs)
DAPDL	Data Processing Action Description Language
WMS	Web Map Server
LAI	Leaf Area Index
MODIS	MODerate resolution Imaging Spectro-radiometer
NCEP	National Center for Environmental Prediction
NOAA	National Oceanic and Atmospheric Administration
NPP	Net Primary Production
NWS	National Weather Service
RAWS	Remote Automated Weather Stations
RHESSys	Regional Hydro-Ecological Simulation System
SCAN	Soil Climate Analysis Network
SNOTEL	SNOWpack TELemetry network
SOGS	Surface Observation and Gridding System
SRTM	Shuttle Radar Topography Mission
SSM/I	Special Sensor Microwave Imager
SST	Sea surface temperature
TOPS	Terrestrial Observation and Prediction System
USFS	United States Forest Service
USGS	United States Geological Survey
XML	Extensible Markup Language

References

Band, L.E. P. Patterson, R.R. Nemani, and S.W. Running. (1993). Forest ecosystem processes at the watershed scale: incorporating hill slope hydrology. *Agricultural and Forest Meteorology*, 63: 93–126.

Barnett, T.P., L. Bengtsson, K. Arpe, M. Flugel, N. Graham, M. Latif, J. Ritchie, E. Roeckner, U. Schlese, U. Schulzweida, and M. Tyree. (1994): Forecasting global ENSO-related climate anomalies. *Tellus*, 46A, 381–397.

Cane, M.A., G. Eshel, and R.W. Buckland. (1994) Forecasting Zimbabwean maize yield using eastern equatorial Pacific sea surface temperature. *Nature*, 370, 204–205.

Canham, C.D., J.J. Cole, and W.K. Lauenroth. (1997) *Models in ecosystem science*. Princeton University press, 456 pp.

Clark, J.S., S.R. Carpenter, M. Barber, S. Collins, A. Dobson, J.A. Foley, D.M. Lodge, M. Pascual, R. Pielke Jr., W. Pizer, C. Pringle, W.V. Reid, K.A. Rose, O. Sala, W.H. Schlesinger, D.H. Wall, and D. Wear. (2001). Ecological forecasts: An emerging imperative. *Science*, 293: 657–660.

Daly, C., R.P. Neilson, and D.L. Phillips. (1994) A Statistical-topographic model for mapping climatological precipitation over mountainous terrain. *Journal of Applied Meteorology*, 33, 140–158.

Goddard, L., S.J. Mason, S.E. Zebiak, C.F. Ropelewski, R. Basher, and M.A. Cane. (2001) Current approaches to climate prediction. *Int. J. Climatology*, 21 (9): 1111–1152.

Golden, K., W. Pang, R. Nemani, and P. Votava. (2003) Automating the processing of Earth observation data. *Proceedings of the 7th International Symposium on Artificial Intelligence, Robotics and Automation for Space (i-SAIRAS 2003)*, Nara, Japan. May 19–23.

Hashimoto, H., R.R. Nemani, M.A. White, W.M. Jolly, S.C. Piper, C.D. Keeling, R.B. Myneni, and S.W. Running. (2004) El Niño–Southern Oscillation–induced variability in terrestrial carbon cycling. *Journal of Geophysical Research-Atmospheres*, 109 (D23): D23110, doi:10.1029/2004JD004959.

Jolly, M.W., J.M. Graham, A. Michaelis, R. Nemani, and S.W. Running. (2004). A flexible, integrated system for generating meteorological surfaces derived from point sources across multiple geographic scales. *Environmental modeling & software*, 15: 112–121.

Johnson, L., D. Roczen, S. Youkhana, R.R. Nemani, and D.F. Bosch. (2003) Mapping vineyard leaf area with multispectral satellite imagery. *Computers & Agriculture*, 38(1): 37–48.

Justice, E. Vermote, J.R.G. Townshend, R. Defries, D.P. Roy, D.K. Hall, V.V. Salomonson, J. Privette, G. Riggs, A. Strahler, W. Lucht, R. Myneni, Y. Knjazihhin, S. Running, R. Nemani, Z. Wan, A. Huete, W. van Leeuwen, R. Wolfe, L. Giglio, J-P. Muller, P. Lewis, and M. Barnsley. (1998) The moderate resolution imaging spectroradiometer (MODIS): land remote sensing for global change research. IEEE *Transactions on Geoscience and Remote Sensing*, 36: 1228–1249.

Milesi, C., H. Hashimoto, S.W. Running and R. Nemani. (2005a). Climate variability, vegetation productivity and people at risk. *Global and Planetary Change*, 47: 221–231.

Milesi, C., C.D. Elvidge, J.B. Dietz, B.J. Tuttle, R.R. Nemani, and S.W. Running. (2005b). Mapping and modeling the biogeochemical cycling of turf grasses in the United States. *Environmental Management*. DOI: 10.1007/s00267-004-0316-2.

Myneni, R.B., S. Hoffman, J. Glassy, Y. Zhang, P. Votava, R. Nemani, and S. Running. (2002) Global products of vegetation leaf area and fraction absorbed PAR from year one of MODIS data. *Remote Sensing of Environment*, 48(4): 329–347.

Nemani, R., S. Running, L. Band, and D. Peterson. (1993) Regional Hydro-Ecological Simulation System: An illustration of the integration of ecosystem models in a GIS. In: M.F. Goodchild, B.O. Parks and L. Steyeart (eds.), *Environmental modeling and GIS*, Oxford press, p. 296–304.

Nemani, R.R., M.A. White, D.R. Cayan, G.V. Jones, S.W. Running, J.C. Coughlan, and D.L. Peterson. (2001) Asymmetric warming along coastal California and its impact on the premium wine industry. *Climate Research*, **19**: 25–34.

Nemani, R.R., C.D. Keeling, H. Hashimoto, W.M. Jolly, S.C. Piper, C.J. Tucker, R.B. Myneni, and S.W. Running. (2003). Climate driven increases in terrestrial net primary production from 1982 to 1999. *Science*, **300**: 1560–1563.

Running, S.W., R.R. Nemani, F.A. Heinsch, M. Zhao, M. Reeves, and H. Hashimoto. (2004) A continuous satellite-derived measure of global terrestrial primary productivity. *Bioscience*, **54(6)**: 547–560.

Roads, S.-C. Chen, J.R.F. Fujioka, H. Juang, and M. Kanamitsu. (1999) ECPC's global to regional fireweather forecast system. Proceedings of the 79th Annual AMS Meeting, Dallas, TX January 10–16.

Robertson, A., U. Lall, S.E. Zebiak, and L. Goddard. (2004). Improved combination of multiple atmospheric general circulation model ensembles for seasonal prediction. *Monthly Weather Review*, **132**, 2732–2744.

Tague, C. and L. Band. (2004). RHESSys: Regional Hydro-ecologic simulation system: An object-oriented approach to spatially distributed modeling of carbon, water and nutrient cycling. *Earth Interactions*, **8(19)**: 1–42.

Thomson, M.C., F.J. Doblas-Reyes, S.J. Mason, R. Hagedorn, S.J. Connor, T. Phindela, A.P. Morse, and T.N. Palmer. (2006): Malaria early warnings based on seasonal climate forecasts from multi-model ensembles. *Nature*, **439**, 576–579.

Thornton, P.E., S.W. Running, and M.A. White. (1997). Generating surfaces of daily meteorological variables over large regions of complex terrain. *Journal of Hydrology*, **190**: 214–251.

Trenberth, K.E. (1996) Short-term climate variations: Recent accomplishments and issues for future progress. *Bull. Amer. Met. Soc.*, **78**: 1081–1096.

Waring, R.H. and S.W. Running. (1998) *Forest Ecosystems: Analysis at Multiple Scales*, 2nd Edition. Academic Press.

White, M.A., P.E. Thornton, S.W. Running, and R.R. Nemani. (2000). Parameterization and sensitivity analysis of the BIOME-BGC terrestrial ecosystem model: net primary production controls. *Earth Interactions*, **4**: 1–85.

White, M.A. and R.R. Nemani. (2004) Soil water forecasting in the continental U.S. *Canadian Journal of Remote Sensing*, **30(5)**: 1–14.

Wood, A.W., E.P. Maurer, P. Edwin, A. Kumar, and D.P. Lettenmaier. (2001) Long-range experimental hydrologic forecasting for the eastern United States, *Journal of Geophysical Research (Atmospheres)*, DOI 10.1029/2001JD000659.

Zebiak, Stephen W. (2003): Research Potential for Improvements in Climate Prediction. *Bull. American Meteor. Society*. **84(12)**, 1692–1696.

Chapter 2

Remote sensing as an aid in arresting soil degradation

R.S. Dwivedi

National Remote Sensing Agency, Department of Space, Govt. of India, Balanagar, Hyderabad, India

2.1 Introduction

The purpose of the chapter is to describe the use of remote sensing techniques to design strategies of arresting soil degradation, and thereby contribute to food security.

Soil degradation is generally be defined as "the deterioration of soil quality through loss of one or more of its functions (production, buffering capacity, etc.)" (http://www/iiasa.ac.at/Research LUC/GIS/giswebpage/documents/rs....). The term, land degradation is used interchangeably for soil degradation. Land degradation, which leads to the development of problem soils, generally, manifests itself in the form of a loss or reduction of land productivity as a result of human activity (UNEP, 1993). Two types of human-induced degradation processes, namely displacement of soil material, and internal soil deterioration are, generally, encountered. The processes included in the first category are water erosion, and wind erosion. Included in the second category of soil degradation are chemical deterioration, physical deterioration, and biological deterioration. The chemical deterioration consists of loss of nutrients, pollution and acidification, salinization and/or alkalinization, discontinuation of flood-induced fertility, and other chemical problems. Sealing or crusting of top soil and subsidence of organic soils, soil compaction, deterioration of soil structure and waterlogging comprise physical deterioration. Biological deterioration includes imbalance of (micro) biological activities. Overgrazing of pasturelands, deforestation, over-intensive annual cropping, mining, etc. are causes of soil degradation.

Soil degradation affects the productivity of land and in turn has significant impact on food security. World-wide, an estimated 305 million ha of land are seriously degraded, and rendered unproductive. About 910 million ha are moderately degraded. An estimated 140 million ha are expected to become seriously degraded in the next 20 years. The world average of 0.28 ha cropland per capita in 1990–1991 would decline to 0.17 ha by the year 2025, as per population projections (World Resources Institute, 1990). Sustainable food production could be achieved by bridging the gap between potential and actual yield in major farming systems by conserving prime farmland for agriculture by (i)Arresting soil degradation and loss of the biological potential of the soils,(ii) Promoting less water use on the basis of agro-ecological, meteorological and marketing factors, and (iii)Launching community centered water harvesting, conservation and use programme to ensure efficient harvest of rainwater and sustainable use of ground water.

(a)

Sheet and rill erosion in parts of Kurnool
district, Andhra Pradesh, Southern India

Rill erosion in part of Belgaum district
of Karnataka, Southern India

(b)

Plate 2.1 Sheet and rill erosion as seen in satellite images. (See *Colour plate 2.1*)

2.2 Role of remote sensing

Because of the advantage of repetitive coverage, and capability of synoptic overview, remote sensing has emerged as a powerful and cost-effective tool to monitor and quantify various manifestations of soil degradation, and assist in decision-making about ecologically-sustainable, economically-viable and people-participatory strategies of reclamation of degraded soils.

2.2.1 *Remote sensing of soil erosion and wind erosion*

Colour plates 2.1, 2.2 and 2.3 show satellite images of sheet and rill erosion, ravines and wind erosion

Dwivedi, Kumar and Tewari (1997) used Landsat-MSS, and TM, and SPOT-MLA data for the detection and delineation of the features associated with soil erosion in a black soil terrain in southern India. An improvement in the level of information on eroded land was observed with improving spatial resolution.

Spaceborne spectral measurements have been used for delineating the reclamative grouping of ravines (a network of gullies). Besides, IRS-1C PAN data have been used for monitoring the spatial extent and distribution of gullied/ravinous lands in western Uttar Pradesh, northern India (Singh, Sharma, and Singh, 1998).

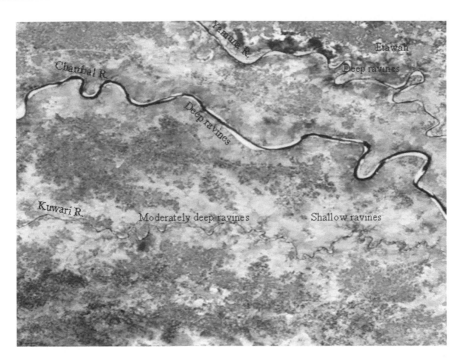

Plate 2.2 Ravines as seen in IRS-1D LISS-III image acquired in February, 2006. (See *Colour plate 2.2*)

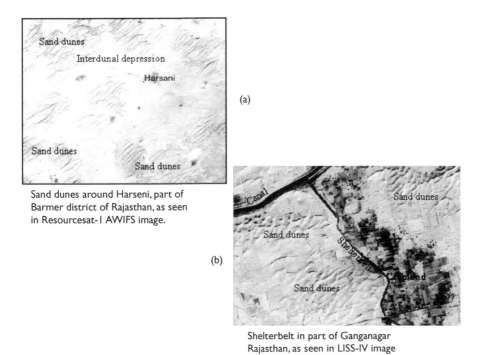

Sand dunes around Harseni, part of Barmer district of Rajasthan, as seen in Resourcesat-1 AWIFS image.

Shelterbelt in part of Ganganagar Rajasthan, as seen in LISS-IV image

Plate 2.3 Wind erosion features as seen in Resourcesat-1 AWIFS and LISS-IV images. (See *Colour plate 2.3*)

Spaceborne multispectral data hold a great promise in deriving information on features associated with the wind erosion. Using brightness and redness indices derived from the Nimbus Coastal Zone Colour Scanner (CZCS) data over northern Africa, Escadafal (1992) distinguished sand seas with different sand types (pale to reddish) apart from other features. In India, the separability of various terrain features related to wind erosion has been studied using earth observation satellite data. Dwivedi *et al.* (1992) used Landsat TM data and its various transforms, namely principal component, Normalized Difference Vegetation Index (NDVI), Soil Brightness Index (SBI) and Perpendicular Vegetation Index (PVI) derived there from over central Luni Basin in Rajasthan, and have shown that the information on sand dunes and other associated features generated by the conjunctive use of the standard data product i.e. false colour composite prints generated from green, red and near-IR spectral bands, and the transforms far exceeds the one derived from the former alone. While working in the Indira Gandhi canal command area in Rajasthan, western India, parabolic dunes could be discriminated from interdunal depressions using Landsat MSS data of 1973 and 1986 coinciding with the before and after commissioning of the canal.

With respect to soil erosion by water only qualitative assessment could be achieved. However, for implementation of reclamation programme quantitative information on soil loss is needed. Soil erosion models viz. empirical as well as process-based models could be used in this endeavour wherein information on vegetation cover and the soil conservation practice followed could be derived from remote sensing data and the information on rest of the parameters could generated conventionally. For deriving quantitative information on wind erosion-related features, namely sand dunes, sand sheets, etc. stereo images with high spatial and 'z'-axis resolution from Cartosat-1 (2.5 m spatial resolution and ~8 m 'z'-axis resolution) and Cartosat-2 (better than 1m spatial resolution and ~3 m 'z'-axis resolution) could be used for generating digital elevation model(DEM) that may enable categorizing dunes based on their reclaimability.

2.2.2 Remote sensing of salt-affected soils

Color plates 2.4 and 2.5 show satellite-images of salt-affected soils.

Salt-affected soils with salt encrustation at the surface are, generally, smoother than non-saline surface and cause high reflectance in the visible and near infrared bands. Quantity and mineralogy of salts together with soil moisture, colour and roughness are the major factors affecting reflectance of salt-affected soils. Salt mineralogy (e.g. carbonates, chlorides and sulphates) determine the presence or otherwise of absorption bands in the electromagnetic spectrum. Based on a laboratory spectra of mixtures of SiO_2 and $NaCl + MgCl_2$, Hick and Russell (1990) observed significant features associated with two of the water absorption bands at around 1.4 and 1.9 μm. Contrastingly, carbonates exhibit absorption features in the thermal region (between 11 μm and 12 μm) due to internal vibration of the CO_3^- group whereas anions have an absorption band near 10.2 μm caused by overtones or combination tonnes of internal vibration of constitutional water molecules. Based on *in situ* spectral measurement, it has been observed that the natural or *in situ* salt-affected soils with salt encrustation at the surface exhibit the maximum spectral response form, followed by sodic or alkali soils formed due to irrigation with high residual sodium carbonate (RSC)

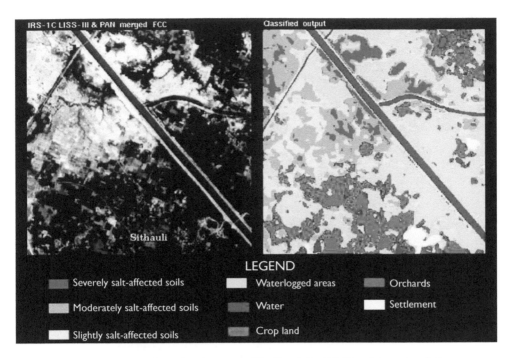

IRS-1C LISS-III & PAN merged FCC   Classified output

Sithauli

LEGEND

Severely salt-affected soils   Waterlogged areas   Orchards

Moderately salt-affected soils   Water   Settlement

Slightly salt-affected soils   Crop land

Plate 2.4 Map showing salt affected soil in part of Rai Bareli dt., U.P. (See *Colour plate 2.4*)

water, natural saline soils, and saline soils formed due to irrigation with saline water in western India. Csillag *et al.* (1993) found the visible (550 to 570 nm), near infrared (900–1030 nm and 1270–1520 nm) and middle infrared bands (1940–2150 nm, 2150–2310 nm and 2330–2400 nm) portion of the spectrum at 20 nm, 40 nm, and 80 nm spectral resolution as the key spectral bands in characterizing the salinity status of soils.

The significant difference in the imaginary part of dielectric constant (I∈) between pure water and saline water at microwave frequencies less than ≥7 GHZ has been used to derive information on soil salinity (Ulaby, Moore and Fung, 1986). Bell *et al.* (2001) have used Small Perturbation Model (SPM), Physical Optics, and Dubius Dielectric Retrieval models for mapping soil salinity in the Alligator River Region of the Northern Territory, Australia using C-(5.33 GHZ), L-(1.25 GHZ) and P-(440 MHZ) bands airborne SAR data.

Owing to a large variation in the surface condition of salt-affected soils on account of the presence of moisture, organic matter, vegetation-both live and dead, and similarity in its spectra with other non-salt affected soils, namely sandy soils and black soils due primarily to similarity in colour, these soils do not exhibit unique spectra. Consequently, the detection and delineation of salt-affected soils using remote sensing data is fraught with several problems. Hyperspectral measurements have been quite promising. Taylor *et al.* (1994) observed absorption features at 1400, 1900 and 2500 nm due to uncombined water in saline soils having $MgCl_2$ while measuring the spectral response from vegetated and bare salt-affected soils with a 24-channel field spectrometer operating in the short-wave infrared region. Szilagyi and Baumgardner

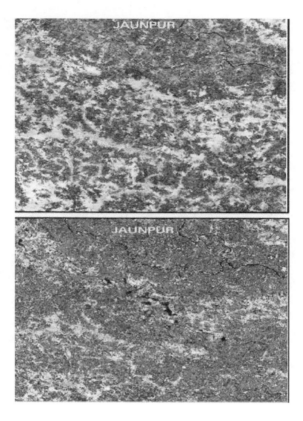

Plate 2.5   Salt-affected soils in part of sharda Sahayak command area (Indo-Gangetic plains), Jaunpur (Uttar Pradesh). (See *Colour plate 2.5*)

(1991) reported the use of high resolution laboratory measurements to identify salinity status of soils.

2.2.3   *Remote sensing of soil salinity*

Multispectral video systems, recording spectral response pattern of vegetation in the visible and near infrared regions of the spectrum have been used for mapping crop variations due to soil salinity. Everett *et al.* (1988) used narrow band videography to detect and estimate the extent of salt-affected soils in Texas, USA. Wiegand, Escobar and Lingle (1994) analyzed and mapped the response of cotton to soil salinity using colour infrared photographs and videography. In addition, other airborne sensors such as digital multispectral video (DMSV) or the digital Multispectral imaging (DMSI) systems have been used for assessment of vegetation degradation due to various factors including land degradation.

Airborne geophysical measurements have also been used to delineate salt-affected soils and a good correlation between the apparent electrical conductivity (ECa) as measured by an electromagnetic induction meter mounted on an aircraft, and electrical conductivity from field samples has been observed. However, the results were not encouraging in areas where rock substratum was covered with a thin soil layer.

The Landsat-MSS data was used for the first time in India at the National Remote Sensing Agency, Hyderabad for mapping salt – affected soils (Singh, Baumgardner, and Kristof, 1977). Subsequently, Landsat-TM, SPOT-MLA and the Indian Remote Sensing Satellite (IRS-1A/-1B), Linear Imaging Self-scanning Sensor (LISS-I and –II) data were used for mapping salt-affected soils. Besides, for mapping salt-affected soils in the Indo-Gangetic alluvial plains Dwivedi and Rao (1992) identified a three-band combination from Landsat-TM data viz. bands-1 (0.45–52 μm), −3 (0.63–0.69 μm) and −5 (1.55–1.75 μm). With the improvement in spatial resolution of 80 m from Landsat MSS to 30 m and 20 m from Landsat TM and SPOT MLA, the level of information that could be derived improved tremendously. While only nature of salt-affected soils in terms of saline, saline-alkali and alkali could be derived from Landsat MSS data, Landsat TM and SPOT MLA data could afford the delineation of the magnitude of soil salinity and/or alkalinity in terms of slight, moderate and strong categories. Apart from inventory of salt-affected soils, temporal behaviour of these soils could be studied using concurrent and historical satellite data/aerial photographs.

With the availability high spatial resolution Linear Imaging Self-scanning Sensor (LISS-III) and Panchromatic (PAN) sensor data from IRS-1C and -1D, salt-affected soils, better delineation of salt-affected soils could be achieved by digitally merging the data from these two sensors' data (Dwivedi et al., 2001). In addition, various image processing techniques, namely band ratioing and principal component analysis, intensity-hue-and -saturation, spectral unmixing and fuzzy logic have also been used for delineation of various categories of salt-affected soils.

2.2.4 Remote sensing of water-logging

Colour plate 2.6 shows the satellite image of water-logging.

Waterlogging refers to a state of soil wherein there is a free water in pores. "An area is said to be waterlogged when the water table rises to an extent that soil pores in the root zone of a crop become saturated, resulting in the restriction of normal circulation of the air, decline in the level of oxygen and increase in the level of carbon dioxide" (Anonymous, 1991).

The presence of moisture in the soil affects its spectral response considerably and a dramatic decrease in the albedo, and other changes related to water and lattice-OH from dry to wet condition of montmorillonite at room temperature has been noted. Further, adding water to montmorillonite sample enhanced the water OH features at 0.94, 1.2, 1.4, and 1.9 μm, due to relatively high surface area and a corresponding high content of adsorbed water. In Western Australia, McFarlane et al. (1992) used portable field spectrometer (PFS) operating in the 0.4 to 2.5 μm region and 256 spectral bands to delineate waterlogged areas of wheat, oat and pastures. For wheat crop there was a good separation between 0.80 and 1.8 μm. Further, there were separations between 1.2 and 1.38, between 1.46 and 1.38, and between 1.94 and 2.5 μm. The PFS spectra for waterlogged and non-waterlogged oat canopies showed that there was slightly better separation in the visible part of the spectrum (0.40 to 0.65 μm) than for wheat. Subsequently Wallace et al. (1993) used a 13-channel airborne multispectral scanner

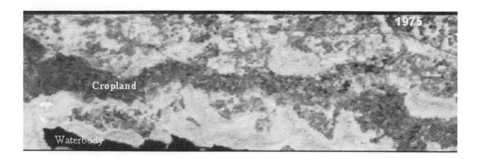

Plate 2.6 Waterlogging around Suratgarh, Ganganagar district, Rajasthan due to seepage from canal. (See *Colour plate 2.6*)

(GEOSCAN Mark1) operating in the visible and near infrared region (0.45–0.97 μm), short-wave (1.98–2.40 μm) and thermal infrared region (8.5–12.5 μm) of the spectrum to detect and map areas in cereal crops where growth has been affected by waterlogging in Western Australia. Near infrared and thermal channels were found to be important in the discrimination between waterlogged and non-waterlogged areas.

Land subject to waterlogging has also been delineated using aerial photographs. Wallace *et al.* (1993) used 1:10,000 scale aerial photographs over part of Western Australia for detection and mapping of waterlogged areas. In India, Sahai and Kalubarme (1985) used black-and-white and colour infrared aerial photographs at 1:30,000 and 1:50,000 scale for delineation of waterlogged areas with water table within 1.5–3.0 m in the Ukai command area in Gujarat, western India.

Landsat TM digital data was used to map waterlogged crops in East Yornaning catchment in Australia (McFarlane *et al.*, 1992). Using space-borne multispectral measurements made in the reflective portion of the spectrum, waterlogged areas with surface ponding or a thin film of water on the surface or wetness of the surface layer could be delineated (Sahai and Kalubarme, 1985). Such a capability has been operationally used for delineating and monitoring water-logged areas apart from salt-affected soils for entire Sharda canal system covering an estimated 6.7 million ha in Uttar Pradesh, northern India (National Remote Sensing Agency, 1995), and in other major command areas of India. However, waterlogging on account of rising ground water table is not easily amenable to be detected using spectral measurements in the reflective portion of the spectrum.

When ground water table rises and reaches very close (within 2 m) to the surface, the growth of most of the mesophytic plants begins to be affected. Being a sub-surface phenomenon, its detection in the optical region of the spectrum is, generally, not feasible. Attempts have been made to detect the shallow ground water tables using ground penetrating radar (GPR) in part of Massachusetts. Additionally, an index based on visible green, near and mid-infra-red region of Landsat-TM bands was used to delineate the degree and extent of water-logging in crop lands in part of Australia. Cialella *et al.*(1997) conjunctively used Normalized Difference Vegetation Index (NDVI) derived from Advanced Visible and Infrared Imaging Spectrometer (AVRIS) and digital elevation model (DEM) over Howland, Maine, USA, to develop a technique for predicting soil drainage classes which are very important from waterlogging point of view.

2.2.5   *Remote sensing of soil crusting*

Soil compaction and crusting are other forms of soil degradation. Two types of crusts, namely lithospheric and biogenic, are normally encountered. A lithospheric crust is the rain crust formed by raindrops. It affects the segregation of the finer particle size at the surface, and its effect is more pronounced in saline soils. Biogenic crust consisting primarily of lower non-vesicular plant (microphytic) cover the upper soil surface in a thin layer.

Higher spectral reflectance values between 0.43 and 0.73 μm has been observed from crusted soil relative to the same soil with the crust broken. In addition, O'Neill (1994) observed the spectral features of microphytic crust between 2.08 and 2.10 μm, and attributed it to the to presence of cellulose. In another study, Karnieli and Tsoar (1994) showed that the microphytic crust caused a decrease in overall albedo in the soils whereas the spectral response related to the biogenic crust permits linear mixing models.

2.3   Conclusion

Remote sensing and allied technologies offer immense potential in deriving information on extent, spatial distribution and dynamics of land subject to various soil degradation processes provided these processes lead to perceptible spectral contrast with the adjoining normal fertile soils. Remote Sensing data in respect of soil degradation can be put to a number of beneficial uses: (i) People-participatory ways of using degraded soils to grow tree crops, such as, biodiesel plantations, and fodder, (ii) Monitoring and quantifying the spread of degradation (such as, soil salinization and waterlogging) to design appropriate measures to roll back the degradation, and (iii) improving water resources, biodiversity, and ecosystem of areas.

Acknowledgements

The author is grateful to Dr K. Radhakrishnan, Director and Dr P. S. Roy, Deputy Director, National Remote Sensing Agency (NRSA) for providing necessary facilities during the preparation of manuscript. Further, technical support provided by Dr K. V. Ramana, Land Resources Group and Dr K.Srinivas, Earth Resources Group, Remote Sensing and GIS Applications Area, NRSA, is gratefully acknowledged.

References

Anonymous (1991) *Waterlogging, soil salinity and alkalinity.* Report of the Working Group on Problem identification in irrigated areas with Suggested Remedial Measures. Ministry of Water Resources, Government of India, New Delhi.

Bell, D., Menges, C., Ahmad, W., & van Zyl., J. J. (2001) The application of dielectric retrieval algorithms for mapping soil salinity in a tropical coastal environment using polarimetric SAR. *Remote Sensing of Environment*, 75, 375–384.

Cilella, A. T., Dubayah, R., Lawrence, W., & Levine, E. (1997) Predicting soil drainage class using remotely sensed and digital elevation data. *Photgrammetric Engineering and Remote Sensing*, 63(2),171–178.

Csillag, F., Pasztor, L., & Biehl, L. L. (1993) Spectral band selection for the Characterization of salinity status of soils. *Remote Sensing of Environment*, 43, 231–242.

Dwivedi, R. S., Ravi Sankar, T., Venkataratnam, L., & Rao, D. P. (1992) Detection and delineation of various desert terrain features using Landsat-TM derived image transforms. *Journal of Arid Environment*, 25: 151–162.

Dwivedi, R. S. & Rao, B. R. M. (1992) The selection of the best possible Landsat-TM band combinations for delineating salt-affected soils. *International Journal of Remote Sensing* 13(11): 2051–2058.

Dwivedi, R. S., Ramana, K. V., Thammappa, S. S., & Singh, A. N. (2001) Mapping salt-affected soils from IRS-1C LISS-III and PAN data. *Photogrammetric Engineering and Remote Sensing*, 67(10), 1167–1175.

Dwivedi, R. S., Kumar, A. B., & Tewari, K. N. (1997) The utility of multi-sensor data for mapping eroded lands. *International Journal of Remote Sensing*, 18(11), 2303–2318.

Escadafal, R. (1992) Remote sensing of north African sand seas with medium resolution satellite imagery (Nimbus-CZCS). *International Symposium on Evolution of Deserts.* Ahmedabad, India, February 11–19, 1992.

Everitt. J., Escobar, D., Gerbermann, A.H., & Alaniz, M. (1988) Detecting saline soils with video imagery. *Photgrammetric Engineering and Remote Sensing*, 54, 1283–1287.

Hick, R. T. & Russell, W. G. R. (1990) Some spectral considerations for remote sensing of soil salinity. *Australian Journal of Soil Research*, 28, 417–431.

Karnieli, A. & H. Tsoar, H. (1994) Spectral reflectance of biogenic crust developed on desert dune sand along the Israel-Egypt border. *International Journal of Remote Sensing*, 16, 369–374.

McFarlane, D. J., Wheaton, G. A., Negus, T. R. & J. F. Wallace, J. F. (1992) Effects of waterlogging on crop and pasture production in the Upper Great Southern, Western Australia. Technical Bulletin No. 86, Department of Agriculture, and Western Australia.

National Remote Sensing Agency (1995) *Study of land degradation problems in Sharda Sahayak command area for sustainable agriculture development.* Project report National Remote Sensing Agency, Department of Space, Government of India.

O'Neill, A. L. (1994) Reflectance spectra of microphytic soil crust in semi-arid Australia. *International Journal of Remote Sensing*, 15, 675–681.

Sahai, B. & Kalubarme, M. H. (1985) Ecological studies in the Ukai command area. *International Journal of Remote Sensing*, 6 (3&4), 401–409.

Singh, A. N., Baumgardner, M. F., & Kristof, S. T. (1977) Delineating salt-affected soils in part of Indo- Gangetic plain by digital analysis of Landsat data, Tech. Report 111477, LARS, Purdue University, West Lafayette, Indiana, USA.

Singh, A. N., Sharma, Y. K., & Singh, S. (1998) Evaluation of IRS-1C PAN data for monitoring gullied and ravinous lands of Western U.P. *Remote sensing and Geographic information system for Natural Resources management.* A joint publication of Indian Society of Remote Sensing and NNRMS, (Department of Space, Government of India, Bangalore) pp149–160.

Szilagyi, A. & Baumgardner, M. F. (1991) Salinity and spectral reflectance of soils. *Proceedings of American Society of Photogrammetry and Remote Sensing Symposium*, Baltimore, 25–29 March, 1991, APRS, Baltimore.

Taylor, G. R., Bennett, B. A., Mah, A. H., & Hewson, R. D. (1994) Spectral properties of salinized land and implications for interpretation of 24-channel imaging spectrometry. *Proceedings of the First International Airborne Remote Sensing Conference and Exhibition.* Strasbourg, France. 11–15 Sept., 1994.

The World Watch Institute (1990) *World Resources 1990–1991.* Chapter 6. Food and Agriculture, Oxford University Press.

U.N.E.P. 1993. *Environmental report 1993–1994.* Blackwell Oxford, U.S.A, Cambridge, UK.

Ulaby, F. T., Moore, R. K., & Fung, A. K. (1986) *Microwave Remote Sensing Active and Passive-From Theory to Applications.*3.Dedham, Massachusetts: Artech House.

Wallace, J., Campbell, N. A., Wheaton, G. A., & McFarlane, D. J. (1993) Spectral discrimination and mapping of waterlogged cereal crops in Western Australia. *International Journal of Remote Sensing* **14(14)**:2731–2743.

Wiegand,C. L., Escobar, D. E., & Lingle, S. E. (1994) Detecting growth variation and salt stress in sugarcane using videography. *In* Proceedings of the Biennial Workshop on Colour Aerial Photography and Videography for Resource Monitoring. American Society of Photogrammetric Engineering and Remote Sensing, p. 185–199.

Chapter 3

Crop management through remote sensing

M.V.R. Sesha Sai, R.S. Dwivedi & D.P. Rao

National Remote Sensing Agency, Department of Space, Government of India, Balanagar, Hyderabad, India

3.1 Introduction

Crop management is an integral part of the sustainable agriculture. Because of the advantage of repetitive coverage and synoptic overview, satellite remote sensing has emerged as a cost-effective tool covering all aspects of soil, water and crop management. A sustainable agricultural system has to be economically viable, agrotechnically feasible, politically and socially acceptable, institutionally manageable and ecologically sound. Satisfying all the requirements of sustainability, fixation of the criteria and estimation of the indicators is a very complex task. Hence, the initial and boundary conditions need to be arrived at after giving due consideration to the prevailing conditions in relation to the expected levels of change. Some of the salient basic tenets of sustainable agriculture are given hereunder:

- Meeting the needs of the Present without compromising the future needs
- Stewardship of both natural and human resources
- Transition to Sustainable Agriculture (SA) is a Process, involving a series of small and realistic steps
- Critical for SA continuum
- Participation of everyone involved is a must

3.2 Systems approach in sustainable agriculture

A 'Systems' perspective is essential to understand the sustainability. The system is envisioned in its broadest sense, from the individual farm to the local ecosystem and to the communities affected by the farming system both locally and globally. An emphasis on systems allows a larger and more thorough view of the consequences of farming on both humans and the environment. The systems approach gives us tools to explore the interconnections between farming and other aspects of the environment. It automatically calls for inter-disciplinary efforts in research and education.

British Columbia Round Table, 1994 on Environment and Economy, has provided some guidelines as operational conditions of sustainability, which are given hereunder:

- Limit the impact on the living within carrying capacity
- Preservation and protection of life supporting systems, biological diversity and renewable resources
- Minimize depletion of non renewable resources
- Promote long term economic development
- Meet basic needs and aim for fair distribution of benefits

- More proactive and participatory governance
- Promote values that support sustainability through information and education

3.3 Advances in space technology

Capabilities of Indian Earth Observation Systems have significantly grown to provide multi-temporal and multi-spectral data of different spatial resolutions that enable generation of information on natural resources and environment at various hierarchical levels from macro to microlevels. Resourcesat-1 with a unique combination of three sensors viz., AWiFS, LISS-III prome and LISS-IV provides data ranging from 5.8 m to 56 m, covering swaths from 23 km to 370X2 km, with a revisit period ranging from 5 to 24 days. Along with the predecessor satellites of Resourcesat-1, valuable information on national natural resources has been generated world over.

3.3.1 *Limitations of space technology*

The limitations in substantial utilization of space-derived remote sensing techniques are two-fold: those due to instrument design and those due to the analytical procedures for extraction of required information. Spaceborne hyperspectral remote sensing technology and techniques have yet to grow to provide timely inputs for implementation of precision farming techniques in real time mode. Current satellite-based sensors have fixed and relative broad spectral bands that may be inappropriate for a given application. Inadequate repeat coverage for intensive agricultural management and long time periods between image acquisition and delivery to user (turn around time), and non-availability of cloud-free optical sensor data especially during *kharif* crop growing season are other noteworthy limitations.

3.4 Geo-spatial information

Towards meeting the objectives envisaged in the concept of sustainable agriculture, spaceborne spectral measurements have been found to be useful in providing information spatially with a regular periodicity. Information on various aspects related to the cropping systems analyses, crop intensification, crop diversification, crop suitability and conformity analyses, etc. could be derived from spaceborne spectral measurements. Monitoring of natural resources, that are of significance to agriculture, such as soil, surface and sub-surface waters, weather, land degradation status and dynamics aid substantially towards evolving as well as evaluating the agricultural systems for sustainability. Two such cases are presented below.

Sustainable agriculture implies (i) improving the productivity of existing cultivated lands (ii) rehabilitation and restoration of fertility of degraded lands; and (iii) adoption of eco-friendly alternate/optimal land use plan. In terms of activities it amounts to intensive agriculture, land degradation studies and watershed development.

3.5 Intensive agriculture

Towards intensification of agriculture remote sensing provides information on existing land use/land cover pattern that could be integrated with information on soils, irrigation facilities and socio-economic data to arrive at the most appropriate crop to be taken on a sustained basis. Two studies viz. delineation of *kharif* (rainy season) and *rabi* (winter) fallow, and suitability for cotton crop were taken up by the Department of Space, Government of India.

3.5.1 Delineation of kharif rice fallows in the rice cropping system

Monsoon fallowing is a common practice in Madhya Pradesh in Central India, presumably evolved as a rational strategy for maximizing household involvement over time. It could be due to: (i) difficulty of soil preparation before monsoon for timely sowing of a rainy season crop; (ii) threat of flooding of the rainy season crops due to heavy rains; (iii) reduction in available soil moisture for the post-rainy season crop due to high transpiration by the rainy season crop, and consequent reduction in yield (Krantz and Quackenbush, 1970).

The International Crops Research Institute for Semi-arid Tropics (ICRISAT) has demonstrated the technical and economic feasibility of double cropping (a short duration monsoon season crop followed by a short duration pulse crop on dry lands with Vertisols, where annual rainfall exceeds 750 mm (ICRISAT, 1987). Timely and reliable information on the extent and spatial distribution of fallow lands during kharif season is a pre-requisite for their effective utilization. Rice is the major food grain cereal crop grown in south Asia and is cultivated mostly during the *kharif* (rainy) season. Since the scope is limited for horizontal expansion, increased cropping intensity on the existing agricultural lands is one of the best crop management options. In this context, post-*kharif* rice fallows offer a considerable scope for achieving sustainable production by introduction of short duration leguminous crops. These crops sustain the crop production system by improving the fertility of the soil through symbiotic nitrogen fixation, checking the soil erosion by serving as cover crops, breaking the build up of pests and diseases, etc.

The Indian Remote Sensing Satellite (IRS) Wide Field Sensor (Waifs) data of 1999 *kharif* season and *rabi* 2000 season were analyzed following total enumeration approach for deriving spatial distribution of *kharif* rice and rice fallows in different states of India and in the neighboring Pakistan, Nepal and Bangladesh nations. Spatial information on the distribution of *kharif* rice and *rabi* fallow lands was logically combined to derive the distribution and area of post kharif rice fallow lands. The analysis indicated that a good potential for utilization of post-*kharif* rice fallows existed in India, Bangladesh and Nepal. However, a systematic soil suitability analysis for raising *rabi* crop needs to be taken up. Spatial distribution of kharif rice and post kharif rice fallows for Orissa state is given in Figure 3.1.

Integration of remote sensing-based spatial information layer of rice fallows with soil and climate layers in GIS domain enabled ICRISAT to identify potential areas for rice-fallow cultivation with a suitable short duration leguminous crop. Thus, this information along with pedo-climatic regimes and the socio-economic conditions in GIS environment enables better utilization of the post *kharif* rice fallow lands towards increasing cropping intensity and enhancing the crop productivity.

3.5.2 Sustainable cotton production

Variations in the cotton productivity and production are attributed to the soil, climatic and socio-economic factors besides the market price during the in and off – seasons of

Kharif rice area	Rabi Fallow area	Fallow land as % of Kharif rice area
3.88 Mha	1.22 Mha	31.4 %

Plate 3.1 Spatial distribution of *kharif* rice and fallow lands in Orissa state. (See *Colour plate 3.1*)

the crop. Best returns on a sustainable basis can only be achieved by cultivating crops in the pedo-climatic regimes that suit them the best. In sub-optimal pedo-climatic conditions, the requirements of externally applied inputs will be large to get sustained net monetary returns. Spatial distribution of cotton crop in relation to its in situ suitability based on the soil characteristics were derived for selected districts of Andhra Pradesh, India. In this conformity analysis, IRS-LISS III data was used for deriving spatial distribution of cotton crop, while soil maps prepared by the National Bureau of Soil Survey and Land Use Planning, India were used to derive the soil suitability regimes for cotton cultivation. Four classes of suitability viz., most suitable, highly suitable, moderately suitable and least suitable, were generated. This information is useful for agricultural extension personnel in identifying the areas most suitable for cotton crop and encouraging farmers to raise this crop towards selection of the districts for encouragement or demotivation of the farmers towards growing cotton crop. Such a representative map for Guntur district, Andhra Pradesh, Southern India is presented as Figure 3.2. In addition, this exercise also helps in discouraging farmers from cultivating cotton crop in areas that are not suitable.

3.6 Sustainable watershed development

In pursuance of the principle of optimal utilization of natural resources based on their potential and limitations for sustainable agriculture, an integrated approach was adopted for generating information on various natural resources, namely soils, surface and ground water, land use/land cover from satellite data and integrating them in GIS environment for arriving at optimal land use plan/interventions for land and water resources development under a national-level project titled 'Integrated Mission for Sustainable Development (IMSD'. Presented here is a case study wherein IRS LISS-II data with 36.25 m spatial resolution was used to generate information

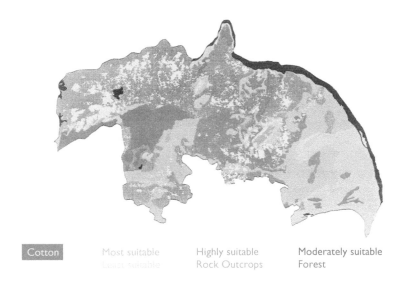

Cotton | Most suitable | Highly suitable | Moderately suitable
Least suitable | Rock Outcrops | Forest

Plate 3.2 Suitability regimes for cotton cultivation in Guntur district, A.P. (See *Colour plate 3.2*)

on hydrogeomorphology, soils land use/land cover, surface water and slope by integrating such information with socio-economic data and other ancillary information optimal land use/developmental for land and water resources has been generated.

Realising the fact that all natural resources co-exist and sectoral approach for their management is not the solution, an integrated approach for natural management including adoption of appropriate intervention on a watershed (hydrological unit) basis has been used based on information derived on these resources from space-borne spectral measurements. Since treatment of all watersheds in a basin does not seem feasible prioritization of watershed is resorted to. In this endeavour too, satellite data hold great promise. Interventions in a watershed lead to improved hydrological regime that helps establishing vegetation cover (either agriculture crops, grasses or other vegetation) that could be monitored using multi-temporal satellite data.

3.7 Land degradation studies

In pursuit of bringing more area under plough, concerted efforts by the Government stake-holders has been made to rehabilitate degraded lands. Remote sensing techniques and GIS offer immense potential in deriving information on the nature, extent, and spatial distribution of degraded lands. Further, in the event of implementation of reclamation programmed remote sensing enables monitoring the progress and success of reclamation programmed.

3.8 Paradigm shift

Previously, production agriculture was empirical and based on 'black box' approaches at field level. With developments in satellite technology, computation and modeling,

Table 3.1 Old concepts and new perspectives in agriculture (Rabbinge, 1997).

Old concept	New perspective	Requirement
White-peg agronomy	Production ecology	Training in systems approaches, Simulation and exploratory studies
Production functions	Target oriented input levels	Development of technical coefficients for production techniques
Heterogeneity as a liability	Heterogeneity as an asset	Fine tuning of measures to specific possibilities using remote sensing and GIS techniques

the traditional approaches of agricultural science towards production agriculture have undergone significant transformation.

The old concepts and new perspectives in agriculture are given in Table 3.1.

The perspective of heterogeneity needs to be looked upon as an asset in the production ecology. Information on heterogeneity needs to be used to further increase the productivity and efficiency of agricultural inputs. Most of the agriculture is performing at efficiency far below its potential. By making use of precision agriculture techniques, more productive agricultural systems that are economically and ecologically efficient, and that can contribute towards sustainable agriculture systems, can be created.

3.9 Precision farming

Precision agriculture aims at adjusting and fine tuning land and crop management to the needs of plants within heterogeneous fields. Thus, the concept behind precision agriculture is based on matching the external inputs such as fertilizers and herbicides, with the variation in local soil conditions in order to reduce costs and negative environmental impacts. Traditionally, farming practices such as fertilization, tillage, and cropping and pest management are defined uniformly for each field on the basis of the data derived from average samples. However, modern techniques of site-specific yield measurements have shown that yield differences within fields commonly vary by a factor of 3 or 4. The new field of 'precision agriculture' takes into account this within – field variability to enable the precise targeting of the interventions such as crop spraying or fertilizer application and other agricultural practices only when and where they are needed. Such practices, therefore, have the dual advantage of maximizing the production and minimizing the environmental damage, and have been made possible by the advances in forming technology, procedures for mapping and interpolating spatial patterns and GIS for overlaying and interpolating several soil, landscape and crop attributes.

Broad delineation of soil-based and/or crop-based management units could be derived from currently operating earth observation missions. The technology of finer delineation and identification of surrogates for diagnostics is still at various stages of research and development. Studies are in progress using crop simulation models, especially to study the crop growth processes and its effect on the variability in the crop yield.

Information Requirement for Precision Agriculture:

Seasonally-stable Conditions

- *Diagnostic Analysis*
- *Soil-based management units, based on soil properties*
- *Real-time Information*

Seasonally-variable Conditions (real-time mode)

- *Crop Phenology*
- *Crop Growth*
- *Soil moisture*
- *Crop Evapotranspiration Rate*
- *Crop Nutrient Deficiency*
- *Weed Infestation*
- *Insect Infestation*
- *Diagnostics of Variability in Crop Production*
- *Mapping Information on Meteorological/Climate Conditions*
- *Addressing Time-Critical Crop Management (TCCM) Applications*

3.10 Role of space technology

Remote sensing holds very good promise in deriving spatio-temporal information required for precision agriculture. Specific role of remote sensing includes enabling generation of field maps on soil attributes and crop variability from point samples, deriving information on stable conditions, variable conditions, and enable determining the causes of variability in crop yield, and ultimately developing management strategies. Applications of GIS in precision agriculture include the storage and management of model input data and the presentation of model results which process model provides. For the dynamic process models, GIS can only partly play a role in storage and management of model input data and presentation of results. Lastly, creation of an operational decision support system (DSS) that GIS technology can provide is a very important element of precision agriculture. Moran *et al.* (1997) have extensively reviewed the role of remote sensing in precision agriculture.

The information on soils and crops derived from the analysis of remotely sensed spatial data could be integrated into GIS environment along with the other relevant spatial and spatial data and could be used for generating valuable information needed for precision farming. These techniques include aid in converting point samples to field maps, mapping soil variability, mapping crop canopy parameters and yield, etc. to provide diagnostics for the yield variability and thus aid in time specific crop management.

3.10.1 *Earth observation systems vs level of spatial information*

Rapid developments in the sensing technologies led to evolution of Earth Observation (EO) systems that are capable of providing high spatial resolution data with regular periodicity to enable generation of information related to natural resources, on a large scale, and their monitoring. The second generation of Indian Remote Sensing satellites, IRS-1C and 1D, carried panchromatic cameras that provided information at 5.8 m

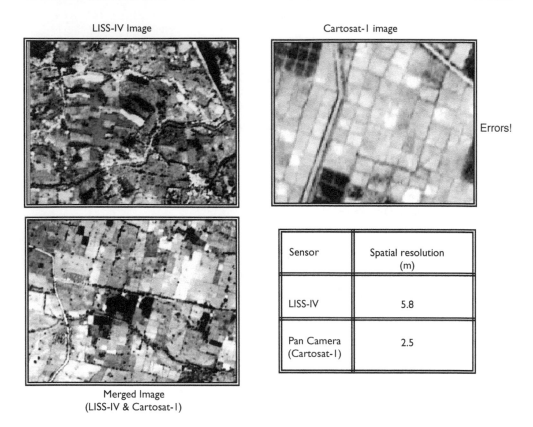

LISS-IV Image

Cartosat-1 image

Errors!

Sensor	Spatial resolution (m)
LISS-IV	5.8
Pan Camera (Cartosat-1)	2.5

Merged Image
(LISS-IV & Cartosat-1)

Plate 3.3 Indian Earth Observation Systems of high spatial resolution. (See *Colour plate 3.3*)

spatial resolution and multispectral data in VNIR region at 23.5 m, besides a Wide Field Sensor (WiFS) at 188 m. The third generation satellite viz., Resourcesat-1 carried LISS-IV sensor that is providing multispectral data at 6 m covering a swath of 23 km or monochromatic data at 6 m covering 70 km swath, in addition to LISS-III at 23.5 m and Advanced WiFS (AWiFS) at 58 m spatial resolution.

These sensors together with their capabilities have enabled generation of information on natural resources at various spatial hierarchical levels. Whilst LISS-IV can provide information upto field level with swath of 23 km AWiFS can provide information at regional level at once with a swath of 380 × 2 km.

In the context of precision farming, LISS-IV data can provide valuable information for precision farming on intra-field variability and LISS-III could provide inter-field variability of soil and crop variables. Data from LISS-IV can be merged with that of Cartosat-1, providing data at 2.5 m enabling better delineation of different land / crop features. Figure 3.3 shows the images of LISS-IV, Cartosat-1 and merged data. Delineation of features is clearly discernible with these datasets. The level of information ranging from a cluster of villages to individual villages and farms can be derived by choosing the appropriate satellite datasets and techniques.

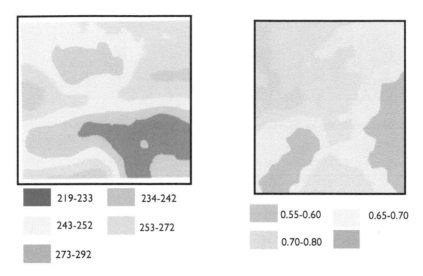

Plate 3.4 Intrafield variablity of soil-N (kg/ha) and yield of soybean (g/m²). (See *Colour plate 3.4*)

3.11 Mapping spatial variability of soil and crop variables

Spectral measurements made from air or spaceborne platforms enable converting point-based samples to continuous soil or land maps using geostatistical methods. Kriging and co-kriging are generally employed to generate surfaces that show a spatial continuity and enable further analyses in GIS environment. The spectral data from sample sites can be extracted and then be related to measured variables at the same sites (yield, available water, salinity, soil nitrogen, etc.). Plate 3.4 given above depicts the surfaces of soil nitrogen and the biomass and yield of soybean fields of the experimental farm at ICRISAT.

3.11.1 Intra field variability of crop condition/vigor

The Normalized Difference Vegetation Index (NDVI) is the surrogate parameter that indicates the crop condition and stage of the crop. It is defined as the ratio of the difference in the spectral reflectance in the near infrared and red regions to their sums. It ranges from -1 to $+1$ during different phenological stages of the crop growth. NDVI typically follows the march of the crop growth, in the sense it increases with the leaf area index of the crops and remains plateau after LAI reaches a peak, generally coinciding with the maximum vegetation stage and starts declining as the crop enters the senescence stage and matures. NDVI indicates the cumulative effect of all factors of crop growth and needs to be interpreted in relation to the prevailing crop growth environment (Wiegand *et al.*, 1991).

Thus, NDVI being an indicator of the crop growth, could also be used for assessing the intra and inter field variability of crop canopy parameters and for identifying and delineating the homogeneous zones for crop management. Further, NDVI derived from high spatial resolution data is useful in detecting within-field subtle anomalies in crop growth/conditions. Such information in conjunction with ground information helps

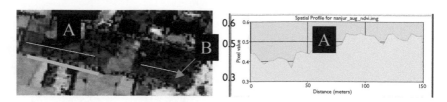

Plate 3.5 Intrafield variability of NDVI in sugarcane. (See *Colour plate 3.5*)

Table 3.2 Response of NDVI in different depth regimes of sorghum plots.

NDVI→ ↓ Soil depth	Max	Min	Mean	Range	Standard deviation	Coefficient of variation%
Deep	0.31	0.19	0.28	0.12	0.05	17.9
Moderately deep	0.28	0.16	0.17	0.12	0.02	11.8
Moderately shallow	0.21	0.10	0.15	0.11	0.02	13.3
Shallow	0.18	0.09	0.13	0.09	0.03	23.1

in decision making for appropriate agrochemical interventions. Plate 3.5 depicts the intra field variability of NDVI in sugarcane plot of 'A', ranging from 0.37 to 0.53 in a sugarcane plot. Ground information revealed that the low NDVI is due to the water logging in that part of the filed of 'A'. Plot 'B' had also shown low NDVI and the cane crop in this plot was mature and about to be harvested.

3.12 Variability in soil depth

Soil depth is an important soil parameter – it influences the crop growth pronouncedly as the depth is an indicator of the capacity of the soil reservoir for soil water and nutrients, etc. Variations in the crop growth due to depth, as manifested in the satellite data by NDVI are mentioned in the Table 3.2. It may be observed from the Table that the maximum and minimum NDVI are the highest in those sorghum plots that are deeper and the lowest in shallow soils. However, the dispersion of these values showed a mixed response with the depth.

3.13 Mapping crop yield

Remote sensing has been used operationally for pre-harvest forecasting of yield . In the simplest approach, final grain yield has been correlated with a single observation of the normalized difference vegetation index (NDVI) or an NDVI time integral at specific times during the season. Regression equation was developed between NDVI and sorghum crop yield and this equation were used for estimation of the variability of crop yield in different sorghum plots. Though detailed diagnostics were not carried out, it was observed that the variations in the crop yield were due to the variations in the crop growth conditions such as soil depth, wetness, slope, clay content, etc.

3.14 Role of information technology

In-season information of crop growth and development on a real time basis enables taking up appropriate crop management practices. Satellite remote sensing data analyzed in association with the ground information helps in the diagnosis of the causes of the variability in crop growth and production. However, this information should reach the farming community in time for decision making. In this regard, information technology plays a pivotal role in disseminating the large scale maps of various natural resources generated using high spatial resolution satellite data. As the resources are optimized in this technology, the cost of the cultivation will also be reduced while improving the production and preservation of the agro-ecosystem. This locale-specific information could reach the farmers through Village Resource Centers planned by the Department of Space and various other knowledge dissemination initiatives of NGOs, etc., to harness the benefits.

3.15 Planning processes & state policies on food and agriculture

Existing federal, state and local government policies often influence the goals of sustainable agriculture. New policies are needed to simultaneously promote environmental health, economic profitability, and social and economic equity. For example, commodity and price support programs could be restructured to allow farmers to realize the full benefits of the productivity gains made possible through alternative practices. Tax and credit policies could be modified to encourage a diverse and decentralized system of family farms rather than corporate concentration and absentee ownership. Government and land grant university research policies could be modified to emphasize the development of sustainable alternatives. Marketing policies and environmental standards could be amended to encourage reduced pesticide use. Coalitions must be created to address these policy concerns at the local, regional, and national level.

Conversion of prime agricultural land to urban uses is a matter of particular concern as rapid growth and escalating land values threaten farming on prime soils. Existing farmland conversion patterns often discourage farmers from adopting sustainable practices and a long-term perspective on the value of land. At the same time, the close proximity of newly developed residential areas to farms is increasing the public demand for environmentally safe farming practices. Comprehensive new policies to protect prime agricultural land and to regulate the urban development are needed, particularly in areas of intense developmental activity. By helping farmers to adopt practices that reduce chemical use and conserve scarce resources, sustainable agriculture research and education can play a key role in building public support for agricultural land preservation. Educating land use planners and decision-makers about sustainable agriculture would further strengthen the endeavors for a economically viable and ecologically sustainable agricultural production system.

3.16 Conclusions

Remote sensing and GIS techniques together can provide valuable information towards sustainable agriculture and precision farming. Best results out of these technologies can be obtained through systems perspective. In order to further advance the entire system towards sustainable agriculture continuum, the transition has to be visualized as a process, involving the farming and consuming communities in all the major important decisions for realizing the objectives of food security. Precision farming is essentially technology driven and the availability of high resolution satellite data providing information on intrafield variability is to be analyzed using advanced techniques like textural segmentation, field classifiers and geo-statistical parameters.

Precision agriculture could be viewed as the information based agriculture, both in space and time domains, for increased agricultural productivity and sustenance of the agro-ecosystem. RS and GIS tools with VRT enable developing decision support systems (DSS) for formulating management strategies. Availability of high resolution satellite data has opened new vistas in generating timely information on intrafield variability of soil and crop variables using advanced techniques such as textural segmentation, field classifiers and geo-statistical parameters. However, finer delineation and surrogates for diagnostics are yet to be derived. Studies are in progress to use crop simulation models to relate crop growth processes and their effects on the variability in the crop yield to evolve advanced crop growth monitoring and management systems.

References

ICRISAT (International Crops Research Institute for the Semi-Arid Tropics) (1987), *Farming system research at ICRISAT*. Proc. Workshop on Farming System Research 17–21 February, 1986, ICRISAT Patancheru Centre, India.

Krantz, B. A. and Quackenbush, T. H. (1970) A proposal for increasing crop production through improved water management and utilization in rainfed agriculture. Unpublished discussion paper, New Delhi, India, The Ford Foundation.

Moran, M.S., Inoue, Y. and Barnes, E.M. (1997) Opportunities and limitations for image-based remote sensing in precision crop management. *Remote Sens. Environ.* 61: 319–36.

Rabbinge, R. (1997) Chairman's introduction In: *Precision agriculture: spatial and temporal variability of environmental quality*. Eds. J.V. Lake, R.B. Gregory, and J.A. Goode. John Wiley, UK, 1–4 pp.

Wiegand, C.L., Richardson, A.J., Escobar, D.E. and Gebermann, A.H. (1991) Vegetation indices in crop assessments. *Remote Sens. Environ.* 35: 105–119.

Chapter 4

Preparation of agro-climatic calendar

M.C. Varshneya[1], Vyas Pandey[2], V.B. Vaidya[2] & B.I. Karande[2]

[1]Anand Agricultural University, Anand, Gujarat, India; [2]Department of Agricultural Meteorology, BACA, Anand Agricultural University, Anand, Gujarat, India

4.1 Introduction

In any economy, forecasting of rainfall and renewable water supplies for the upcoming water year are critically important in water management process for crop production, industrial use, hydropower generation, domestic use as well as sustenance of aquatic species. The agricultural production in India is largely dependent on the monsoon rainfall. The timely onset, well distributed and sufficient monsoon rainfall is the key for better agricultural production in the country which directly influences rural poverty alleviation. The total irrigated area in India is around 32%. Therefore, major thrust for increasing agricultural production will have to be in the rainfed area, whose productivity is dependent on the prediction of temporal and spatial distribution of rainfall.

4.2 Crop calendar

Prediction of weather is the prerequisite for the preparation of crop calendars. There are several methods of rainfall prediction both modern as well as traditional as discussed below.

4.2.1 Long range forecasts of the Indian Meteorological Department (IMD)

There is lot of spatial as well as temporal variation in rainfall in India. There are various methods, modern as well as traditional, for prediction of rainfall. Modern method of long range prediction is based on 16 parameters, which IMD was using from 1988–2002. Then they have updated it to 10 parameter model in two stages of prediction i.e. first stage of forecast is issued in the month of April by using 8 parameters and second stage of forecast is issued in the month of July by using another two parameters (Varshneya, 2007).

The power regression model:

The different parameters used in the model are
Arabian Sea Surface Temperature (January + February)
Eurasian Snow Cover (December of previous year)
Northwest Europe Mean Temperature (January)
Nino-3 SST (July + August + September of previous year)
South Indian Ocean SST Index (March)

East Asia Pressure (February + March)
Europe pressure Gradient (January) and
50 hPa Wind Pattern (January + February)

Probabilistic Model:

The model estimates the probability of South-west Monsoon season (June–September) rainfall for the country as a whole in 5 predefined categories viz.,

Deficient (less than 90% of LPA)
Below normal (90% to 98% of LPA)
Near normal (98 to 102% of LPA)
Above normal (102 to 110% of LPA) and
Excess rainfall (more than 110% of LPA)

National Centre for Medium Range Weather Forecasting (NCMRWF), DST, Government of India – Medium range weather forecast:

National Centre for Medium range Weather Forecasting (NCMRWF) is predicting location specific rainfall and other five parameters for four days twice a week. NCMRWF is giving prediction based on Numerical Weather Prediction (NWP) technique with the help of super computer to 107 Agromet Advisory Field Units (AMFU) throughout the country. The average accuracy of rainfall forecast on Yes/No basis for Pune was 70% and for Anand in Gujarat it was 68–76% in monsoon season (Varshneya, 2007).

4.2.2 AAU's monsoon research almanac

Based on traditional approaches, Anand Agricultural University, Anand . prepared in 2005 Nakshtra-Charan wise forecast for eight agro climatic zones of Gujarat. The validation of this forecast on Yes/No basis indicated that accuracy varied between 42% to 73% for various zones.

Validation of AAU almanac rainfall forecast:

Validation of rainfall forecast, for the year 2006, was done on Yes/No skill score (%) basis, for different zones of Gujarat state, India. The average accuracy of rainfall forecast for state as a whole for June 40.4%, for July 62.5%, for August 52.5% and for September it was 50.9%. In June, forecast accuracy ranged between 30 to 57% for different zones. There was less number of rain events, in June, due to late onset of monsoon. During July accuracy was between 42 to 74% for different zones. In August, accuracy ranged between 48–58%, while in September it was between 40–57%. In August and September rainfall was well distributed spatially over Gujarat, hence, range of accuracy was less.

4.2.3 Artificial neural network (ANN)

Artificial neural networks (ANN) are class of models, inspired by biological nervous systems, denoted by neurons, and working in parallel. The elements are connected by synoptic weights, which are allowed to adapt through a learning process. Now-a-days, neural networks are applied in hydrological prediction (Kingston et al., 2003), pattern recognition, vision, speech recognition, classification, and control systems.

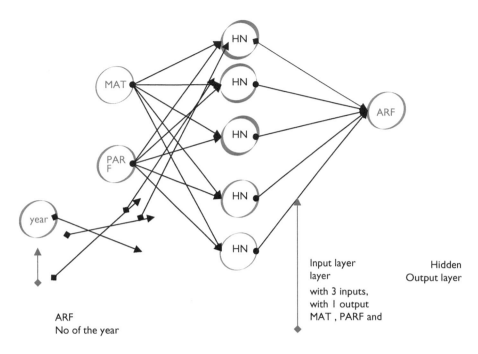

Input layer
layer
with 3 inputs,
with 1 output
MAT , PARF and

Hidden
Output layer

ARF
No of the year

Plate 4.1 Artificial neural network. (See *Colour plate 4.1*)

Hall *et al.* (1993) and Hsu *et al.* (1993) have applied artificial neural network for rainfall- runoff modeling. Goswami *et al.* (1990) have used ANNs with three layers, namely, input layer, hidden layer and output layer for experimental forecasts of all India Summer Monsoon Rainfall for 2002 and 2003. Here an attempt to represent the rainfall process in terms of a single-hidden layer feed forward Neural Network is made (Fig. 4.1) (Kulshrestha, 2006).

Network architecture

ANN consists of an input layer of two neurons, one hidden layer and an output layer with one neuron. Each neuron is connected by feed forward network (Fig. 4.1). Number of nodes in the input and output layer is equal to the number of variables of inputs and output respectively. Figure 4.1 shows three nodes are in the input layer that has known values and one is in output layer to be predicted.

In the present analysis numbers of neurons in input layer, hidden layer and output layer are 2, 147 and 1 respectively (Table 4.1).

The results of actual and model predicted rainfall for the year 2002 to 2006 are depicted in Fig. 4.2. From the student's t-test for two tails at 95% of confidence level predicted ARF by ANN, is found to be significantly close to the actual rainfall except for the year 2005. For the outliers like 1693.0 mm, in the year of 2005, in the data series, ANN is unable to predict the ARF accurately. Predicted Annual Rainfall (ARF) of the year 2002 is 520.9 mm and actual 479.2 mm. Difference between these two ARF is 41.7 mm only. RMSE and PAE are computed for the four years (Table 4.2). Predicted Annual Rainfall (ARF) of the year 2005 is 1319.5 mm. Actual Annual Rainfall (ARF)

Table 4.1 Details of the parameter values used in ANN training.

Sr. No.	Predicting year	Number of epochs used	Learning rate	Momentum	No. of neurons	Error goal
1	2002	22230	0.001	0.3	147	0.0007
2	2003	237540	0.001	0.5	147	0.00001
3	2004	103684	0.001	0.3	147	0.00008
4	2005	18362	0.001	0.5	160	0.00089
5	2006	5942	0.001	0.5	147	0.0075

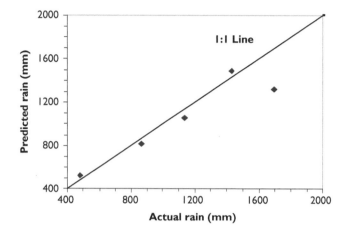

Plate 4.2 Actual vs. predicted rain for different years by ANN. (See *Colour plate 4.2*)

Table 4.2 Details of the ANN results.

Sr. No.	Predicting year (mm)	Predicted ARF I (mm)	Actual ARF II (mm)	Difference | I − II | (mm)	RMSE	PAE (%)
1	2002	0520.9	0479.2	41.7		
2	2003	1054.0	1135.4	081.4		
3	2004	0816.4	0866.0	049.6	63.01	6.5
4	2005	*1319.5	1693.0	*373.5		
5	2006	1485.1	1428.7	56.4		

*This value is not included in the computation of RMSE.

is 1693.0 mm, which is very high. Difference between these two ARF is 373.5 mm. This year's ARF is an outlier and therefore, ANN method can not predict this accurately. From the predicted ARF found RMSE is 63.01 and PAE is 6.5%. For the year 2006, predicted ARF is 1485.1mm against the actual ARF 1428.7 mm.

4.2.4 Sea surface temperatures

Sea Surface Temperatures (SSTs) measured from AVHRR (Advanced Very High Resolution Rasdiometer) and MODIS (Moderate Resolution Imaging Spectrometer) satellite systems have proved most useful as climate predictors, by making it possible to predict the temporal and spatial distribution of rainfall on adjacent land. The three primary modes of Pacific sea surface temperature (SST) variability – the El Niño–Southern Oscillation (ENSO), the Pacific decadal oscillation, and the North Pacific mode – have proved useful in predicting the U.S. warm season hydroclimate in terms of precipitation, drought and stream flow (Barlow *et al.*, 2001). Rajagopalan *et al.*, (2005) developed a scheme of water resources management by identifying large scale climate predictors via climate diagnostics, and then use the identified predictors to generate ensemble of streamflow forecast. On the basis of the Sea Surface Temperatures, other atmospheric phenomena, and statistical analysis of climate data, the Meteorological Department of South Africa has started providing as a service customized high-resolution hydroclimatic calendars for large farms, resorts, ranches, wildlife parks, etc.

References

Barlow, M., S. Nigam and E.H. Berberg, (2001) ENSO, Pacific Decadal variability, and U.S. Summer time precipitation, drought and stream flow, *J. Climate*, **14** (9), 2105–2128.

Goswami, P., G. Crasta, V. Sreekanth, K.V. Shobha and K.M. Naik. (1999) Experimental forecast of all India monthly and summer monsoon rainfall using neural network. *Curr. Sci.*, **38** (76): 1481–1483.

Hall, M.J. and A.W. Minns. (1993) Rainfall – runoff modeling as a problem in artificial intelligence experience with a neural network. Proc. 4th National Hydrology Symposium, *British Hydrological Soc. Credit.* pp. 5.51–5.57.

Hsu, K.L., H.V. Gupta and S. Sorooshian. (1993) Artificial Neural Network modeling of the rainfall-runoff process. *Water Resources Research*, **29** (4):1185–1194.

Kingston,G., M. Lambert and H. Maier. (2003) *Development of stochastic artificial neural networks for hydrological prediction.* Center for Applied Modelling in Water Engineering, University of Adelaide, Australia.

Kulshrestha, M.S.(2006) Prediction of weather parameters using harmonic analysis and artificial neural network. Ph. D. thesis (Applied Mathematics) submitted to M S University of Baroda, Vadodara. P. 1–295 (unpublished).

Rajagoplan, B., S. Regonda, K. Grantz, M. Clark, and E. Zagona. (2005) Ensemble Streamflow Forecasting: Methods and Applications. In *Advances in Water Science Methodologies* (Ed. U. Aswathanarayana), p. 97–116. Leiden, The Netherlands: A.A. Balkema.

Varshneya, M.C. 2007. Climate and Weather Forecast for Agriculture. A lecture delivered at Indian Science Congress held at Chidambaram (Tamilnadu) on 06-01-2007, published at *www.aau.in* website, pp. 1–18.

Chapter 5

How to use soil microorganisms to optimize soil productivity

Bhavdish N. Johri & Devendra K. Choudhary
Department of Biotechnology, Barkatullah University, Bhopal, India

Subhendu Chaudhuri
Bidhan Chandra Krishi Vishwavidyalaya, Kalyani, West Bengal, India

5.1 Introduction

The purpose of the Chapter is to elucidate the ways and means of using the soil microorganisms to optimize the productivity of the soil, through an understanding of the soil microbial processes related to Below Ground Bio Diversity (BGBD).

Soil, the complex geomorphological unit over the earth's crust that sustains the biosphere, contains a largely unknown microbial universe. With 10^{16} prokaryotes in a tonne of soil compared to mere 10^{11} stars in the galaxy, soil has been described as the 'final frontier'. The soil or the belowground microbial communities are central to most of the planet's geochemical cycles, agriculture and plant production and a host of other functions that in essence sustain the diverse aboveground life forms. A microbial community of consequence can harbour anywhere between 10^{10} to 10^{17} bacteria, representing more than 10^7 different taxonomic entities and near countless number of functional groups.

While the microbiological processes occurring in the top few centimeters of the earth's surface determine the existence of life, the opacity of soil has traditionally precluded knowing the functional attributes of many resident communities that make up the belowground biodiversity (BGBD). The scenario has recently changed since one can now see the opaque material through lasers and can study the communities *in situ* utilizing the tools of 'new biology'.

5.2 Soil microbial diversity and land-soil productivity

The diverse 'process-functional' relations of the of terrestrial ecosystems flow from the diverse 'population-community' structures of soil microorganisms. As drivers and managers of natural soil processes the community structure of belowground microorganisms or the biodiversity provide for the biological basis of sustainable aboveground production. The important productivity sustaining natural processes that are linked to and are derived from the functions of BGBD are the following.

- Nutrient cycling
- Regulation of soil organic matter dynamics and sequestration of atmospheric carbon
- Modification of soil physical structure and water regimes

- Enhancement of nutrient acquisition by rhizospheric and endophytic association – the mutualists (mycorrhizas, rhizobia), P-solubilizers etc.
- Enhancement of plant health by interacting with pathogens and pests, predators and parasites
- Promotion of abiotic stress tolerance, bioaccumulation and bioremediation of xenobiotics
- Enhancement of plant defense through induced systemic resistance and other mechanisms

5.2.1 The BGBD – AGBD sustainability link

According to the tenets of ecology or the ecological code a biodiversity driven functionally stable ecosystem represents a sustainable production system. In this milieu, BGBD acting as an integrator of ecosystem process-function relations help maintaining plant biodiversity – structure, composition, productivity and stability of terrestrial ecosystems. In the context of sustainability of agroecosystems it is thus imperative to add value to the BGBD that sustains land-soil productivity. The illustration given below brings the point into further focus.

5.3 Sustaining biological basis of agriculture – the case of Indo-Gangetic Plains

'The Indo-Gangetic Plains (IGP) of India with about 13% of geographical area coverage produces about 50% of food grains for 40% of population of the country. There has been a general decline in soil fertility throughout the IGP, soil organic matter content has gone down and natural soil chemical degradation processes have set in. Wheat or rice productivity remains static or tends to decline even after proper fertilization and water regime. *The biological component of these soils is gradually eroding resulting in reduced productive efficiency of the inputs with the resultant fall in near-time productivity and portents of far-time productivity loss'.*

The valid question that arises from the above is 'can we sustain the biological basis of agriculture, i.e., land-soil productivity vis a vis crop production'? To answer some of the biophysical tenets of sustainability of such and other stressed agroecosystems answers to the following questions within the stated framework of process-function and population-community relations of BGBD become important.

- What are the links between crop productivity and the BGBD driven land-soil productivity?
- How does BGBD work to sustain productivity in spatially and temporally variable production systems?
- What are the methodologies to assess and evaluate the biological basis in its exact contributory dimensions in different agroecosystems?
- What interventions can help to improve the biological basis?

5.4 Rhizosphere, the hotspot of belowground biodiversity – aboveground productivity link

Root exudates, sloughed-off mucilage, dead cells, lysates etc. are responsible for changed composition of the soil biota in the root-soil interface – the rhizosphere.

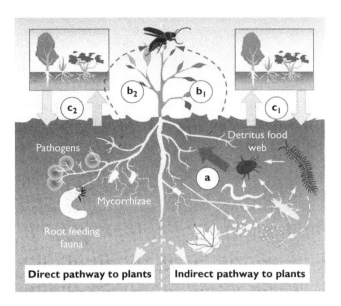

Figure 5.1 Soil-rhizosphere interplay. Aboveground communities are affected by both direct and indirect consequences of soil food web organisms.
From Wardle et al. (2004) 'Ecological linkages between aboveground and belowground biota'. Copyright Science (AAAS, USA). Reproduction with permission.

The hallmark of rhizosphere lies in the competitive and non-competitive interactions between the constituent microbial communities on one hand and the biophysical and biochemical interactions between the roots and the biotic communities (Fig. 5.1) on the other. Liberation of no less than 1000 different low molecular weight metabolites from the plant roots through exocytosis, diffusion, leaching, wounding, ion fluxes and exudation leads to aggregation of soil particles, chelation of lowly available metal ions, allelopathic influences on neighbouring roots, antagonism through antimicrobials of various kinds, formation of biofilms, and processes leading to detoxification of undesirable molecules etc. On account of this, the rhizosphere is considered as a hot spot of belowground biodiversity in so far as it relates to above ground productivity (Bais et al., 2006).

Since the major functional attributes of BGBD are better expressed in the rhizosphere than in bulk soil, extensive researches of rhizosphere biodiversity of crops and other economic plants have been undertaken in different agroecologies. Modern molecular tools such as SSCP, DGGE, T-RFLP have permitted the study of community level changes in the rhizosphere of crop plants grown with normal management practices as also under stressed conditions. This has permitted preparation of very accurate rhizosphre microbial fingerprints of several crops, particularly with reference to beneficial communities comprising of arbuscular mycorrhizal fungi, rhizobia, plant growth promoting pseudomonads, bacilli and others and has resulted in search of effective microbial gene pool for rhizosphere manipulation. Use of species-specific probes coupled to DNA sequencing and now, microarrays permits functional assessment of the rhizosphere activity under *in situ* conditions. This has uplifted rhizosphere biology from an empirical science to exactitude and the same is now at the heart of attempts for biological management of production systems for sustainability.

5.5 Dissecting the rhizosphere micro-habitat for functional biodiversity

5.5.1 The positive and negative roles of the plant communities

The rhizosphere is the connecting link between the plant and soil microbial diversity at functional level. Rhizosphere microhabitats are dominated by a variety of microbial communities, which by their presence and intimate associations with the roots introduce many different changed characters that make the habitat a functional zone of the plant's physiology. The community members occupy different trophic hierarchies by developing a variety of saprotrophic and biotrophic relations with the roots that could be detrimental or beneficial for the plant. Detrimental relations are formed by the specific groups of pathogenic bacteria and fungi, nematodes and protozoa, and the not so specific, non-pathogenic but deleterious microorganisms that are damaging for the roots. The detrimental rhizospheric microorganisms form a separate domain of study of soil microbial diversity vis-à-vis plant pathogenesis. The beneficial microorganisms include microbial groups that are able to promote plant growth and health, performance and yield by a variety of nutritional and non-nutritional mechanisms. These include the broad functional groups like the (i) plant growth promoting rhizosphere microorganisms (PGPRMs), (ii) decomposers of organic detritus, and (iii) fungal and bacterial antagonists of root pathogens. Needless to say that there are overlaps in functions of these somewhat arbitrary and symptomatic groups whereby some of the PGPRMs may occupy many different functions (Barea et al., 2005).

The many different rhizobacteria and rhizofungi that solubilize insoluble or sparingly soluble inorganic phosphate complexes by releasing organic acids are examples of PGPRMs preferentially colonizing the rhizosphere and benefiting plant by mobilizing scarcely available nutrients. Such nutrient transformation for greater plant availability is but one mechanism behind plant growth promotion by the passive colonizers of the soil-root interface. Many bacterial species strains colonize the rhizoplane of desert and rock-weathering plants and release minerals (P, K, Mg, Cu, Zn etc) for the plant's benefit (Puente et al., 2004). The asymbiotic root associative nitrogen fixers – Azosprillum and the like, that are effective mobilizers of nitrogen for the plants, are examples of growth promoting microorganisms that intimately associate with the roots. Many of these bacteria and fungi can become endophytic and help in plant growth promotion by varieties of ways including nitrogen fixation, nutrient transport etc. The symbiotic N_2 – fixing rhizobia and the mycorrhizal fungi are well-known examples of the colonizers of the rhizosphere that promote plant growth by endophytic association. More common among the PGPRMs are the plant growth promoting rhizobacteria (PGPR).

5.5.2 Biological nitrogen fixation

Diverse groups of free-living soil bacteria can fix nitrogen using soil carbon as the energy source. Major portion of the fixed nitrogen adds to the soil available pool and in time becomes available to the plant and other microorganisms. Rhizobium is the foremost genus of eubacteria that acts as the primary symbiotic fixer of N_2 with leguminous plants. These bacteria lead to the formation of root nodules where N_2 fixation takes place. *Azospirillum* spp., *Klebsiella* spp., and *Enterobacter* spp. have been reported as ubiquitous associative rhizobacteria in diverse habitats of grasses and cereal crops in both tropical and temperate regions. The design of universal nif H primer requires a high degree of DNA sequence degeneracy.

5.5.3 *Microbial allelopathy or inter-microbial antagonism in the rhizosphere*

Allelopathic potential of biocontrol microorganisms is manifest in a number of complex interactions between the plant, the pathogen, the biocontrol agent, the microbial community in and around the roots, and the physico-chemical soil environment. Plants release metabolically active cells from their roots and deposit about 20% of the photosynthetic C to the roots effecting a highly evolved and balanced relationship between the plant and specific groups of microorganisms. Events like rain and drought lead to fluctuations in salt concentration, pH, osmotic and water potential and soil particle structure. Such dynamicity of rhizosphere introduces typified interactions between the communities leading to disease and also suppression of diseases.

Some soils are naturally suppressive against some soil-borne plant pathogens e.g., *Fusarium, Gaeumannomyces, Rhizoctonia, Macrophomina, Pythium*, and *Phytophthora* due to the presence of microorganisms that are parasites, predators and non-specific antagonists. Certain biocontrol fungi such as *Trichoderma* spp suppress plant diseases by parasitizing the pathogens. Hyphal branches of the mycoparasitic fungi, such as *Trichoderma* spp. extend towards and coil round the hyphae of the pathogenic fungi. Following attachment, many lytic enzymes such as proteases, chitinases, and glucanases take part in digestion of the host cell wall. Then there are the nematophagous fungi that are predators of several plant pathogenic nematodes. A diverse group of root-associated prokaryotes and eukaryotes have antagonistic properties against root pathogens and deleterious microorganisms in the rhizosphere. *Agrobacterium, Bacillus* spp., *Streptomyces, Burkholderia*, and *Pseudomonas* spp. are effective bacterial antagonists whereas *Trichoderama* spp. dominates among the eukaryotic antagonists in the rhizosphere (Whipps, 2001).

Biological control of pathogenic and other deleterious microorganisms in soil-root interface is often attributed to antibiosis where antibiotics produced by Gram −ve and Gram +ve antagonistic bacteria play direct role in disease suppression. The highly effective disease-suppressive fluorescent pseudomonad species strains owe their suppressive potential to antibiotics produced in the rhizosphere. Phenazine derivatives were the first antibiotics implicated in biocontrol produced by fluorescent pseudomonads, such as *Pseudomonas fluorescens* and *P. aureofaciens*. Several other antibiotics and antimicrobials are produced in the rhizosphere by *P. fluorescens*, including HCN, 2, 4-DAPG, and pyoluteorin, which directly interfere with growth of various pathogens and contribute to disease suppression (Dwivedi and Johri, 2003). The Gram +ve bacilli have also received attention as potential biocontrol agents. Because of the stable endospores produced by the bacilli that survive under high soil temperature and desiccation, bacilli are preferred biocontrol agents for tropical soils. Among the fungi, *Trichoderma* and *Gliocladium* play important roles in disease suppression. Both produce antimicrobial compounds and suppress disease by different mechanisms, which include production of the structurally complex antibiotics viz., gliovirin and gliotoxin (Harman *et al.*, 2004).

5.5.4 Iron competition

Biological control also involves suppression of the pathogenic microorganisms by competition for and deprivation of nutrients. The best-studied example of this mechanism is competition for iron. Iron is present in soil in highly insoluble form and is only available to organisms at concentrations at or below 10^{-18} M in soil solution at pH 7. This poses a challenge for the bacteria, which require iron at very low concentration for growth. Many rhizobacteria, such as fluorescent pseudomonads have evolved high-affinity iron uptake system to shuttle lowly available iron into the cell. It involves various classes of siderophores, which are iron binding protein ligands that transport iron into the cell. High-affinity iron uptake system of the PGPRs that have promiscuous siderophores may interfere with iron nutrition of many pathogenic bacteria and fungi leading to their suppression (Sharma *et al.*, 2003).

5.5.5 Microbial signals molecules and the role of volatiles

The role of volatiles of microbial origin as signal molecules for plant defense has come to light recently. Plants have the ability to acquire enhanced level of resistance to pathogens after exposure to biotic stimuli provided by many different plant growth-promoting rhizobacteria (PGPR). Several PGPR e.g., *Pseudomonas* spp., *Bacillus pumilus* and *Serratia marcescens*, have the capacity to colonize root system if applied to the seeds and protect plants against foliar diseases by triggering the plant's inherent defense metabolism through their association. These in association with plant roots elicit a steady state of defense or induced systemic resistance (ISR) in plants. This is often referred to as rhizobacteria-mediated ISR. A network of interconnected signaling pathways regulates induced defenses of plants against pathogens. The primary components of the network are plant signal molecules – salicylic acid (SA), jasmonic acid (JA), ethylene (ET), and probably nitric oxide (NO). Exogneous application of these often results in higher level of plant resistance to pathogens. Signal transduction leading to ISR has been seen to be triggered by several low molecular weight volatile compounds of microbial origin in the rhizosphere that may even include any of the above molecules. It has been suggested that signal transduction leading to ISR requires responsiveness to both jasmonic acid (JA) and ethylene. Methyl jasmonate (MeJA) and the ethylene precursor i.e., ACC are effective in inducing resistance against phytopathogenic microflora. It is postulated that ethylene signaling is required at the site of application of inducers which are involved in the generation or translocation of the systematically transported ISR signals. JA and ethylene act in concert in activating defense responses. JA and derivatives induces the expression of genes encoding defense-related proteins e.g., thionins and proteinase inhibitors whereas ethylene activates several members of the pathogenesis-related (PR) gene super family. They also act synergistically in stimulating elicitor-induced PR gene expression and systematically induce defense responses.

Volatile organic compounds (VOCs) provide resistance to plants by having different mechanisms. The phenyl propanoid pathway component, SA, appears to be a critical plant messenger of pathogen recognition and disease resistance. Jasmonic acid, a product of lipoxygenase pathway is a potent regulator that mediates plant responses to mechanical damage and pathogenesis. Recently, rhizobacterial volatile compounds have been observed to serve as agents for triggering growth promotion in *Arabidopsis*. In particular, the volatile components 2,3-butanediol and acetoin were released exclusively from two bacterial strains that triggered high level of growth promotion. Furthermore, application of 2,3-butanediol enhanced plant growth whereas bacterial mutants blocked in 2,3-butanediol and acetoin synthesis were devoid of growth-promotion capacity. Accordingly, several genera of PGPR strains have been assessed for eliciting ISR and growth promotion by volatiles under *in vitro* conditions (Ryu *et al.*, 2004).

VOCs occur in the biosphere over a range of concentrations and many are microbial in origin. Many chemically mediated interactions have been reported in the biosphere, e.g., in the insect world, between plants and mammals. The compounds involved in these interactions are termed 'infochemicals'. Frequently, changes in microbial process rates cannot be explained by corresponding changes in inputs and the environment. It is possible that such phenomena result from infochemical mediated interactions in the microbial facet of the biosphere; VOCs are ideal candidates for this role (Wheately, 2002).

5.5.6 *Microbial cooperation in the rhizosphere*

Soil microbial populations are immersed in a framework of interactions to affect plant fitness and soil health quality. Cooperative inter-community interaction between the different types of microorganisms in the rhizosphere is a facet of rhizosphere biodiversity that seems to have profound functional implications. Direct cooperation between the members of different microbial communities contributes to the promotion of key processes benefiting plant growth and health. The two most well documented cases of cooperation include, (i) the co-operation between PGPR and *Rhizobium* for improved N_2-fixation, and (ii) interaction between rhizobacteria and AM fungi to establish a functional myco-rhizosphere.

(i) *PGPR-Rhizobium co-operation to improve N_2 fixation*

During root colonization, interaction takes place between the rhizobia and PGPR in the root-soil interface. Several PGPR improve nodulation and N_2 fixation in legume plants. Results of several studies applying ^{15}N-based techniques under field conditions reinforce evidences of such beneficial co-operative effects between the PGPR and rhizobia. Several *Pseudomonas* strains increase nodule number and acetylene reduction in soybean plants inoculated with *Bradyrhizobium japonicum*. Enhanced nodule formation is one among the many co-inoculation benefits, which seems to arise from production of plant hormones by the PGPR. Inoculation of phosphate solubilizing *Pseudomonas* enhances nodulation and N_2 fixation with parallel increase in P-content in alfalfa plant tissues. PGPR isolated from a Cd-contaminated soil increased nodulation of clover growing in the same soil. Cd uptake by the plants and rhizobia was reduced, preventing Cd toxicity and enabling nodulation as Cd accumulation by the PGPR reduced its concentration in soil solution. An increase in soil enzymatic activities (e.g., phosphatase, β-glucosidase, and dehydrogenase) and of auxin production around PGPR inoculated roots may also be involved in the PGPR effect on nodulation (Vivas *et al.*, 2005).

(ii)   Cooperation between AM fungi and rhizosphere bacteria to establish a functional mycorrhizosphere

Many species of rhizosphere bacteria are known to stimulate mycelial growth of the mycorrhizal fungi and enhance mycorrhiza formation. These are known to produce compounds that increase the rates and quality of root exudation stimulating morphogenetic development of AM fungal mycelia in the rhizosphere that facilitates root penetration. Hormones and microbial metabolites produced by soil microflora may also affect AM establishment (Gaur *et al.*, 2004). The rhizosphere of the mycorrhizal plants, termed mycorrhizosphere, may functionally differ from that of the non-mycorrhizal plants. Some AM fungi have established a specific type of symbiosis with endosymbiotic bacterium, *Burkholderia*, recently reassigned to a new taxon named candidatus *Glomeribacter gigasporarum*. These bacteria have specific metabolic genes that influence AM function. AM mycelium releases energy-rich organic compounds which reflect increased growth and activity of microbial saprophytes in the mycorrhizosphere. The establishment of PGPR inoculants in the rhizosphere is affected by AM fungal co-inoculation where AM inoculation improves the establishment of both inoculated and indigenous phosphate-solubilizing rhizobacteria (Bianciotto and Bonfante, 2002).

5.5.7   Rhizosphere bacterial signaling or quorum sensing

It has been widely recognized that bacteria do not live in the rhizosphere as isolated entities but instead exist as communities which exploit elaborate system of intercellular communication to facilitate their adaptation to changing environmental conditions. This intercellular communication is called quorum sensing (QS). Quorum sensing system employs N-acyl-homoserine lactones (Acyl HLs) as the signaling molecule. It has now been evident that acyl HLs produced by a wide variety of Gram +ve and Gram −ve rhizospheric bacteria control a diverse range of cell density dependent reactions (Sharma *et al.*, 2003).

Many important plant pathogens use QS to regulate virulence. The disruption of signaling system prevents the pathogens from responding to the signal and thereby prevents the expression of virulence factors. Multitrophic interactions mediate the ability of fungal pathogens to cause plant disease and the ability of bacterial antagonists to suppress disease. Antibiotic production (e.g., DAPG, Phenazines, Pyoluteorin, and Pyrollnitrine) by the antagonists is modulated by abiotic and host plant factors. Pyoluteorin production by *P. fluorescens* in the rhizosphere influences gene expression by neighbouring bacterial cells and this can be considered as signaling molecule influencing the expression of traits that contribute to biological control by *Pseudomonas* spp. in the rhizosphere. Acyl-HL influences the expression of phenazine biosynthesis gene by *P. fluorescens* CHA0, whereas siderophores influence the expression of pyoverdine biosynthesis gene. These examples provide evidence that the expression of biocontrol traits by rhizobacteria is influenced profoundly by signal molecules produced by their co-inhabitants in the rhizosphere.

5.6   Soil microbial diversity and agro-ecosystem management for sustainability

The paradigm of sustainable agriculture rests on a large interconnected conceptual framework of biophysical, social and economic processes of which maintenance of the biological basis of land-soil productivity is an integral part. The ecological tenets of structure-function relationship of ecosystems as empirically applied to soil microorganisms and discussed above would show that (i) maintenance of a diverse pool of functional microbial community is essential for land-soil sustainability, and (ii) the loss of belowground biodiversity makes the production systems more vulnerable and less resilient to environmental stresses, such as those introduced by intensive soil and crop management practices.

Soil biodiversity management for sustainable productivity is attempted at three different scales:

1. Keystone biota level management – biological nitrogen fixation, mycorrhizas, earthworms etc.
2. Soil level management – organic matter inputs, balanced mineral fertilizers and amendments, tillage and irrigation
3. Cropping system level management – cropping system design in space and time, choice crops and varieties, genetic manipulation, microsysmbioses, rhizosphere microbial dynamic etc.

5.6.1   Keystone indicators of soil biological health and their management

Simple positive relationships exist between soil biological quality, soil functions and productivity in undisturbed natural ecosystems. Recent experimental evidence from biologically driven alternative agricultural production systems have provided empirical information about changes and improvements in soil biological processes and quality traits. In this context, it is important to delineate keystone BGBD markers that satisfy reliability and reproducibility. Against the current state of our knowledge it is difficult to envisage that any single indicator will be valid for the different production systems. Instead, system and function specific biotic indicators which satisfy the stringent requirements of functionality, reliability and reproducibility will help advance the cause of BGBD search for sustainable management of land soil productivity.

Keystone biota level management aims at optimizing the goods and services provided by the diverse groups of soil organisms independently and by their interactions. BNF is a dominant feature of the terrestrial ecosystems. Considering the low turn over – highly mineralized state of nitrogen in tropical soils, BNF holds high promise for managing nitrogen supply to crops in sustainable manner. The mycorrhizal fungi function as one main carbon sink in soil with direct links to the atmosphere. The biodiversity of the ecto- and arbuscular mycorrhizal fungi as they are distributed in climatically determined ecological zones satisfy the prime requirement of a biological indicator of soil sustainability – universality of distribution and direct link with the plants. In the context of search for function specific biotic indicators it is imperative to enquire whether the biodiversity of mycorrhizal fungi can be used as indicators of carbon cycling and the interconnected processes vis a vis elevated carbon dioxide in the atmosphere.

Arbuscular mycorhizal fungi (AMF) is one keystone biotic component of the nitrogen mineralized, low phosphorus containing tropical soils. With mycorrhizas, C:N and C:P ratios decline to slow down carbon loss by microbial oxidation. The residual 'slow' decomposing organic carbon helps formation of water stable aggregates to improve the soil physical properties. Glomalin, the extrametrical mycelial protein of the AM fungi is seen to play a significant role in this context. Sharing of the common genetic pathways of symbiosis development between AM and *Rhizobium* nodulation in legumes, as has become evident now, will show that there is a hidden link between the two processes than that is superficially noticed by the synergistic or cooperative actions between the two (Sastry *et al.*, 2000). As a keystone biotic component of tropical soils AM diversity can offer avenues for sustainable management.

The role played by earthworms in promoting physical structure and chemical properties of soils as well as other indirect effects, such as promoting nodulation, dispersal of AM fungal propagules, suppression of diseases etc make them key species groups of soil biodiversity management for sustainability.

5.6.2 Soil level management

The nexus between system productivity and soil organic matter (SOM, especially the more labile part of it, through aggradations of soil fertility and soil physical structure underwrites the SOM-BGBD and soil sustainability linkage. The labile fraction of soil organic matter in the rhizosphere and in the rhizoplane serves as ready source of energy rich carbon compounds necessary to support a high level of microbial diversity and activity to drive the natural soil processes that sustain nutrition, growth and health of plants. The less labile SOM fraction together with the polysaccharides and other fractions from the roots and microbes (e.g. glomalin of the extrametrical AM mycelium) provide the cementing material for soil aggregation.

Management of soil organic matter is clearly a key factor for long-term maintenance and improvement of agricultural productivity as well as restoration of degraded agricultural lands. Along with off-site incorporation of plant residues, on-site management of crop and fallow vegetation and soil biology using naturally occurring microorganisms and crop mixtures offer short term opportunities. Plant breeding and modern biotechnology can also offer the scope for managing plant microbe interactions to improve soil organic matter. Balanced mineral nutrients and amendments imply, reduced and conservation tillage, improved water management and such other soil management practices that are incidental to maintenance of rich soil biodiversity. High phosphate fertilization is known to select less efficient AM mutualists that fail to transfer phosphorus to plants (Sharma *et al.*, 1999). Tillage affects the fate of organic C and nitrogen in soil. Decay of crop residues is delayed under reduced tillage making slow steady turn over of residue nitrogen to the soil. Such management examples for biological sustenance of crop production are many and together with SOM offer scope for soil level management of belowground biodiversity for sustaining crop production.

5.6.3 *Cropping system level management*

Cropping system level management of soil biodiversity for sustainability has many overt and covert dimensions where indigenous technological knowledge has come to be of great value. Crops lose their AM association in soils intercropped with non-host crops, sequenced with long bare fallows. In continuous cropping monoculture, AM species diversity may shift from those that are good mutualists to those that are good as competitive colonizers but give less benefit to hosts. Polyculture of two or more crops in row, alley or intercropping in close proximity provide a mixture of hosts with differing AM species relations to maintain more effective community of AM mutualists. The use of AM building hosts in the pre-cropping and fallow cover periods help selecting effective AM mutualists and active inocula. Fertlizers, especially at high doses, as are used in production agriculture have negative selection effect on the AMF, nitrogen fixing organisms etc.

A variety or genotype specific differences in root associative BNF potential, arbuscular mycorrhiza, *Rhizobium* nodulation are well known and can be exploited for achieving larger benefit from the existing soil community. Choice of crops in sequence and varieties in combination over space and time go a long way to conserve soil biodiversity for sustainable production. Crop genotype specific differences in supporting soil biological processes have opened up a new area of plant breeding and biotechnology for exploiting soil biology for production management of crops and cropping systems in sustainable manner. The genetic pathway of micro-symbioses getting known fast offer scope for manipulation of crops and varieties for increased mycorrhiza or nodulation response, much in the same way as breeding varieties for resistance to pests and diseases. Modern plant breeding by making selection under high nutrient and water supply has largely dispensed with the plants ability to fetch for itself through a large efficient pool of microorganisms at the root-soil interface, the rhizosphere. Rhizosphere biotechnology in all its qualitative and quantitative dimensions now offers chances of rehabilitating the microorganisms.

References

Bais, H.P., Weir, T.L., Perry, L.G., Gilroy, S., and Vivanco, J.M. (2006) The role of root exudates in rhizosphere interactions with plants and other organisms. *Annu. Rev. Plant Biol.* **57**, 233–266.

Barea, J-M., Pozo, M.J., Azćon, R., and Azćon-Aguilar, C. (2005) Microbial co-operation in the rhizosphere. *J. Exp. Bot.* **56**, 1761–1778.

Bianciotto, V., and Bonfante, P. (2002) Arbuscular mycorrhizal fungi: a specialized niche for rhizospheric and endocellular bacteria. Ant. Van Leeu., *Int J. Gen. and Mol. Microbiol.* **81**, 365–371.

Dwivedi, D., and Johri, B.N. (2003) Antifungals from fluorescent pseudomonads: Biosynthesis and Regulation. *Curr. Sci.* **85**, 1693–1703.

Gaur, R., Shani, N., Kawaljeet, Johri, B.N., Rossi, P., and Aragno, M. (2004) Diacetylphloroglucinaol-producing pseudomonads do not influence AM fungi in wheat rhizosphere. *Curr. Sci.* **86**, 453–457.

Harman, G.E., Howell, C.R., Viterbo, A., Chet, I., and Lorito, M. (2004) Trichoderma species-opportunistic, avirulent plant symbionts. *Nature Rev. Microbiol.* **2**, 43–56.

Puente, M.E., Bashan, Y., Li, C.Y., and Lebsky, V.K. (2004) Microbial populations and activities in the rhizosphere of rock-weathering desert plants. I. Root colonization and weathering of igneous rocks. *Plant Biol.* **6**, 629–642.

Ryu, C.M., Farag, M.A., Hu, C-H., Reddy, M.S., Kloepper, J.W., and Pare, P.W. (2004) Bacterial volatiles induce systemic resistance in *Arabidopsis*. *Plant Physiol.* **134**, 1017–1026.

Sastry, M.S.R., Sharma, A.K., and Johri, B.N. (2000) Effect of an AM fungal consortium and Pseudomonas on the growth and nutrient uptake of *Eucalyptus* hybrid. *Mycorrhiza,* **10**, 55–61.

Sharma, A., Johri, B.N., Sharma, A.K., and Glick, B.R. (2003a) Plant growth-promoting bacterium Pseudomonas sp. strain GRP$_3$ influences iron acquisition in Mung bean (Vigna radiate L. Wilzeck). *Soil Biol. Biochem.* **35**, 887–894.

Sharma, A., Sahgal, M., and Johri, B.N. (2003b) Microbial communication in the rhizosphere: Operation of quorum sensing. *Curr. Sci.* **85**, 43–48.

Sharma, A.K., Srivastava, P.C., and Johri, B.N (1999) Effect of VAM on uptake and translocation of P in *Sesbania aculeate*. *Curr. Sci.* **77**, 1351–1355.

Vivas, A., Barea, J.M., and Azćon, R. (2005) Interactive effect of *Brevibacillus brevis* and *Glomus mosseae*, both isolated from Cd-contaminated soil, on plant growth, physiological mycorrhizal fungal characteristics and soil enzymatic activities in Cd polluted soil. *Environ. Poll.* **134**, 257–266.

Wardle, D.A., Bardgett, R.D., Klironomos, J.N., Setälä, H., van der Putten, W.H., and Wall, D.H. (2004) Ecological linkages between aboveground and belowground biota. *Science.* **304**, 1629–1633.

Wheately, R.E. (2002) The consequences of volatile organic compound mediated bacterial and fungal interactions. *Ant.Van. Leeu.* **81**, 357–364.

Whipps, J.M. (2001) Microbial interactions and biocontrol in the rhizosphere. *J. Exp. Bot.* **52**, 487–511.

Chapter 6

Pathway of arsenic from water to food, West Bengal, India

D. Chandrasekharam

Department of Earth Sciences, Indian Institute of Technology Bombay, Mumbai, India

6.1 Introduction

Arseniasis (manifested in the form of skin lesions, vascular damage, cancers of the bladder, lung, liver and kidney, etc.) arises from the ingestion of excessive quantities of arsenic, through drinking water (as in Bangladesh, West Bengal (India), Inner Mongolia, Shaanxi, Xinjiang (China), Taiwan, western USA, Argentina, Chile, etc – vide comprehensive reviews by Aswathanarayana, 2001; Chappell *et al.*, 2002), diet (Norra *et al.*, 2005; Mukherjee *et al.*, 2006) and inhalation of arsenic-containing aerosols (arising from the burning of high-As coal for cooking, keeping warm and drying of grains, as in Guizhou province of China; Sun, 1999). In consonance with the theme of the volume, the chapter traces the pathway of arsenic to man through food, in relation to the content of the element in irrigation water.

The large agrarian population of West Bengal drinks groundwater with arsenic content anywhere between 0.05 and 3.7 mg/L. More than 44% of this population suffers from arsenic related diseases like conjunctivitis, melanosis, hyperkeratosis, and hyper pigmentation. In certain areas gangrene in the limb, malignant neoplasm and even skin cancer have also been observed. The worst affected are children below the age of 12 years. Recent work reveals that arsenic is geogenic and anthropogenic input is small (Chandrasekharam *et al.*, 2001; Chandrasekharam, 2005; Stüben *et al.*, 2003). Several mechanisms of arsenic release into groundwater have been documented from various parts of the world (Smedley and Kinniburgh, 2002). Examples of reducing (e.g. Bengal Basin), oxidizing (Argentina and Chile) and both reducing and oxidizing mechanisms (USA) have been well documented (Smedley and Kinniburgh, 2002). Adsorption and desorption/dissolution by iron and manganese oxides are the commonly recognized processes of arsenic release into the groundwater (Smedley and Kinniburgh, 2002; Stüben *et al.*, 2003). Besides naturally occurring redox conditions, West Bengal experiences "irrigation controlled" redox conditions, especially in areas where rice cultivation through tube-well irrigation is practiced. Irrigation practices have contributed a large quantity of arsenic in to the shallow aquifers of West Bengal.

6.2 Bioaccumulation of arsenic in food crops

It is well known that the soluble inorganic arsenicals are more toxic than the organic ones, and the trivalent forms (AsIII) are more toxic than the pentavalent ones (AsV).

Arsenic is a deadly poison in large doses to humans, but the arsenic poisoning in West Bengal and Bangladesh arose from slow, prolonged poisoning. Arsenic poisoning in mammals is less prevalent, as most of the mammals have a built-in mechanism to detoxify arsenic. They do so through the process of methylating inorganic arsenic to methylarsonic acid or dimethylarsonic acid. The methylated arsenites are less reactive, less toxic, and more readily excreted in urine. About 60–70% of arsenic is thus excreted within 48 hours (Aswathanarayana, 2001). In the case of humans, this mechanism does not appear to exist and hence arsenic gets accumulated in several parts of the body. Accumulation of arsenic in human hair (180–20340 µg/kg) and nails (380–44890 µg/kg; Mukherjee et al., 2006) in West Bengal and Bangladesh, is indicative of chronic arsenic toxicity. Urine analysis can be made use of to monitor the *ongoing* exposure, and arsenic content of nails and hair is a good indicator of *chronic* arsenic toxicity. Recent work indicates that, besides groundwater, food is a major pathway of arsenic in to human system (Norra et al., 2005, Huq and Naidu, 2005).

In West Bengal, groundwater used for irrigation contains 0.05 and 3.7 mg/L of Arsenic (Stüben et al., 2003; Norra et al., 2005). In the light of this, detailed investigations have been carried out in Malda district in West Bengal on the distribution, speciation and mobility of arsenic in the soils of paddy and wheat fields and also on the concentration of arsenic in different parts of rice (Boro and Aman variety) and wheat plants cultivated in such soils (Norra et al., 2005). The arsenic content of groundwater used for irrigation in this area varies from 519 to 782 µg/L. Boro is cultivated using groundwater from December to May while Aman is cultivated from July to December using rain water. Arsenic content in different parts of rice and wheat plants and in respective soils is given in Table 6.1 (Norra et al., 2005).

Similarly, high arsenic content in vegetables grown using groundwater with high arsenic content in West Bengal and Bangladesh have been reported by several workers (Roychowdhury et al., 2003, Huq and Naidu, 2005). Arsenic content in vegetables and cereals grown using groundwater with 85 to 108 µg/L of arsenic content in several districts in West Bengal varies from 20–21 µg/kg, and 130–179 µg/kg, respectively (Mukherjee et al., 2006). This concentration is 300% greater compared to the mean concentration generally reported in vegetables and cereals elsewhere in the world (Dabeka et al., 1993). Due to repeated irrigation with such groundwater, soils, supporting the cultivation of vegetables and cereals, also registered high arsenic content 10.7 mg/kg; (Roychowdhury et al., 2002). It has also been reported that the arsenic uptake by plants is influenced by the amount of arsenic absorbed by soils and the plant species (Huq and Naidu, 2005). For example, "*arum*" a leafy vegetable, is widely consumed by the locals as a source of vitamins, "A" and "C", and iron. Uptake of arsenic by this plant varies with in the region. In certain areas "*arum*" has recorded arsenic content as high as 138 mg/kg while in certain other areas the concentration is as low as

Table 6.1 Arsenic content in the soils of rice and wheat and in different parts of rice and wheat plants.

	As (mg/kg) in rice plant	As (mg/kg) in wheat plant
Soil	7 to 10	10 to 17
Root	169–178	0.3–0.7
Stem	6 to 7	0.4–0.7
Husk	1	
Grain	0.3	0.7

0.21 mg/kg (Huq and Naidu, 2005). Thus people consuming 100 gm of "*arum*" that contains 0.22 mg/kg of arsenic will ingest the maximum daily allowable limit of arsenic by eating this leafy vegetable alone. According to provisional tolerable intake value of arsenic set by WHO (1993), an adult male can consume 9–11 µg/kg body weight/day while an adult female can consume 11–13 µg/kg body weight/day of arsenic. Children, below 10 years, can consume 12–15 µg/kg body weight/day of arsenic. Thus consuming even small quantity of arum with 138 mg/kg of arsenic will far exceed the limit set by WHO. This gives an idea of the amount of arsenic ingested by people daily through vegetables alone. Even food cooked with arsenic contaminated groundwater showed high values (0.12–1.45 mg/kg; Huq and Naidu, 2005) which is well above the limit prescribed by WHO. Thus it is clear that bioaccumulation of arsenic (in food crops) is strongly influenced by the irrigated water, soil type, its chemical and physical characteristics, and micro-organisms (Mukherjee *et al.*, 2006). Accumulation of arsenic by food crops is a consequence of the irrigation practices adopted by the local population for long periods.

6.3   Irrigation practices in West Bengal

Increasing demand for food changed the entire system of agriculture in rural West Bengal. Nearly 72% of West Bengal population (88 million as on 2001) live in rural areas and are agrarian by profession. More than 90% of the land in West Bengal is under irrigation and paddy is widely cultivated. Since groundwater is easily accessible, agrarian community as well as the government agencies practiced tube well irrigation since 1959, by extracting large quantities of groundwater from shallow as well as deep aquifers. Arsenic content in both deep and shallow aquifers varies from 0.05 to 560 µg/L (Stüben *et al.*, 2003; Norra *et al.*, 2005; Mukherjee *et al.*, 2006). These aquifers are located within the 3.5 km thick sedimentary sequences of different geological ages (Fig 6.1; Chandrasekharam, 2006). Though, on a local scale, three prominent aquifers are identifiable (Stüben *et al.*, 2003) due to the presence of clay horizons between the aquifers, on a regional scale this distinction is not applicable due to swelling and pinching of clay layers.

This practice of tube well irrigation started in 1959 and by 1976, 20,000 tube wells were drilled providing water for cultivation through out the year. By 2001 this number has gone up to 5,50,000 bringing over 47,650 km^2 of land under rice cultivation. Farmers are able to cultivate their land through out the year and a system of multiple cropping method emerged as a lucrative farming practice. Thus the same soil is ploughed several times to accommodate different food crops (Chandrasekharam, 2005). This system of irrigated farming supported ancillary industries like animal husbandry, fertilizer and other agro-based industries and created huge employment for the rural population. Thus millions of people were benefited by this method of farming. What was not realized by the agrarian population is the dark cloud lurking behind the "feel good" life pattern. Tube well irrigation has completely eclipsed the lift irrigation system that was practiced since 1976 in West Bengal. For example in the year 1976, 20,000 shallow and deep tube wells were drilled as against 700 lift irrigation units from rivers. Subsequently after a period of ten years, the number of lift irrigation units fell to about 69 while tube well irrigation units went up by an addition of 5634 tube wells. Considering the surface drainage system of West Bengal (Fig. 6.1), the local

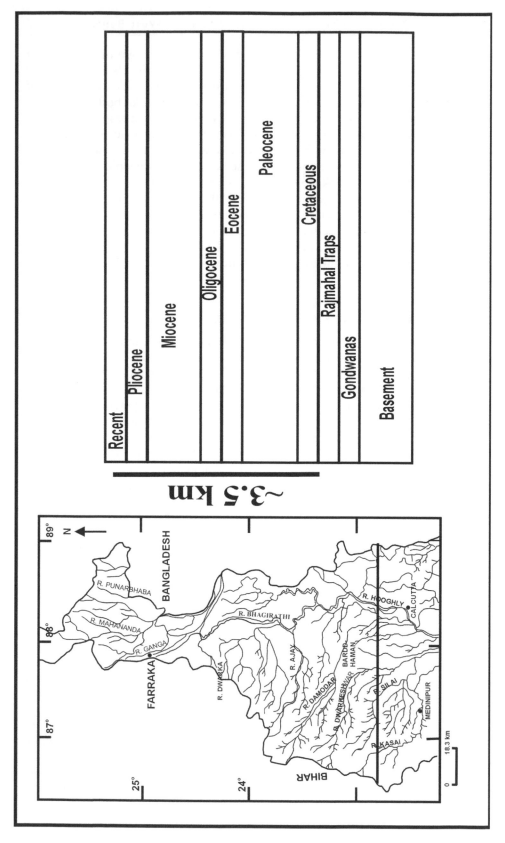

Figure 6.1 Map of West Bengal showing the drainage pattern and the sedimentary sequence across (indicated by thick line) the southern part of the basin.

government should have encouraged lift irrigation and canal irrigation schemes instead of supporting tube well irrigation (Chandrasekharam, 2005). Investigation on rice plants from an area cultivated using surface water has shown that the arsenic content in the roots is about 20 mg/kg compared to those cultivated using tube wells reported in Table 6.1.

6.4 Tube well irrigation and its impact on arsenic content

In large parts of West Bengal both deep and shallow aquifers contain arsenic beyond the limit ($>10\,\mu g/L$) prescribed for drinking water standards by WHO (Chandrasekharam et al., 2001; Stueben et al., 2003; Smedley and Kinniburgh, 2002). Because of heavy rainfall and frequent floods, arsenic content in the shallow aquifers tends to be low due to high base flow. It would be interesting to know the arsenic content in the shallow aquifers prior to 1959, the year which triggered tube well irrigation practice. However this data is not available for comparison.

West Bengal experiences floods every year due to south-west and north-east monsoons thus extending the rain fall from June to December. About 43% of the total area of 89,000 km^2 gets flooded during monsoon causing considerable loss of life, property and physical damage to more than 30% of population (total population as in 2001 is 88 million). Further, rice cultivation requires 5–10 cm standing water in the field and thus cultivable and non cultivable lands are under reducing environment through out the year.

According to Masscheleyn et al. (1991), at higher soil redox levels ($\sim$500 mV) arsenic solubility is low while at low redox levels (-200 mV) the soluble arsenic content increases thirteen-fold as compared to 500 mV. Maximum conversion of As (V) to As (III) takes place under redox potential of $+100$ mV and below. Around $+150$ mV, iron and manganese also get mobilized into the aqueous phase releasing arsenic into the solution. This process has been recognized in the field conditions also in several areas in West Bengal, where dissolution of both iron and manganese oxides are responsible for increasing the arsenic levels in groundwater (Stüben et al., 2003, Norra et al., 2005). Onken and Hossner (1996) reported a most interesting experimental finding, which is very relevant to the conditions prevailing in West Bengal. According to this experiment, soil solution under flooding conditions recorded maximum concentration of arsenic compared to the soils under non-flooding conditions. Release of soluble arsenic under flooding conditions takes place within few hours of flooding. The work reported by Masscheleyn et al. (1991) and Onken and Hossner (1996) suggests a strong concordance between experimental work and flooding of rice fields in West Bengal.

Indian farmers plough the roots of the rice plants back into the soil after harvesting. This practice is prevalent in the Indian agrarian community for centuries. Due to change in redox conditions during the subsequent crops, the arsenic in rice roots gets mobilized and infiltrates in to the shallow aquifers. Thus arsenic from groundwater, after entering the food chain, gets concentrated in the roots and enters the shallow aquifer thus establishing an *"arsenic flow cycle"* in the rice fields. If this practice continues, the probability of contaminating the entire shallow (and deeper aquifers) aquifers is very high in the very near future and entire West Bengal population may not get safe drinking water from the shallow as well as deep aquifers.

6.5 Arsenic removal techniques for domestic water

Over the past two decades, research laboratories across the world developed arsenic removal techniques to provide safe drinking water to the rural population. Oxidation, coagulation and ion exchange process are some of the methods commonly recommended to remove arsenic from drinking water (Driehaus, 2005). In oxidation technique, As(III) is oxidized to As (V) by using oxidants like potassium permanganate and Fenton's reagent. Arsenic precipitate is removed by filtration. Alumina and Fe^{2+} salts are used to remove arsenic through coagulation process. This is most frequently applied for arsenic removal at pH below 7.5. In the case of ion exchange method, a resin with chelating groups saturated with ferric ions is applied for removal of As (V) and As (III). Both redox forms effectively removed in optimum pH level (pH 3–6 for As (V) and pH 8–9 for As (III)). Though such methods are very effective in obtaining arsenic-free drinking water, they are not widely practiced due to several socio-economic impediments. Rooftop rainwater harvesting for drinking water is a sensible way to get over the problem.

6.6 Arsenic removal techniques for irrigation water

Rott and Friedle (1999) proposed a novel technique for the subterranean removal of arsenic from groundwater *in situ*. Oxygen-enriched water is injected under pressure into the aquifer. The introduced oxygen activates the autotrophic micro-organisms, oxidizes As (III) to As (V), and facilitates the precipitation of arsenic and its absorption on iron hydroxide and manganese oxide. Field experiments validate the technical viability of the method, but the socio-economic viability of this technique is yet to be established.

Perhaps the best way to solve this problem of arsenic is to use surface water resources, especially in regions like West Bengal that has excellent drainage net-work (Fig. 6.1).

6.7 Remedial solution

The government of West Bengal realized the effect of irrigation on the quality of groundwater and is advocating rainwater harvesting to improve the quality and quantity of groundwater in the state. The Central Groundwater Board (CGWB) has been entrusted with this task for three districts. From the drainage net work (Fig. 6.1), it is apparent that the entire Bengal Basin has sufficient surface water to recharge the aquifers. How far CGWB will succeed in reducing the incidence of arsenic in groundwater through rainwater harvesting is matter to be examined over a period time. Perhaps adopting alternate methods of irrigation may be a viable solution to control arsenic levels in the shallow aquifers as well as in the food crops.

According to Indian Council of Agricultural Research (ICAR, 1992), West Bengal can be divided in to six major agro-ecological sub-regions. The four most important sub-regions are: 1. Hot dry sub-humid; 2. Hot moist sub-humid (Bengal Basin); 3. Hot per-humid; 4. Hot moist sub-humid (Gangetic delta).

The hot moist sub-humid regions (e.g. Baharampur in Murshidabad district) have different type of soil and experiences flooding while hot dry sub-humid regions (e.g. Ahmadpur in Birbhum district) are free from flooding with soil type different from the former region (Chandrasekharam, 2005). The most interesting fact is that in the hot moist sub-humid regions, agriculture is based on irrigation and crops are grown through-out the year while in the hot dry sub-humid regions rain/canal-fed cultivation is adopted. In both these regions rice is cultivated extensively. As described above, arsenic accumulation in rice roots in canal/rain cultivated areas is far less compared to the areas where tube wells are used for irrigation. The arsenic content in the rivers draining the Bengal Basin is less than 1.9 μg/L.

Thus in general, groundwater from deep and shallow aquifers is extensively used in central and eastern regions while surface water irrigation is common along the western region. Though the mitigation of the reducing conditions due to floods is not possible, recycling of arsenic-containing groundwater for irrigation can be avoided. Since rice is cultivated through rain/canal irrigation in large part of West Bengal, attempts should be made to promote similar irrigation culture in the moist sub-humid regions of West Bengal as well. Considering the drainage system in the Bengal Basin (Fig. 6.1), this is possible by creating a net work of surface irrigation canal system through out the Bengal basin. This will provide arsenic- free water to the rice crop and prevents accumulation of arsenic in root zone of the plants. Both groundwater in the shallow aquifers and the soil solutions can thus be made free from arsenic. The local government should take initiative to educate the farmers of the advantages of such system for the benefit of the large population. In the long run this will provide safe drinking water as well arsenic free food to the population of West Bengal.

References

Aswathanarayana, U. (2001). *Water Resources Management and the Environment*, Lisse (The Netherlands), A.A. Balkema, p. 314–328.

Chappell, W.R., Abernathy, C.O., Calderon, R.L. and Thomas, D.J. 2002. (Eds). Proceedings, 5th Conference on arsenic exposure and health effects, July 14–18, 2002, San Diego, California, Elsevier Pub. 400 p.

Chandrasekharam, D., Julie Karmakar., Berner, Z. and Stueben, D. (2001). Arsenic contamination in groundwater, Murshidabad district, West Bengal. *Proceed. Water-Rock Interaction 1.* (ed) A.Cidu, 1051–1058 A.A.Balkema, The Netherlands.

Chandrasekharam, D. (2005). Arsenic pollution in groundwater of West Bengal, India: where we stand. In *"Natural arsenic in groundwater: occurrence, remediation and management"* (eds) J. Bundschuh, P. Bhattacharya and D. Chandrasekharam), AA. Balkema, Taylor & Francis Group, London: 25–29.

Chandrasekharam, D. (2006). Geogenic arsenic pollution of groundwater: West Bengal. In *"Groundwater Flow and Mass Transport Modeling"* (ed) T. Vinoda Rao, Allied Pub. Pvt.Ltd., New Delhi, 2006, 132–144.

Dabeka, R.W., Mckenzie, A.D., Lacroix, G.M.A., Cleroux, C., Bowe, S., Graham, R.A., Conacher, H.B.S. and Verdier, P. (1993). Survey of arsenic in total diet food composites and estimation of the dietary intake of arsenic by Canadian adults and children. *J. AOAC Int.* 76: 14–25.

Driehaus, W. (2005). Technologies for arsenic removal from potable water. In *"Natural arsenic in groundwater: Occurrence, Remediation and Management"*, (Eds). J. Bundschuh, P. Battacharya and D. Chandrasekharam, Taylor and Francis Group Pub., London, 189–203.

Huq, S.M.I. and Ravi Naidu (2005). Arsenic in groundwater and contamination of the food chain: Bangladesh Scenario. In *"Natural arsenic in groundwater: Occurrence, Remediation and Management"*, (Eds). J. Bundschuh, P. Battacharya and D. Chandrasekharam, Taylor and Francis Group Pub., London, 95–101.

ICAR, 1992. Indian Council of Agricultural Research, Report 48.

Masscheleyn, P., Delaune, R.D. and Patrick Jr., W.H. (1991). Effect of redox potential and pH on arsenic speciation and solubility in a contaminated soil. *Environ. Sci. Tech.*, **25**: 1414–1419.

Mukherjee, A.B., Bhattacharya, P., Jacks, G. Banerjee, D.M., Ramanathan, A.L., Mahanta, C. Chandrasekharam, D. and Naidu, R. (2006). Groundwater Arsenic Contamination in India: Extent and severity. *In "Managing Arsenic in the Environment: From soil to human health"* (Eds) R. Naidu, E. Smith, G. Owens, P. Nadebaum, and P. Bhattacharya. CSIRO Publishing, Melbourne, Australia, 664 p.

Norra, S., Berner, Z.A., Agarwala, P., Wagner, F., Chandrasekharam, D. and Stüben. D. (2005). Impact of irrigation with As rich groundwater on soil and crops: a geochemical case study in Malda District, West Bengal. *App. Geochem, 20*, 1890–1906.

Onken, B.M. and Hossner, L.R. (1996). Determination of Arsenic species in soil solution under flooded conditions. *Jour. Soil Sci. Soc. Am.*, **60**: 1385–1392.

Rott, U. and Friedle, M. (1999). Subterrenean removal of arsenic from groundwater. In Chappell, W.R., Abernathy, C.O., and Calderon, R.L. (Eds.). *Arsenic Exposure and Health Effects*. Amsterdam : Elsevier, p. 389–396.

Roychowdhury, T., Tokunaga, T. & Ando, M. (2003). Survey of arsenic and other heavy metals in food composites and drinking water and estimate of dietary intake by the villagers from an arsenic-affected area of West Bengal, India. *Sci. Total Environ.* 308: 15–35.

Smedley, P.L. and Kinniburgh, D.G. (2002). A review of the source, behaviour and distribution of arsenic in natural waters. *App. Geochem.*, **17**: 517–568.

Stüben, D., Berner, Z., Chandrasekharam, D. and Julie Karmakar (2003). Arsenic pollution in groundwater of West Bengal, India: Geochemical evidences for mobilization of As under reducing conditions. *App. Geochem*, 18: 1417–1434.

WHO (1993). World Health Organization. *Guidelines for drinking water quality*, vol. 1, Recommendations, 2nd edn, WHO, Geneva.

Chapter 7

How to do with less water

U. Aswathanarayana

Mahadevan International Centre for Water Resources Management, Hyderabad, India

7.1 Introduction

There have been two significant advances, which need to be incorporated in water resources management methodologies. The first area of advance is conceptual – there is widespread recognition that it is only through synergy between the earth (including the atmospheric and oceanic realms), space and information sciences, is it possible to reduce the predictive uncertainty in hydrological sciences, and address the complex issues involved in the management of the four kinds of waters (rain water, surface water, groundwater and soil water) and generate employment in the process of utilizing them. The second advance is in the area of tools – because of the advantages of repetitive coverage and capability for synoptic overview, satellite remote sensing has emerged as a powerful and cost-effective tool covering all aspects of water resources management.

We have to develop new ways and means of using remote sensing data to underpin the study of interactions between the atmospheric, oceanic and hydrological processes, for various applications. Different water-related applications, such as water resources management, environmental monitoring, climate prediction, agriculture, preparation for and mitigation of extreme weather events, etc, are characterized by widely varying requirements of spatial, temporal, and spectral resolutions, and need different data assimilation methodologies and technology transfer practices. We have to cover various dimensions and potential growth areas of remote sensing, such as the existing and projected satellite sensors, image analysis, data assimilation methods, technology transfer modalities, agriculture-related applications (involving evapotranspiration and soil moisture), prediction of runoff and flood risk, management of water resources in watersheds through linkages with geomorphology, etc.

There is need to promote Research & Development and training in regard to ways and means by which the scope of remote sensing applications is enlarged and made commercially viable, through steps such as, launching of dedicated satellite systems, developing new retrieval algorithms for remote sensing data and formatting them for ingestion into GIS packages, interaction with stakeholders, training of cadres, public policy, etc., customized to the biophysical and socioeconomic situations in different countries.

The new concepts and tools are to be applied to the following specific water issues.

7.2 How to do with less water

In the past, when we needed water, we simply looked for it on the surface or underground, and tapped it. It is this kind of thinking that led to the destruction of the Aral Sea, and the dewatering of the Colorado Basin in Mexico. During the last 30 years, the number of tube wells in India went up from 2 million to 23 million, with the consequence that the water-table has been going down by 1–3 m per year. The water-table beneath Beijing, China, went down by about 60 m. during the last 20 years. The present situation is evidently unsustainable, and the only way to avoid the impending catastrophe, is to cut down our water use and bring about changes in the economy to enable us to use water more efficiently.

The use and reuse of water have to be based on the following considerations: (i) The highest quality water (at the rate of, say, 2–4 L per capita per day) should be reserved for drinking purposes (as different from water for other domestic uses, such as bathing, washing clothes and utensils and sanitation, which can use slightly inferior water). In many Indian homes, the traditional practice has been to keep drinking water separately from water to be used for washing purposes. It is no doubt technologically feasible but prohibitively expensive to purify contaminated water (say, municipal waste water) to a level of purity acceptable for domestic purposes (ii) Municipal waste water which contains (say) pathogens can be used in most industries, except food industries, (iii) Sewage which contains organic matter and nitrates is not only acceptable but even be desirable for use in irrigation.

The quantity of water used in industries in the industrialized countries is about 30 times more than in the developing countries.

In most industrialized countries, water use is becoming less. For instance, USA uses far less water per person, and less water in total, than it did twenty-five years ago. In some cases, water use is reduced because of scarcity, but in most cases, countries deliberately changed their economies, to enable water to be used more efficiently. In the case of Japan, during the period, 1965–89, the amount of water needed to produce a million dollars' worth of goods went down from fifty million litres to thirteen million litres. A similar pattern of decrease in per capita consumption of water has been observed in Finland, parts of Australia, much of Europe, and even Hong Kong. The Nobel Prize-winning economist Simon Kuznets is of the view that in the case of the industrialized countries, as technologies mature and efficiency improves, they become more conscious of the importance of the preservation of the quality of environment, which has the consequence of using the natural resources, including water, in a more sustainable manner.

In countries with monsoon climate (e.g. India), the rainfall is restricted to a few (typically 3–4) months of the year. It is also erratic. Instances are known where the annual precipitation in an area occurred in a matter of a few days of intense downpour. Under these circumstances, harvesting and storage of surface water is a geophysical necessity. The runoff in a watershed can be harvested both through a high dam on a river, *plus* a series of small check dams on the streams, *plus* recharge of groundwater. It should be noted that it is not a case of *either/or* but *and*. The stakeholder communities have to determine as to what combination of the three approaches is the most sensible.

About one-third of the precipitation ends up as "Blue" water in rivers, lakes and aquifers, which is presently over-used. About two-thirds of the precipitation ends up as "Green" water (soil moisture and evapotranspiration). These have great potential to increase food production.

The water storage per capita varies greatly from country to country. In USA and Australia, it is $5000 \, m^3$. In South Africa, Mexico, Morocco and China, it is $1000 \, m^3$. It is very low in India ($200 \, m^3$). There is evident need to increase the water storage capacity in India, in the form of surface water reservoirs, irrigation ponds in farms, groundwater reservoirs, etc.

The waste water problem has indeed a win-win solution. Ecological treatment of waste water has a number of merits: besides removing the pollutants effectively, it is low-cost, energy-saving, resource-recovering, easy operation and maintenance. The reuse of wastewater not only avoids contamination of the hydrological system but the treated water can be put to beneficial uses such as irrigation, industry, recreation, etc.

7.3 Integrated use of the four kinds of water: The hydroclimatic calendar

The four kinds of water (i.e. rainwater, surface water, groundwater, and soil water) are interlinked, and should be treated as a continuum.

The issues that need to be coordinated between those who generate relevant water science (e.g. hydrometeorologists, and geohydrologists,), and those who make decisions on the basis of that information (e.g. administrators, water managers) have become so complex, with several levels of uncertainty, that a new class of professional consultants have come into the picture. For instance, on the basis of the analysis of satellite-based, climate-related information.(including ENSO impacts), the Meteorological Department of South Africa offers high resolution, customized advice to companies and organizations for specific areas, on agricultural applications, reservoir control, power generation, wild life, tourism, transportation, habitat protection, etc. Thus, a water manager can get a hydroclimatic decision calendar, to enable him to plan his operations. Similarly, farmers can get customized precision agricultural advice taking into account the soil moisture, and selective application of N-fertilizer and herbicide to minimize costs, while reducing nitrogen contamination to the atmosphere and to the watershed. Display of virtual images of various possible scenarios is an effective way to sensitize the administrators, farmers and citizens to the implications of various techno-socio-economic options of water use and water-related natural hazards (such as, floods and drought).

7.4 Techno-socio-economic dimensions of water use

Since all economic activities involve the use of water in some manner, control of water supply is closely linked to political power and generation of wealth (incidentally, the Chinese character for "political order" is based on the symbol for "water," which implies that those who control water control people). No wonder, most of the water supply decisions made tend to have a political bias. Science has a great potential to help the policy makers and water managers to take informed, rational decisions on all aspects of water resources management (such as, storage, allocation, delivery, treatment, use and reuse, etc. of water), to promote economic development, ecological sustainability and social equity. Healthy economies can coexist with healthy ecosystems. This process is best accomplished through a dialogue between the scientists, policy makers and stakeholders, covering all aspects of water use and water rights.

Since many combinations of use and reuse of water are possible, simultaneously and in sequence, it is critically important to know how the various uses of water combine and interact in space and time. A knowledge of the value addition through the use of an acre-ft (a acre-ft $= 1235\,\text{m}^3$) of water for alternative uses permit informed decision about the directions in which the economy could move in order to make the best use of all its sources, as the scarcity of water increases. A critically important input in this study is the quality of water at a particular place and particular time in a watershed.

Crops cannot be grown without water, and agriculture has been and will always be the principal user of water. Irrigation accounts for about 85% of the water use in India. The increasing demand for food for the growing population necessitated the development of irrigation on a large scale, as irrigated agriculture is far more productive than rainfed agriculture. Thus, though only 15% of the lands in the world are irrigated, they account for almost half of the total crop production in the world in terms of value.

Rice is the staple food in many parts of Asia – Pacific region, and it will continue to be cultivated despite its high water demand. Irrigated rice is a highly water-intensive crop (3000–5000 L /kg of rice). In countries such as India and China, rice crop alone uses more than 50% of all the total water consumed in the country. Any attempt to reduce the quantity of water used in agriculture should address the problem of ways and means of reducing the water needs of rice. Two approaches are being attempted in China and India: (i) developing low water-need and drought-tolerant rice paddy strains, (ii) agronomic methods, such as SRI, which need less water. Compared to the water requirement of lowland flooded rice, the aerobic rice system can save about 45 per cent of water .Aerobic rice is both a concept of growing rice and appropriate genotypes suited for such a growth. In the present set-up of nominal water charges, and guaranteed price for paddy and in some states, free electricity, the farmer has no incentive to conserve water. Two administrative steps are recommended which will result in the conservation of water: (i) Irrigation water should be available on demand (just as we switch on the light in the house only when needed, unlike in the office where people switch on the lights even when not needed, just because they do not have to pay for it!), (ii) Irrigation water should be priced in proportion to the quantity used per unit area of land – a farmer irrigating one ha. of paddy should pay water-cess 8 times more than a farmer who irrigates one ha of bajra. Also, if a farmer uses lesser amount of water per unit of land, he should get water at a cheaper rate. *All this can be achieved without imposing any major economic burden on the farmer.*

A fruitful approach that is being used extensively in Madagascar and Cambodia, and is becoming popular in India is SRI – System of Rice Intensification. Irrespective of the strain of rice used, it makes do with lesser amount of water, and smaller quantity of seed, and yields a larger harvest. Though it is 25–50% more labour intensive, the expenditure against this head is more than made up by the high profitability of SRI.

"On the same amount of land that Chinese farmers grow four thousand kilograms of rice each year, Indians grow no more than sixteen hundred, and they use ten times more water to do it than is necessary" (Michael Specter, *New Yorker*, Oct. 23, 2006). Although there is much controversy about genetically modified crops, several strains of rice have been developed that require only a fraction of the water most farmers use today.

India has 20% of the world's population, and 4% of the water resources. Also, India has 30% of the farm animals in the world. Animals need large quantities of drinking water, though not necessarily of high potable quality: 40 L/d for cattle, horses and camels, 15 L/d for hogs, and 10 L/d for sheep. In South Africa, families give the wash water to the pigs, and wild animals such as impala, which need only 1.5 L/d of drinking water, and ostrich, which does not drink water at all, are raised in commercial farms for meat and eggs (fifteen omelets can be made from one ostrich egg!). It is not an accident that ostrich farms have come up in the arid areas of SW USA, and Australia. Generally, animals survive drought better than crops. Hence in chronically drought-stricken areas, there is need to promote animal husbandry by landless labour families, and cold-storage facilities for the storage and marketing of meat.

Water rights are by far the most difficult issues in water resources management, associated as they are with custom, tradition and culture. Water is a human heritage. Hinduism is the only religion in which water is worshipped and is used in worship. We have long been accustomed to thinking of water as a free good. It is no longer possible to maintain the traditional approach. Water is a human right not only for the farmers, *but all citizens*. A farmer cannot decide to grow only rice in his farm and demand irrigation water needed for his farm.

Just as primary education is a human right, but MBA education is not, drinking water alone is a human right, but not irrigation water or industrial water. This perception has profound implications for water management. A family of ten members can demand larger quantity of drinking water than a family with two members, *as a matter of right*. But a farmer with 20 acres of land cannot demand ten times more irrigation water than a farmer with two acres of land, *as a matter of right*. This is so because that according to UN Principles, water should be recognized as an economic good for all purposes, other than drinking water.

Prevention of mining of groundwater is the key for the sustainable and socially equitable groundwater management

Advances in remote sensing and information technologies have made it possible on one hand to monitor the aquifer depletion, and on the other hand, to enforce compliance with water rights. There is a clear correlation between evapotranspiration (ET) and groundwater pumpage in an area. Landsat data is processed through software called METRIC to develop seasonal evapotranspiration levels for an area, and thereby monitor the aquifer depletion. The water rights entitlement for an individual farmer is determined depending upon the area of the farm, rainfall, aquifer characteristics, etc. The compliance with water entitlement is monitored remotely for each farmer through GPS location of a well, capacity of the pump, measurement of power consumption, and well flow, and the extent of groundwater pumpage If it has been found that a particular farmer is withdrawing groundwater beyond his entitlement, he is heavily penalized, and the electricity connection to the water pump could be cut off.

Industries are using either less and less water per unit of production or switching over to dry technologies, for the simple reason that the less the quantity of water used in processing, the less would be the effluents that need to be disposed of. For instance, about 35 m^3 of water was used for refining one tonne of crude. Research has brought down the water requirement steadily. Modern refineries currently use less than 0.12 m^3/t.

Singapore which imports virtually all its water requirements sensitizes its citizens about ways and means of using less water for domestic purposes, such as bathing and washing.

7.5 Conclusions

There is little doubt that almost any water use could be accomplished with less water than currently used. Various combinations of techno-socio-economic approaches should be attempted, in combination with administrative incentives and disincentives, to achieve this.

Chapter 8

Irrigated agriculture for food security

V.V.N. Murty

Irrigation Engineering and Management, Asian Institute of Technology, Bangkok

8.1 Introduction

Based on the source of water for crop water use, agricultural production systems are broadly classified as rainfed agriculture systems wherein the entire crop water use is met from rainfall and irrigated agriculture systems wherein the available rainfall is supplemented with external water supplies. It is estimated that at present 277 million hectares i.e., about 18% of world's arable land is under irrigation and is responsible for about 40% of the crop output (International Commission on Irrigation and Drainage – ICID, 2006). Even though at present contribution of rainfed agriculture to the world's food production is more than the irrigated agriculture, there is a considerable effort to shift to irrigated agriculture where ever possible. This is for the reason that the relative immunity which irrigation offers from an erratic or inadequate rainfall. With assured water supply from irrigation farmers are able to invest in other inputs like fertilizers, improved tillage practices etc., in order to obtain higher crop yields. Figure 8.1 shows in general the yield potentials of rainfed and irrigated agricultural systems.

Agricultural crops, livestock and fisheries are the three principal sources of food supply for human needs. Agricultural crops accounts for more than three quarters of the food consumed (77.5%), with livestock (15.9%) and fishery products (6.6%) making up the balance (van Hofwegen and Svendsen, 2000). The world population is expected to increase at the rate of about 1.5% during 2005–2010, stabilizing thereafter. The

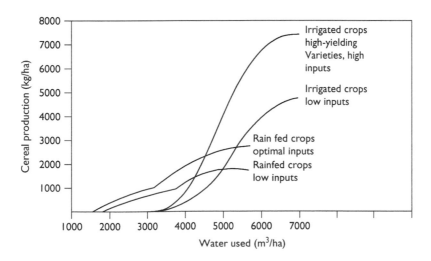

Figure 8.1 Yield potentials of rainfed and irrigated agriculture systems (adapted from Chitale, 2000).

world needs additional food production not only to meet the demands of the increasing population but also to meet the deficit food supplies to the existing population.

The role of water availability for agriculture and consequently to food security are crucial in several countries of the Asia-Pacific region and especially in India. At present, the Asia-Pacific region is the home of the world's majority population. After the Second World War, several countries in this region faced acute food shortages. To meet this situation as well as to meet the demands of the increasing population all most all countries in this region have embarked on the construction of irrigation systems. China, India, Thailand, Indonesia, Sri Lanka, Philippines are some of the countries that have seen large scale development of irrigation facilities. No doubt these irrigation systems have helped to a great extent to meet the increasing demands for food but their operation and management particularly of the larger ones have come in for considerable criticism. It is felt that the irrigation efficiencies are generally below the acceptable values, equity in water distribution is not satisfactory and irrigation systems have caused environmental concerns like public health issues and deterioration of water quality of the surface and groundwater resources. There is considerable scope for improving the agricultural productivity in the existing irrigation systems. The Food and Agriculture Organization of the U.N. (Rome) and the International Commission on Irrigation and Drainage (New Delhi) are disseminating information about the improved management of irrigated agriculture systems around the world.

In this chapter we will examine the role of the irrigated agriculture systems and suggest measures for improving their productivity keeping in view their sustainability over a period of time. The role of irrigation in increasing agricultural production has been conclusively demonstrated in several countries around the world. In the State of Andhra Pradesh in India irrigation systems have helped in bringing large semi-arid areas under irrigation and substantially improved the agricultural productivity (Government of A.P., 2003).

8.2 Irrigated agricultural systems

Irrigation water is drawn from a wide variety of sources like storage on rivers, lifting from rivers, natural lakes, constructed ponds and groundwater reservoirs. Sometimes the projects are classified as major (large), medium and minor (small) projects depending one the area they irrigate or the investments made. A large irrigation system usually consists of one or more storage reservoirs or diversion structures, or pumping plants and delivers water to the command areas through a network of canals and related structures. Water is delivered to several farmers with individual land holdings. In case of the smaller systems the source is very near the area to be irrigated. Groundwater usage is through wells and they are usually located in the area to be irrigated.

Operating and managing the irrigation systems whatever may be their size require a good knowledge base. The engineering aspects involve the development of the source, conveyance, land preparation and designing the application method. Knowledge about crop water requirements, irrigation schedules, water quality management and agronomical aspects of crop production are required for obtaining maximum productivity from the system.

The available water supplies in an irrigation system are not the same every year and the structures constructed could deteriorate with time. The system operation should be

Table 8.1 Water requirements of different crops.

Crop	Growing season	Total ETcrop (mm)
Rice	Kharif/Rabi	1100–1300
Sorghum	Kharif	400–550
	Rabi	450–600
Maize	Kharif	400–550
	Rabi	450–600
Wheat	Rabi	450–550
Finger millet	Kharif	350–450
	Rabi	400–550
Green gram and	Kharif	300–400
Black gram	Rabi	350–450
Bengal gram	Rabi	350–450
Groundnut	Kharif	300–450
	Rabi	350–650
Sunflower	Kharif	300–450
	Rabi	350–500
Cotton	Part Kharif and Rabi	350–850
Red gram	Part Kharif and Rabi or Rabi	350–800
Sugarcane	Perennial	1400–2000 (per year)
Banana	Perennial	1600–2250 (per year)
Plantation crops	Perennial	1250–1800 (per year)

(Kharif and Rabi refer to the two crop growing seasons, Viz, rainy and non-rainy seasons).

planned in such a way that the available supplies are equitably distributed, the delivery systems are properly maintained and finally the environmental hazards are kept to a minimum. The system should have plans for the years when water supplies available are below normal. The design and operation of each irrigation system is specific to the location of the project and the command area. Strictly uniform policies may not be applicable to all irrigation systems, but certain guiding principles applicable to all systems could be developed. Based on available studies and experiences, guidelines for the management of a particular irrigation system need to be developed.

8.3 Crop water requirements

Crop water requirements for each crop are basically determined using the climatic and soil parameters. There is a fair understanding of the crop water requirements for the crops grown under irrigated conditions. Doorenbos (1992) outlines procedures for determining crop water requirements under different climatic conditions. As an example, Table 8.1 indicates the crop water requirements which are used in the State of Andhra Pradesh in India (Govt. of A. P., 2003).

The crop water requirements given in Table 1 include the amount of effective rainfall even though it is not indicated clearly. The crop water requirements are based on the determination of the evapotranspiration (ET) values based on climatic parameters. Other quantities like deep percolation losses and water required for land preparations are added to the ET values. Values of crop water requirements will help in selecting the suitable cropping sequences at a given location.

8.4 Irrigation schedules

The crop water requirements provide an idea of the total water needed for the life cycle of the crop. However for maximizing crop production water has to be supplied to the crops at different growth stages depending on its physiological nature as well as soil moisture depletion allowed in the root zone. The relation between the water applied and the yield is discussed for various crops in Doorenbos and Kassam (1979). If water is not supplied or supplied in lesser amounts than required there will be a reduction in the yield of the crop. In some situations when there are water shortages, water supply may have to be reduced deliberately in a planned manner and such a practice is known as deficit irrigation. In deficit irrigation the crop yields may not reach the maximum levels but some acceptable level of yields will be obtained.

For each situation, irrigation schedules based on climate, soil and crop variety should be worked out and attempts should be made to meet the schedules with adequate amounts of water.

8.5 Methods of irrigation

In irrigation practice water can be applied below the surface, at the surface or over the surface of the soil. The terms subsurface, surface and overhead or sprinkler are used to describe these three methods of irrigation respectively. The surface irrigation methods are easy to adopt and therefore are most commonly adopted. A wide variety of surface irrigation methods like the check basin, border and furrow methods are used to suit the crop and soil requirements. These methods need to be carefully designed to suit the soil, crop and available water flows. Guide lines for using these methods should be developed for each region for use of the farmers. In adopting the surface irrigation methods land development is a prerequisite – an aspect not fully recognized.

Drip or trickle method of irrigation is a system of applying water on to or in to the soil very near the root-zone of the plants. Water is conveyed from the source through a network of pipes usually of PVC and delivered to the plants. While drip systems are suitable for a wide variety of crops, they are well suited for horticultural and root crops Drip systems have a good number of advantages like efficient use of the water, concurrent use of water and fertilizer, less weed problem, and possibility of using poor quality of irrigation water. Difficulties in adopting this system apart from the higher initial cost, are the durability of the components and blockage of the outlets and consequent improper functioning of the system.

Sprinkler irrigation is a method of applying water above the ground surface some-what resembling rainfall. The spray is obtained by the flow of water under pressure through small orifices or nozzles referred to as sprinklers. A pump is required to develop the pressure but in some situations when the source of water is high enough from the area to be irrigated, the required pressure may be developed by gravity alone. Sprinkler irrigation is suitable for areas with porous soils, areas which are unsuitable or uneconomical for land leveling, when the rate of flow is too small to distribute water efficiently by surface irrigation methods and where frequent light applications are needed. It also eliminates the field channels required as in case of surface irrigation methods. Limitations include apart from higher initial costs as compared to surface irrigation, uneven water distribution due to high winds and evaporation losses when operating under high temperatures.

In a given situation the method of irrigation to be adopted should be properly selected and designed to suit the crop, land topography and soil. Once the system is established proper guidelines may be provided for its operation to the farmer.

8.6 Water balance in irrigation systems

In any irrigation system, water balance calculations consisting of water availability and demands will be required to understand the functioning of the system as well as planning water deliveries. While the water availabilities can not exactly be known, the water demands should be worked out based on the command area and crop water requirements. In case of minor and medium projects water balance components could be calculated for the entire system, where as in case of major projects it could be for each canal unit. Figures about effective rainfall and irrigation efficiencies have to be obtained from available research information. The water balance components could be approximate in the beginning but could be improved year after year as more accurate information is available.

A simplified water balance model for an irrigated area could be written as:

$$Q_r + Q_i = Q_c + Q_{nc} + Q_{sm} + Q_{gw} + Q_{sf} + Q_{rf}$$

Where,

Q_r = Amount of rainfall received in the area.
Q_i = Amount of irrigation water applied.
Q_c = Amount of water consumed by the crops in the area (evapotranspiration).
Q_{nc} = Amount of water consumed by the other vegetation in the non cropped areas.
Q_{sm} = Amount of water stored in the soil profile as soil moisture.
Q_{gw} = Amount of water joining the ground water.
Q_{sf} = Amount of water flowing as surface runoff outside the area.
Q_{rf} = Amount of return flows to the drain/stream. (both from surface and subsurface).

Even though this is very approximate it gives an idea of the water use in the system. Better approximations can be made when a canal unit or individual fields are considered.

8.7 Water deliveries

The most important aspect of the irrigation system management is the delivery of water from the storage or diversion to the farmer's fields. In smaller systems, water might be conveyed through a single channel and directly delivered to the fields. In the larger systems, water is conveyed through a network of main and secondary canals with several controls and finally delivered to the tertiary units.

The two main issues concerned with the water deliveries are the rate and duration of flow. Beyond the project outlet, the delivered water is disturbed to farmer fields containing usually more than one crop. Ideally, the total water deliveries should match the crop requirements. Mismatch of deliveries and requirements will result either in excess or deficit supply. In a given situation, it is desirable to keep either the excesses or the deficits to a minimum.

In terms of the water deliveries, the following points need to be attended to in each irrigation system:

1. Assess the water requirements and adequacy of water deliveries at various levels (e.g. tertiary and secondary), from the perspectives of both the water suppliers and water users.
2. Identify the underlying principles and procedures for decisions on the timing and amounts of water allocated at various levels within an irrigation system, both before the irrigation season begins and during the season, when unexpected water shortages may occur.
3. Evolve policies of water deliveries at different levels in collaboration with water users associations wherever possible. External interference in water delivery policies could occur in some situations. The irrigation system manager has to show some tact and fair play in dealing with such situations.

8.8 Improving irrigation efficiencies

In irrigation water management, it is necessary to evaluate the irrigation practices from the time water leaves the source till it is utilized by the crops. The concept of efficiency which is an input-output relationship is applied to irrigation practice. There are several ways in which irrigation efficiencies are expressed, but two important approaches viz., the conveyance efficiency and the project efficiency need to be considered for quick appraisal. The conveyance efficiency refers to the ratio of the water delivered to the point of use to the water diverted from the source while the project efficiency refers to the ratio of the water used (or required) by the crops to the water diverted at the project head works. The project efficiency reflects the overall efficiency of the system. In large irrigation systems delivery of water to an individual farmer involves conveyance over a large distance and involving passing through several structures and finally delivered to the farmer's field from the field channel. Because of these factors project efficiencies in large irrigation systems cannot be very high. Generally a value of 50% is considered as satisfactory in large irrigation systems.

The term water use efficiency is used to relate crop yield with water use. Two terms are commonly used for this purpose. One is the crop water use efficiency which is the ratio of the crop yield to the amount of water depleted by the crop in the process of evapotranspiration. The other term is the field water use efficiency which is the ratio of the crop yield to the total amount of water used in the yield. The yield may be expressed as kilograms or quintals per hectare and evapotranspiration or water use expressed in terms of hectare-centimeters.

Irrigation efficiencies can be improved by following water saving as well as efficient water use measures. Canals with large seepage losses are to be identified and lined. Lining could later be extended to field channels, particularly in areas with large seepage losses. Field improvements like land leveling, improving field layouts and adoption of drip or sprinkler systems etc. will improve the application efficiencies and ultimately the overall efficiency of the system. Selection of crop sequences suiting the available water resources is also important so that higher water use efficiencies are achieved.

8.9   Equity of water distribution in large irrigation systems

In irrigation systems especially in the larger ones, water should be made available to all the farmers irrespective of the location of their fields. In India there is no uniform policy of water distribution in the irrigation systems in different States. For example in the state of Andhra Pradesh (A.P.), India, in large irrigation systems where water is delivered to a group of farmers, the concept of equity was not built in the project plans. After the irrigation system has come into operation, farmers have started drawing water as per the availability. There is an element of indiscipline and farmers at the head reaches tend to draw more water than required. In the command area development projects, a rotational system of water allocation has been introduced.

In the irrigation systems of Punjab, time for which water can be drawn from the field channels is decided depending on the area owned by the farmer. All farmers commanded by the outlet get water, but may not be the full crop water requirement. The low water allowance forced the farmers to adopt conjunctive use of groundwater and also helped in better water management and higher water use efficiency on the farm.

In the irrigation systems of along with water users associations, it is desirable to introduce rotational supply of water in the command areas. Conjunctive use of ground-water and use of drainage water wherever possible should also be introduced to help in equitable water distribution to the tail end farmers as well.

8.10   Conjunctive use of surface and groundwater

In irrigated areas, because of the continued deep percolation from applied irrigation water, watertables rise over a period of time and tend to come near the land surface. Areas wherein watertables come near the land surface and affect crop growth are known as water logged lands. Rising watertables may also bring up harmful salts to the land surface. These situations could partly be remedied by pumping the ground water so that the watertables are controlled. The pumped water is also used for irrigation conjunctively with the surface water already available.

8.11   Drainage in irrigation systems

Drainage refers to the removal and disposal of excess water not required by the crops. Continued irrigation over a large number of years without adequate drainage facilities is resulting in large tracts of irrigated areas becoming unproductive due to water logging or soil salinity or both the adverse conditions. In many irrigation systems, open drains are provided to drain the surface runoff especially during the monsoon period. The need for field level drainage and subsurface drainage is not well recognized.

In the new irrigation projects that are being formulated drainage should form a part of the project proposals. In the existing projects, the need for drainage should be investigated as a case by case basis and steps need to be taken for implementing the drainage measures. Drainage water of satisfactory quality could be used for irrigation, or could be blended with irrigation water before use. Such a step may help the tail-end farmers in large irrigation systems.

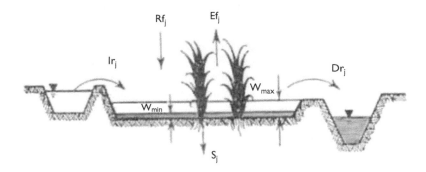

Plate 8.1 Components of the water balance in an individual rice field. (Ir = Irrigation; Rf = Rainfall; Et = Evapotranspiration; S = Seepage and Percolation; Dr = Drainage; Wmax = Maximum depth of water; Wmin = Minimum depth of water required; j = Time period considered). (See *Colour plate 8.1*)

8.12   Water management in rice

Rice is the most important irrigated crop in most of the countries in the Asia-Pacific region. In areas where irrigation water is available, farmers tend to grow rice. It is estimated that out of the total water required by the crop, 3% (40 mm) is used for the nursery, 16% (~200 mm) for land preparation and 81% (~1000 mm) for field irrigation of the crop. The percolation losses are found to be variable from field to field and it is estimated that up to 50 to 60% of the water applied can be lost through percolation depending on the soil conditions.

Plate 8.1 shows the various components in the water balance of an individual rice field.

Research studies indicate that water requirement of rice varies at different growth periods and consequently the depth of water to be maintained also varies with the growth stage of the crop. It has also been observed that drainage for a day or two during the tillering stage (mid-season drainage) helps to stimulate the growth of roots and checks the development of non-productive tillers.

In canal command areas the activities of land preparation and transplanting have to be suitably staggered so that water deliveries could be planned to meet the demands.

In terms of applied research relating to water management in rice the following aspects need to be considered for field level studies.

(i)   The water balance components need to be observed under different soil types and steps to be initiated for managing them effectively.

(ii)   Agronomical practices like weed control, direct seeding, age of the seedlings at the time of transplanting, puddling etc. should be studied as to develop practices suitable for reducing water requirements.

(iii)   Land leveling in rice fields, embankments and field layout need to be studied for providing conditions for efficient water use, maximizing effective rainfall and providing irrigation and drainage.

Agronomic practices, such as SRI (System of Rice Intensification) and aerobic rice, which use less water, are described elsewhere in the volume.

8.13 Environmental hazards in irrigation systems

At present environmental hazards are causing concern in major irrigation systems all over the world. Irrigation systems cause both favourable and unfavourable changes in the irrigated areas over a period of time. Some environmental hazards that affect the irrigation systems are listed below (Murty and Takeuchi, 1996):

- Soil degradation
- Waterlogging, formation of stagnant water pools and vector breeding places
- Stagnant drainage channels that promote the growth of aquatic weeds
- Spread of certain vector borne diseases
- Irrigation return flows causing water quality problems for aquatic habitats
- Contamination of irrigation water with urban wastes.

The agencies responsible for irrigation systems management should be aware of these and any other possible hazards in the system and adopt corrective measures.

8.14 Participatory irrigation management

In large irrigation systems it is now realized that improvements are possible only when the beneficiaries also participate in the management of the system. The Andhra Pradesh Farmers Management of Irrigation Systems Act enacted in 1997 (Government of A.P., 2003), is an example of a legislation supporting the establishment of water users associations in irrigation systems. It is expected that the implementation of this Act will address such problems as inadequate water availability at the lowest of the outlets, poor maintenance of the system at the field level and inequitable distribution of water at the farmers' level. It is also felt that when farmers manage the system themselves they will have full understanding and knowledge of the system and hence will be more willing to pay water charges.

Participatory irrigation management (PIM) through the Water Users Association coupled with the availability of technical information for on-farm water management are likely to bring significant improvements in water management at the field level in large irrigation systems.

8.15 Monitoring and evaluation

Each irrigation system should have regular program of monitoring and evaluation which should lead to better management of the system. Monitoring should consist of both physical and economic parameters and should include climatic parameters at selected points in the command area, cropped areas, water deliveries at different levels, groundwater levels, drainage conditions and condition of the water delivery system. Evaluation of the irrigation system should consist of a set of performance indicators which may include: area of land irrigated in relation to water supply, crop yields, profitability to the farmer, comparison between head and tail-end regions and performance of the water users associations.

8.16 Computer applications

With the availability of high speed personal computers, softwares have been developed to assist in the management of irrigated agriculture systems. These include softwares for estimating crop water requirements, designing surface irrigation methods, designing sprinkler and drip irrigation systems, designing drainage systems, estimating irrigation return flows, water delivery estimations in large irrigation systems etc. Geological information systems (GIS) which are useful in mapping the irrigation command areas or watersheds also assist in efficient management of the irrigated agricultural systems.

8.17 Institutional arrangements and capacity building

Irrigation systems in most of the situations are managed by the engineers who were responsible for their construction. Management of irrigation systems need a different knowledge and approach than the constructional aspects. Management of irrigation systems requires a good knowledge of the agricultural aspects as well. Existing organizations need therefore to be strengthened and reformatted to include personnel with knowledge in the hydraulics of the system, agronomical practices and environmental issues. Training and awareness programs need to be conducted for people at different levels like senior officers, field level officers, technicians and farmers. To assist the farmers in taking up improved practices an extension agency devoted to agricultural water management may also be necessary. When farmers associations are involved in the management of irrigation systems such organizations have to be sustained with suitable administrative support. Non-governmental organizations associated with irrigation systems should have enough technical capabilities. Legal provisions like prevention of unauthorized tapping of water from irrigation canals, polluting irrigation canals etc., will be required as per the location specific situations. Institutional arrangements and capacity building have to be dynamic processes to meet the system requirements.

8.18 Prospect and retrospect

As irrigated agriculture contributes to increased crop production, a number of irrigation systems have been established and more will continue to be established in the future at possible locations. It is estimated that more than 70 percent of the fresh water resources are used for agriculture. Municipal and industrial needs are demanding more water and their future demands may have to come from the water hitherto being used by agriculture. Agriculture and irrigation systems have therefore use water as efficiently as possible. It may also be noted that irrigated agricultural systems also contribute towards live stock development and aquaculture development thus contributing to overall food production. Irrigated agriculture systems cannot operate under isolation. Irrigation systems in addition to supplying water for crops at some locations supply water for municipal as well as industrial needs. Irrigation systems need to be developed keeping in view the overall water resources development of a region (Das, 2005). Chitale (2000) also argues for a comprehensive approach for water resources development and suggests several steps to bring in the so called 'blue revolution'.

In regions or countries where additional food production and food security is needed irrigated agricultural systems offer good investment possibilities. In a given region or country only a part of the agricultural lands can be brought under irrigation with the available water resources. The irrigated areas can be made productive on a continuous basis when they are managed with efficient water management practices along with the required agricultural practices and adequately addressing the environmental concerns. Institutional arrangements that are responsive to the needs of the system and the stakeholders are equally important.

References

Chitale, M.A.(2000). *A Blue Revolution*, Bharatiya Vidya Bhawan, Pune.

Das, Sandeep (2005). *Water – Need for a Comprehensive Management Policy*. Working Paper Series on Agriculture and the Poor, No.33. Bazaar Chintan, New Delhi.

Doorenbos, J. (1992). *Crop Water Requirements. Irrigation and Drainage*. Paper 24. Food and Agriculture Organization of the U.N. (Rome).

Doorenbos, J. and A.H. Kassam (1979). *Yield Response to Water*. FAO Irrigation and Drainage Paper 33, Food and Agriculture Organization of the United Nations (Rome).

Government of Andhra Pradesh (2003). *Andhra Pradesh Water Vision – A Shared Water Vision*, Volume I. – Methods, Position Papers and District Reports. Volume II, Government. Insurance Building, Tilak Road, Hyderabad, Andhra Pradesh: Water Conservation Mission.

International Commission of Irrigation and Drainage (ICID) (2006). *Water for Food*. Paper presented at the World Water Forum.

Murty, V.V.N. and K. Takeuchi (1996). *Land and Water Development for Agriculture in the Asia-Pacific Region*. Oxford and IBH Publishing Co. Ltd. New Delhi.

van Hofwegen, P. and Svendsen, M. (2000). *A Vision of Water for Food and Rural Development*. World Water Forum, The Hague.

Chapter 9

Improved livelihoods and food security through unlocking the potential of rainfed agriculture

Suhas P. Wani[1], T.K. Sreedevi[1], Johan Rockstrom[2],
Thawilkal Wangkahart[3], Y.S. Ramakrishna[4], Yin Dixin[5],
A.V.R. Kesava Rao[1] & Zhong Li[6]

[1] International Crops Research Institute for the Semi-Arid Tropics (ICRISAT), Patancheru, Andhra Pradesh, India;
[2] Stockholm Environment Institute (SEI), Stockholm, Sweden; [3] c/o Agricultural Research and Development,
Region 3, Muang, Khon Kaen, Thailand; [4] Central Research Institute for Dryland Agriculture (CRIDA),
Santoshnagar, Hyderabad, Andhra Pradesh, India; [5] Guizhou Academy of Agricultural Sciences (GAAS),
Integrated Rural Development Center, Guiyang, Guizhou Province, China; [6] Yunnan Academy of
Agricultural Sciences (YAAS), Kunming, Yunnan Province, China

9.1 Importance of rain-fed agriculture

Eighty per cent of the world's agricultural land area is rainfed and generates 58% of the world's staple foods (SIWI, 2001). The importance of rainfed agriculture varies regionally, but produces most food for poor communities in developing countries. In sub-Saharan Africa (SSA) more than 95% of the farmed land is rainfed, while the corresponding figure for Latin America is almost 90%, for South Asia about 60%, for East Asia 65% and for Near East and North Africa 75%. Farming systems in sub-Saharan Africa and Latin America are almost exclusively rainfed, while a predominant blue water dependence in irrigation is concentrated in the West Asian (>80% dependence) and North African regions (>60% dependence) (Rockström, 2003). In South and East Asia the picture is mixed, with countries depending in varying degrees on both rainfed and irrigated agriculture (e.g., India where 60% of water use in agriculture are estimated to originate from directly infiltrated rainfall, while 40% originates from extraction of river and groundwater for irrigation). A survey of "irrigation schemes" in Tanzania has shown that over 80% of them are supplementary irrigated systems where the bulk of water for crops is supplied by direct rainfall (MAFS, Tanzania Ministry of Agriculture, 2003).

Most of 852 million hungry and malnourished people in the world are in Asia, particularly in India (221 million) and in China (142 million). In Asia 75% of the poor are in rural areas and they depend on agriculture for their livelihood. About half of the hungry live in smallholder farming households, while two-tenths are landless. About 10% are pastoralists, fish folk and forest users (Sanchez et al., 2005). Hungry people are highly vulnerable to crises and hazards. The crises may be caused by natural disasters, such as major droughts or floods. Water (freshwater) is a limiting natural resource and plays an important role in providing livelihood support for rural populations where agriculture is the key occupation. Water scarcity is a significant problem for farmers in Africa, Asia, and the near East where 80–90 per cent of water

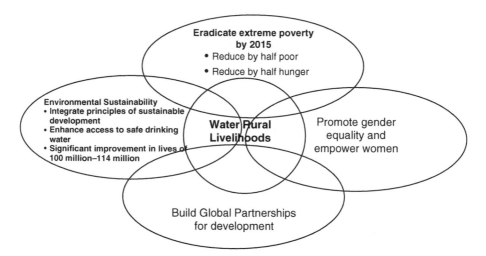

Figure 9.1 Water an important driver for the millennium development goals.

withdrawals are used for agriculture (FAO 2000, Rosegrant et al., 2002). Water a finite resource, the very basis of life and the single most important feature of our planet, is the most threatened natural resource today. Water is most important driver for four of the millennium development goals (MDGs) as shown in the Figure 9.1. In the context of four MDGs contribution of water resources management through direct interventions are suggested to achieve the milestones by 2015.

Improving social capital investments in water infrastructure as a catalyst for regional development and pivotal role of community-based organizations (CBOs) in water management is highlighted by the task force on the MDG. Rain-fed agriculture that constitutes the livelihood base for the vast majority of rural inhabitants (about 75 per cent of the poor in South Asia, and about 80 per cent of the population in east Africa) in the developing countries is a source of food security, employment and cash income (Rockstrom *et al.*, 2003).

9.2   Constraints in rainfed agriculture areas

An insight into the rain-fed regions shows a grim picture of water-scarcity, fragile environments, drought and land degradation due to soil erosion by wind and water, low rainwater use efficiency (35–45%), high population pressure, poverty, low investments in water use efficiency measures, poor infrastructure and inappropriate policies (Wani *et al.*, 2003 a&c, Rockstrom *et al.*, 2007). Drought and land degradation are interlinked in a cause and effect relationship and both in turn are the causes of poverty. This unholy nexus between drought, poverty and land degradation has to be broken to meet the MDG of halving the number of food insecure poor by 2015. A global assessment of the extent and form of land degradation showed that 57% of the total area of drylands occurring in two major Asian countries, namely China (178.9 m ha) and India (108.6 m ha), are degraded (UNEP, 1997). Accelerated erosion resulting in

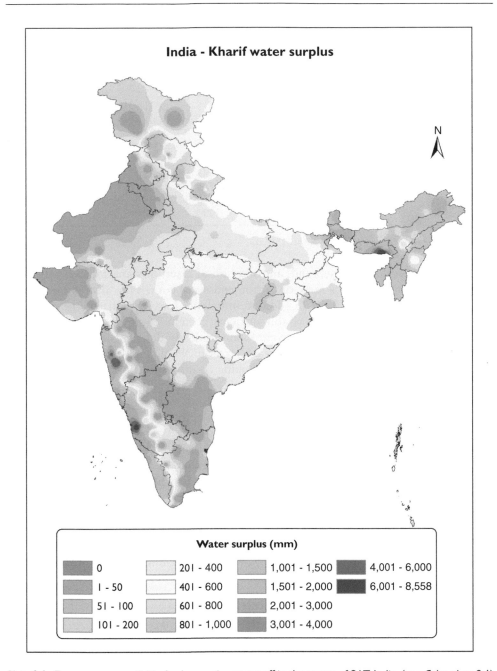

Plate 9.1 Excess water available for harvesting as runoff in the states of SAT India. (see *Color plate 9.1*)

the loss of nutrient rich top fertile soil however, occurs nearly everywhere where agriculture is practiced and is irreversible. The torrential character of the seasonal rainfall creates high risk for the cultivated lands. In India, alone some 150 million ha are affected by water erosion and 18 m ha by wind erosion. Thus, erosion leaves behind

Table 9.1 Annual water balance characters (all values in mm).

Country	Location	Rainfall	PET	AET	WS	WD
China	Xiaoxingcun	641	1464	641	Nil	815
	Lucheba	1284	891	831	384	60
Thailand	Wang Chai	1171	1315	1031	138	284
	Tad Fa	1220	1511	1081	147	430
Vietnam	Chine	2028	1246	1124	907	122
	Vinh Phuc	1585	1138	1076	508	62
India	Bundi	755	1641	570	186	1071
	Guna	1091	1643	681	396	962
	Junagadh	868	1764	524	354	1240
	Nemmikal	816	1740	735	89	1001
	Tirunelveli	568	1890	542	Nil	1347

an impoverished soil on one hand, and siltation of reservoirs and tanks on the other. This degradation induced source of carbon emissions contribute also to far reaching global warming consequences. In addition imbalanced use of nutrients in agriculture by the farmers results in mining of soil nutrients. Recent studies in India revealed that 80 to 100% of the farmers' fields were found critically deficient in zinc, boron, and sulphur in addition to nitrogen and organic carbon (Wani et al., 2006a). If the current production practices are continued, developing countries in Asia and Africa will face a serious food shortage in the very near future.

Weekly water balances of selected watersheds in China, Thailand and Vietnam were completed based on long-term agrometeorological data and soil type. The water balance components included potential evapotranspiration (PET), actual evapotranspiration (AET), water surplus (WS) and water deficit (WD). PET varied from about 890 mm at Lucheba in China to 1890 mm at Tirunelveli in South India (Table 9.1). AET values are relatively lower at the watersheds in China and India compared to those in Thailand and Vietnam. Varying levels of water surplus and water deficit occur at the watersheds. Among all the locations, Tirunelveli in India has the largest water deficit (1347 mm) and no water surplus. Chine in Vietnam has the largest water surplus of 907 mm. These analyses defined the dependability for moisture availability for crop production and opportunities for water harvesting and groundwater recharge.

9.3   Potential of rainfed agriculture

In several regions of the world rainfed agriculture generates among the world's highest yields. These are predominantly temperate regions, with relatively reliable rainfall and inherently productive soils. Even in tropical regions, particularly in the sub-humid and humid zones, agricultural yields in commercial rainfed agriculture exceed 5–6 t ha^{-1} (Rockström and Falkenmark, 2000; Wani et al., 2003a, b). At the same time, the dry sub-humid and semi-arid regions have experienced the lowest yields and the weakest productivity improvements. Here, yields oscillate in the region of 0.5–2 t ha^{-1}, with an average of 1 t ha^{-1}, in sub-Saharan Africa, and 1–1.5 t ha^{-1}, in the SAT Asia and Central and West Asia and North Africa (CWANA) for rainfed

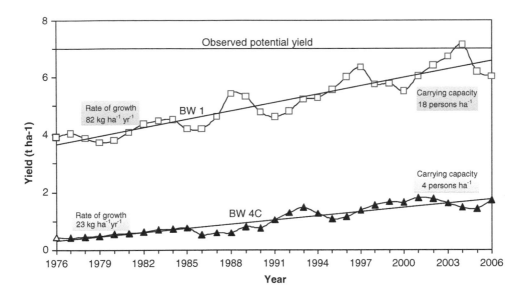

Plate 9.2 Three-year moving average of sorghum and pigeon pea grain yield under improved management and on farmers' fields in a deep Vertisol catchment, Patancheru, India. (See *Colour plate 9.2*)

agriculture (Rockström, and Falkenmark 2000; Wani *et al.*, 2003a, b, Rockstrom *et al.*, 2007). Evidence from long-term experiments at ICRISAT, Patancheru, India since 1976, demonstrated the virtuous cycle of persistent yield increase through improved land, water, and nutrient management in rainfed agriculture. Improved systems of sorghum/pigeonpea intercrops produced higher mean grain yields (5.1 t ha^{-1} per yr) compared to 1.1 t ha^{-1} per yr, average yield of sole sorghum in the traditional (farmers') post-rainy system where crops are grown on stored soil moisture (Plate 9.2). The annual gain in grain yield in the improved system was 82 kg ha^{-1} per yr compared with 23 kg ha^{-1} per yr in the traditional system. The large yield gap between attainable yield and farmers' practice as well as between the attainable yield of 5.1 t ha^{-1} and potential yield of 7 t ha^{-1} shows that a large potential of rainfed agriculture remains to be tapped. Moreover, the improved management system is still gaining in productivity as well as improved soil quality (physical, chemical, and biological parameters) along with increased carbon sequestration of 300 kg C ha^{-1} per year (Wani *et al.*, 2003b). Yield gap analyses, undertaken by the Comprehensive Assessment, for major rainfed crops in semi-arid regions in Asia (Fig 9.4) and Africa and rainfed wheat in West Asia and North Africa (WANA), reveal large yield gaps, with farmers' yields being a factor of 2–4 lower than achievable yields for major rainfed crops grown in Asia and Africa (Rockstrom *et al.*, 2007).

Farmers' yields continue to be very low compared to the experimental yields (attainable yields) as well as simulated crop yields (potential yields), resulting in a very significant yield gap between actual and attainable rainfed yields. The difference is largely explained by inappropriate soil, water, and crop management options at the farm level, combined with persistent land degradation.

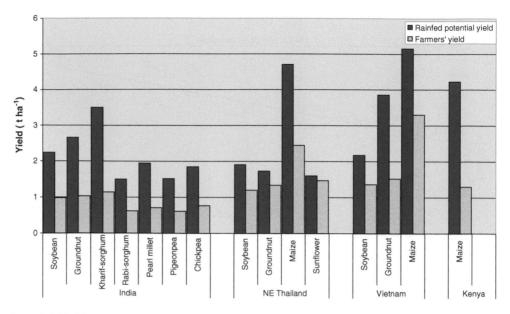

Figure 9.2 Yield gap analysis of important rainfed crops in different countries.

The vast potential of the rainfed agriculture need to be unlocked through knowledge-based management of natural resources for increasing the productivity and incomes to achieve food secured developing world.

9.4   Need for a new paradigm for water management in rainfed agriculture

For enhancing rainwater use efficiency in rainfed agriculture, management of water alone can not result in enhanced water productivity as in these areas crop yields are limited by more than water limitation. ICRISAT's experience in rainfed areas has clearly demonstrated that more than water quantity *per se*, management of water resources is the limitation in the SAT (Wani *et al.*, 2006a).

Based on the Policy on water resource management for agriculture remains focused on irrigation, and the framework for integrated water resource management (IWRM) at catchment and basin scales are primarily concentrated on allocation and management of blue water in rivers, groundwater and lakes. The evidence from the comprehensive assessment of water for food and poverty reduction indicated water for agriculture is larger than irrigation, and there is an urgent need for a widening of the policy scope to include explicit strategies for water management in rainfed agriculture including grazing and forest systems. However, what is needed is effective integration so as to have a focus on the investments options on water management across the continuum from rainfed to irrigated agriculture. This is the time to abandon the obsolete sectoral divide between irrigated and rainfed agriculture, which would place water resource management and planning more centrally in the policy domain of agriculture at large, and not as today, as a part of water resource policy (Molden *et al.*, 2007).

Furthermore, the current focus on water resource planning at the river basin scale is not appropriate for water management in rainfed agriculture, which overwhelmingly occurs on farms of <5 ha at the scale of small catchments, below the river basin scale. Therefore, focus should be to manage water at the catchment scale (or small tributary scale of a river basin), opening for much needed investments in water resource management also in rainfed agriculture (Rockström et al., 2007).

In several countries, central and state governments have emphasised management of rainfed agriculture under various programmes. Important efforts for example have been made under the watershed development programmes in India. Originally, these programmes were implemented by different ministries such as the Ministry of Agriculture, the Ministry of Rural Development and the Ministry of Forestry, causing difficulties for integrated water management. Recently, steps were taken to unify the programme according to the "Hariyali Guidelines" (Wani et al., 2006a). Detailed meta analysis of 311 watershed case studies in India revealed that watershed programs are silently revolutionizing rainfed areas with positive impacts (B:C ratio of 1:2.14, IRR of 22%, increased cropping intensity by 63%, increased irrigated areas by 34%, reduced run off by 13% and increased employment by 181 person days per year per ha). However, 65% of the watersheds were performing below average performance as they lacked community participation, programs were supply driven, equity and sustainability issues were eluding and compartmental approach was adopted (Joshi et al., 2004).

Based on detailed studies and synthesis of the results, impacts, shortcomings, learnings from large number of watershed programs and on-farm experiences gained, ICRISAT-led consortium developed an innovative farmers' participatory consortium model for integrated watershed management (Wani et al., 2002, 2003a,c). ICRISAT-led watershed espouses the Integrated Genetic Natural Resources Management (IGNRM) approach where activities are implemented at landscape level. Research and development (R&D) interventions at landscape level were conducted at benchmark sites representing the different SAT agroecoregions. The entire process revolves around the four E's (empowerment, equity, efficiency and environment), which are addressed by adopting specific strategies prescribed by the four C's (consortium, convergence, cooperation and capacity building). The consortium strategy brings together institutions from the scientific, non-government, government, and farmers group for knowledge management. Convergence allows integration and negotiation of ideas among actors. Cooperation enjoins all stakeholders to harness the power of collective actions. Capacity building engages in empowerment for sustainability (Wani et al., 2003b).

In 2005, the National Commission on Farmers adopted a holistic integrated watershed management approach, with focus on rainwater harvesting and improving soil health for sustainable development of drought prone rainfed areas (Government of India, 2005). Recently, Government of India has established National Authority for Development of Rainfed Areas (NADORA) with the mandate to converge various programmes for integrated development of rainfed agriculture in the country. These are welcome developments where policy makers have realised the need to develop rainfed areas for reducing poverty and increasing agricultural production. However, it is just a beginning and lot more still needs to be done to provide institutional and policy support for development of rainfed areas. Thus, it has become increasingly clear that water management for rainfed agriculture requires a landscape perspective, and

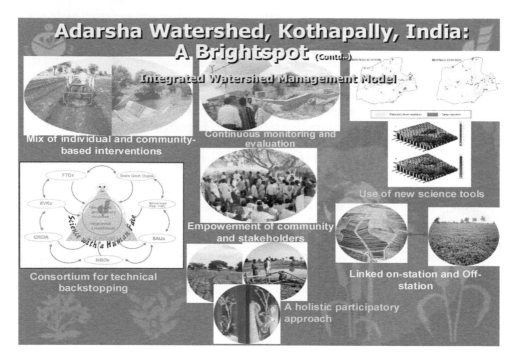

Plate 9.3 An innovative consortium model for integrated watershed management. (See *Colour plate 9.3*)

involves cross-scale interactions from farm household scale to watershed/catchment scale and upstream-down stream linkages.

9.5 Shifting non-productive evaporation to productive transpiration

Rainwater use efficiency in arid and SAT is 35 to 50% and up to 50% of the rainwater falling on crop or pasture fields is lost as non-productive evaporation. This is a key window for improvement of green water productivity, as it entails shifting non-productive evaporation to productive transpiration, with no downstream water trade-off. This *vapour shift* (or transfer), where management of soil physical conditions, soil fertility, crop varieties and agronomy are combined to shift the evaporative loss into useful transpiration by plants, is a particular opportunity in arid, semi-arid and dry-subhumid regions (Rockstrom *et al.*, 2007).

Field measurements of rainfed grain yields and actual green water flows indicate that when doubling yields from 1 to $2\,t\,ha^{-1}$ in semi-arid tropical agro-ecosystems, green water productivity may improve from approximately $3500\,m^3/t^{-1}$ to less than $2000\,m^3/t^{-1}$. This is a result of the dynamic nature of water productivity improvements when moving from very low yields to higher yields. At low yields, crop water uptake is low and evaporative losses high, as the leaf area coverage of the soil is low, which together results in high losses of rainwater as evaporation from soil. When yield levels increase, shading of soil improves, and when yields reach $4–5\,t/ha^{-1}$ and above, the canopy density is so high that the opportunity to reduce evaporation in favour of increased transpiration reduces, lowering the relative improvement of water productivity. This indicates that large opportunities of improving water productivity are found in low-yielding farming systems (Rockström, 2003; Oweis *et al.*, 1998), i.e., particularly in rainfed agriculture as compared to irrigated agriculture where water productivity already is higher due to better yields.

9.6 Investments in rainfed areas produce multiple benefits

Through the use of new science tools (i.e. remote sensing, GIS, and simulation modeling) twinned with an understanding of the entire food production-utilization system (i.e. food quality and market) and genuine involvement of stakeholders, ICRISAT-led watersheds effected remarkable impacts to SAT resource-poor farm households.

9.6.1 Reducing rural poverty

Reducing rural poverty in the watershed communities is evident in the transformation of their economies. The ICRISAT model ensured improved productivity with the adoption of cost-efficient water harvesting structures as an entry point for improving livelihoods. Crop intensification and diversification with high-value crops is one leading example that allowed households to achieve production of basic staples and surplus for modest incomes. Provision for improving the capacity of farm households through training and networking and for alleviating livelihood enhanced participation most especially of the most vulnerable groups like women and the landless. The self-help groups (SHGs) common in the watershed villages of India and an improved initiative in China provided income and empowerment of women. The environmental clubs whose conceptualization is traced from Bundi watershed of Rajasthan, India inculcated environmental protection, sanitation and hygiene among the children.

Building on social capital made the huge difference in addressing rural poverty of watershed communities. A case in point is Kothapally watershed. Today, it is a prosperous village on the path of long-term sustainability and has become a beacon for science-led rural development. In 2001, the average village income from agriculture, livestock and non-farming sources was US$795 compared with the neighboring non-watershed village with US$622 (Fig. 9.3). The villagers proudly professed "*We did not face any difficulty for water even during the drought year of 2002. When surrounding villages had no drinking water, our wells had sufficient water*".

To date, the village prides itself with households owning 5 tractors, 7 lorries and 30 auto rickshaws. People from surrounding villages come to Kothapally for on-farm employment. There were evidences to suggest that with more training on livelihood and enterprise development, migration is bound to cease. Between 2000 and 2003, investments in new livelihood enterprises such as seed oil mill, tree nursery, and worm composting increased average income by 77% in Powerguda, a tribal village in Andhra Pradesh.

Crop-Livestock integration is another facet harnessed for poverty reduction. The Lucheba watershed, Guizhou province of southern China has transformed its economy through modest injection of capital-allied contributions of labor and finance, to create basic infrastructures like access road and drinking water supply. With technical

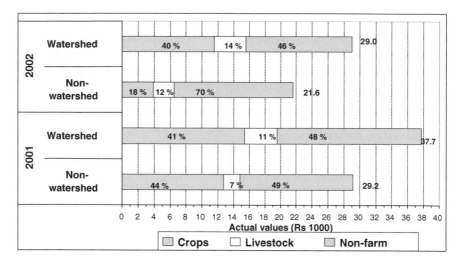

Figure 9.3 Income stability and resilience effects during drought year (2002) in Adarsha watershed, Kothapally, AP, India.

support from the consortium, the farming system was intensified from rice and rape seed to tending livestock (pig raising) and horticultural crops (fruit trees like *Zizipus*; vegetables like beans, peas, sweetpotato) and groundnuts. Forage production specifically wild buckwheat as an alley crop was a good forage grass for pigs. This cropping technology was also effective in controlling erosion and increasing farm income in sloping lands. This holds true in many watersheds of India where the improvement in fodder production have intensified livestock activities like breed improvement (artificial insemination and natural means) and livestock center/health camp establishment (Wani *et al.*, 2006b).

In Tad Fa and Wang Chai watersheds in Thailand, there was a 45% increase in farm income within three years. Farmers earned an average net income of US$1195 per cropping season. A complete turnaround in livelihood system of farm households was inevitable in ICRISAT-led watersheds.

9.6.2 *Increasing crop productivity*

Increasing crop productivity is common in all the watersheds and evident in so short period from the inception of watershed interventions. To cite few cases, in benchmark watersheds of Andhra Pradesh, improved crop management technologies increased maize yield by 2.5 times and sorghum by 3 times (Wani *et al.*, 2006a). Over-all, in 65 community watersheds (each measuring approximately 500 ha), implementing best-bet practices resulted in significant yield advantages in sorghum (35–270%), maize (30–174%), pearl millet (72–242%), groundnut (28–179%), sole pigeonpea (97–204%) and as an intercrop (40–110%). In Thanh Ha watershed of Vietnam, yields of soybean, groundnut and mungbean increased by three to four folds (2.8–3.5 t ha^{-1}) as compared with baseline yields (0.5 to 1.0 t ha^{-1}) reducing the yield gaps between potential farmers' yields. A reduction in N fertilizer (90–120 kg urea ha^{-1}) by 38% increased maize yield by 18%. In Tad Fa watershed

Table 9.2 Crop yields in Adarsha watershed, Kothapally, during 1999–2005.

Crop	1998 Baseline	Yield (kg ha^{-1})						
		1999	2000	2001	2002	2003	2004	2005
Sole maize	1500	3250	3750	3300	3480	3920	3420	3920
Intercropped maize	–	2700	2790	2800	3080	3130	2950	3360
(Traditional)		700	1600	1600	1800	1950	2025	2275
Intercropped pigeonpea	190	640	940	800	720	950	680	925
(Traditional)		200	180	–	–	–	–	–
Sole sorghum	1070	3050	3170	2600	2425	2290	2325	2250
Intercropped sorghum	–	1770	1940	2200	–	2110	1980	1960

of northeastern Thailand, maize yield increased by 27–34% with improved crop management.

9.6.3 *Improving water availability*

Improving water availability in the watersheds was attributed to efficient management of rainwater and *in-situ* conservation; establishing water harvesting structures (WHS) improved groundwater levels. Findings in most of the watershed sites reveal that open wells located near WHS have significantly higher water levels compared to those away from the WHS. Even after the rainy season, the water level in wells nearer to WHS sustained good groundwater yield. In the various watersheds of India like Lalatora, treated area registered a groundwater level rise by 7.3 m. At Bundi, the average rise was at 5.7 m and the irrigated area increased from 207 ha to 343 ha. In Kothapally watershed, the groundwater level rise was at 4.2 m in open wells (Fig. 9.4). The various WHS resulted in an additional groundwater recharge per year of approximately 4,28,000 m^3 on the average. With this improvement in groundwater availability, the supply of clean drinking water was guaranteed. In Lucheba watershed, a drinking water project, which constitutes a water storage tank and pipelines to farm households, was a joint effort of the community and the watershed project. This solved the drinking water problem for 62 households and more than 300 livestock. Earlier every farmer's household used to spend 2–3 hours per day fetching drinking water. This was the main motivation for the excellent farmers' participation in the project. On the other hand, collective pumping of well water out establishing efficient water distribution system enabled farmers group to earn more income by growing watermelon with reduced drudgery for women who had to carry on head from a long distance, pumping of water from the river as a means to irrigate watermelon has provided maximum income for households in Thanh Ha watershed (Wani *et al.*, 2006b).

9.6.4 *Sustaining development and protecting the environment*

Sustaining development and protecting the environment are the two-pronged achievements of the watersheds. The effectiveness of improved watershed technologies was evident in reducing run-off volume, peak run-off rate and soil loss and improving groundwater recharge. This is particularly significant in Tad Fa watershed where

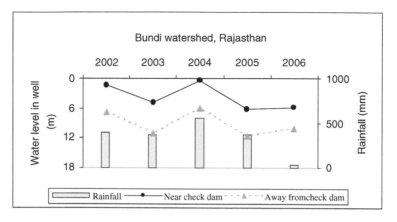

Estimated additional groundwater recharge due to watershed interventions =
6,75,000 m³ per year.

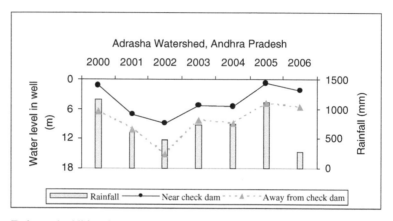

Estimated additional groundwater recharge due to watershed interventions =
4,27,800 m³ per year

Figure 9.4 The impact of watershed interventions on groundwater levels at two benchmark sites in India.

interventions such as contour cultivation at midslopes, vegetative bunds planted with *Vetiver*, fruit trees grown on steep slopes and relay cropping with rice bean reduced seasonal run-off to less than half (194 mm) and soil loss less than 1/7th (4.21 t ha⁻¹) as compared to the conventional system (473 mm run-off and soil loss 31.2 t ha⁻¹). This holds true with peak run-off rate where the reduction is approximately one-third (Table 9.3).

Large number of fields (80–100%) in the SAT were found severely deficient in Zn, B, and S along with N and P. Amendment of the deficient micro- and secondary nutrients increased crop yields by 30 to 70% resulting in overall increase in water and nutrient use efficiency. Introduction of integrated pest management (IPM) and improved cropping systems decreased the use of pesticides worth US$44–66 ha⁻¹. Crop rotation using legumes in Wang Chai watershed substantially reduced N requirement for

Table 9.3 Seasonal rainfall, runoff and soil loss from different benchmark watersheds in India and Thailand.

Watershed	Seasonal rainfall (mm)	Runoff (mm)		Soil loss (t ha^{-1})	
		Treated	Untreated	Treated	Untreated
Tad Fa, Khon Kaen, NE Thailand	1284	169	364	4.21	31.2
Kothapally, Andhra Pradesh, India	743	44	67	0.82	1.90
Ringnodia, Madhya Pradesh, India	764	21	66	0.75	2.2
Lalatora, Madhya Pradesh, India	1046	70	273	0.63	3.2

rainfed sugarcane. IPM practices which brought into use local knowledge using insect traps of molasses, light traps and tobacco waste led to extensive vegetable production in Xiaoxingcun (China) and Wang Chai (Thailand) watersheds.

Improved land and water management practices along with integrated nutrient management (INM) comprising of applications of inorganic fertilizers and organic amendments such as crop residues, vermicompost, farm manures, *Gliricidia* loppings as well as crop diversification with legumes not only enhanced productivity but also improved soil quality. Increased carbon sequestration of 7.4 t ha^{-1} in 24 years was observed with improved management options in a long-term watershed experiment at ICRISAT. By adopting fuel-switch for carbon, women SHGs in Powerguda (a remote village of Andhra Pradesh, India) have pioneered the sale of carbon units (147 t CO_2 C) to the World Bank from their 4,500 *Pongamia* trees, seeds of which are collected for producing saplings for distribution/promotion of biodiesel plantation. Normalized difference vegetation index (NDVI) estimation from the satellite images showed that within four years, vegetation cover could increase by 35% in Kothapally. The IGNRM options in the watersheds reduced loss of NO_3-N in run off water (8 vs 14 kg N ha^{-1}). Introduction of IPM in cotton and pigoenpea substantially reduced the number of chemical insecticidal sprays during the season and use of pesticides reduced the pollution of water bodies with harmful chemicals. Reduced runoff and erosion reduced risk of downstream flooding and siltation of water bodies that directly improved environmental quality in the watersheds.

9.6.5 Conserving biodiversity

Conserving biodiversity in the watersheds was engendered through participatory NRM. The index of surface percentage of crops (ISPC), crop agro-biodiversity factor (CAF), and surface variability of main crops changed as a result of integrated watershed management (IWM) interventions. Pronounced agro-biodiversity impacts were observed in Kothapally watershed where farmers now grow 22 crops in a season with a remarkable shift in cropping pattern from cotton (200 ha in 1998 to 100 ha in 2002) to a maize/pigeonpea intercrop system (40 ha to 180 ha); thereby changing the CAF from 0.41 in 1998 to 0.73 in 2002. In Thanh Ha, Vietnam the CAF changed from 0.25 in 1998 to 0.6 in 2002 with the introduction of legumes. Similarly, rehabilitation of the common property resource land in Bundi watershed through the collective action of the community ensured the availability of fodder for all the households and income of US $1670 y^{-1} for the SHG through sale of grass to the surrounding villages. Aboveground diversity of plants (54 plant species belonging to 35 families) as well as belowground diversity of microorganisms (21 bacterial isolates, 31 fungal species and 1.6 times higher biomass C) was evident in rehabilitated CPR as compared to the degraded CPR land (9 plant species, 18 bacterial isolates and 20 fungal isolates of which 75% belong to *Aspergillus* genus).

9.6.6   Promoting natural resource management (NRM) at landscape level

Promoting natural resource management (NRM) at landscape level is the scale of work done by the ICRISAT consortium. Benefiting from data obtained from using new science tools like remote sensing, a comprehensive understanding of the effects of the changes (i.e. vegetation cover on degraded lands) in the watersheds is made. This in turn has provided the indicators to assess agricultural productivity. Promoting NRM at the landscape level by using tools that provide the needed database is anticipated to have better impact because of the possible integration of all the factors (natural resources with the ancillary information).

While there were some interventions at plot to farm level, the impact factors of NRM such as sustainability of production, soil and water quality, and other environment resources have been looked at from a landscape perspective. This accounts for some successes in addressing concerns on equity issue like benefits for the poorest people such as the landless who are unable to take advantage of improved soil/water conditions and expansion of water intensive crops triggering renewed water stress. These remain as legitimate challenges of a holistic thinking, which can be better unraveled from a landscape scale. To date, the articulation of this recognition is seen in policy recommendations for serious attention to capacity building and not just for construction activities.

Equal concern was made on on-site and off-site impacts. The effect of water conservation at the upper ridge to downstream communities has been factored in. Water harvesting structures specifically the rehabilitation of the *nala* (drain) bund at the upper portion in Bundi watershed allowed irrigation of 6.6 ha at the downstream part. Another case is the Aniyala watershed located at the lower topo-sequence of Rajasamadhiyala watershed. Excess water flows of the 21 water harvesting structures in Rajasamadhiyala cascades into Aniyala. This has increased groundwater recharge by 25% and improved the groundwater source by 50% in a normal rainfall year. Because of this, there was an increase in crop production by 25–30% (Sreedevi *et al.*, 2006). The quality and number of livestock in the village improved because of water and fodder availability. Off-site effects of watershed specifically equity issues is one area that needs to be strengthened for enhanced impact.

9.6.7 Enhancing partnerships and institutional innovations

Enhancing partnerships and institutional innovations through the consortium approach was the major impetus for harnessing watershed's potential to reduce households' poverty. The underlying element of the consortium approach adapted in ICRISAT-led watersheds is engaging a range of actors with the locales as the primary implementing unit. Complex issues were effectively addressed by the joint efforts of ICRISAT and with key partners namely the national agricultural research systems (NARSs), non-government organizations (NGOs), government organizations (GOs), agricultural universities, community-based organization and other private interest groups with farm households as the key decision-makers. In SHGs, like village seed-banks, these were established not just to provide timely and quality seeds. These created the venue for receiving technical support and building the capacity of members like women for the management of conservation and livelihood development activities. Incorporating knowledge-based entry point in the approach led to the facilitation of rapport and at the same time enabled the community to take rational decisions for their own development. As demonstrated by ICRISAT, the strongest merit of consortium approach is in capacity building where farm households are not the sole beneficiaries. Researchers, development workers and students of various disciplines are also trained, and policymakers from the NARSs sensitized on the entire gamut of community watershed activities. Private-public partnership (PPP) has provided the means for increased investments not only for enhancing productivity but also for building institutions as engines for people-led natural resource management.

From another aspect, the consortium approach has contributed to scaling through the nucleus-satellite scheme and building productive alliances for further research and technical backstopping. With cooperation, a balanced R.& D. was implemented rather than a 'purist model' of participation or blind adherence to government guidelines. A balanced R&D in community watersheds has encouraged scientific debate and at the same time promoted development through tangible economic benefits

The contributions of other international agricultural research centers (IARCs) like the International Water Management Institute (IWMI), International Livestock Research Institute (ILRI) and World Wildlife Fund (WWF) have become allies because of common denominators like goal (poverty reduction) and subject (water resources). It must be reckoned that while centers have their own mandates, these will have to be addressed from a holistic perspective seeking the assistance and contributions of other centers; their technical expertise and findings. This not only maximized the use of resources but the problem situation in watersheds allowed for an integrated approach requiring the alliance of institutions and stakeholders. Similarly, the various networks like the Association for Strengthening Agricultural Research in Eastern and Central Africa (ASARECA) and Cereals and Legumes Asia Network (CLAN) have provided an added venue for exchange and collaboration. This led to a strong south-south partnership.

9.7 Conclusion

Rainfed areas which constitute about 80% of cultivated areas worldwide, are also where 65 million poor people reside in the SAT. Along with water scarcity, land degradation, poverty, malnutrition and demographic pressure are important constraints, which need urgent attention. In dry sub-humid and SAT areas yields of rainfed agriculture oscillate between 1 to $1.5\,t\,ha^{-1}$ as against the potential of $5\,t\,ha^{-1}$ in the SAT. There is a need to have a new paradigm for water resource management in rainfed areas where at catchment scale water need to be managed in integrated manner in a continuum from rainfed to supplemented irrigation using harvested run-off water or recharged groundwater. Evidence clearly demonstrated that water alone cannot do the job of increasing productivity as other limiting factors such as nutrients, pests, low quality seeds infrastructure and lack of knowledge held back the potential. Investments in rainfed areas produce multiple benefits such as reducing poverty, developing social capital, community-empowerment, building institutions, protecting environment, reducing land degradation, conserving biodiversity, sequestering carbon and provide environmental services.

References

FAO (Food and Agriculture Organization of the United Nations) (2000). *Crops and Drops.* Rome, Italy.

Govt. of India (2005). "Serving Farmers and saving farming – 2006: Year of Agricultural Renewal, Third Report". National Commission on farmers, Ministry of Agriculture, Govt. of India, New Delhi, December 2005. 307pp

Joshi, P.K., Vasudha Pangare, Shiferaw, B., Wani, S.P., Bouma, J., and Scott, C. (2004). Socioeconomic and policy research on watershed management in India: Synthesis of past experiences and needs for future research. Global Theme on Agroecosystems Report no.7. Patancheru 502324, Andhra Pradesh, India: International Crops Research Institute for the Semi-Arid Tropics. 88pp.

MAFS. (2003). "Study on Irrigation Master Plan". *Tanzania Ministry of Agriculture and Food Security (MAFS)*, Dar es Salaam, Tanzania.

Molden, D. (2007) . Comprehensive Assessment of Water Management in Agriculture. "Water for Food: Water for life – A Comprehensive Assessment of Water Management in Agriculture", London: Earthscan, and Colombo: International Water Management Institute. 40 pp.

Oweis, T., Pala, M., and Ryan, J. (1998). "Stabilizing rain-fed wheat yields with supplemental irrigation and nitrogen in a Mediterranean-type climate". *Agronomy Journal* 90, 672–681.

Rockström, J., and Falkenmark, M. (2000). "Semiarid crop production from a hydrological perspective: Gap between potential and actual yields". *Critical Reviews in Plant Science* 19(4), 319–346

Rockström, J. (2003). "Water for Food and Nature in the Tropics: Vapour Shift in Rainfed Agriculture". Invited paper to the *Special issue 2003 of Royal Society Transactions B Biology, Theme Water cycle as Life Support Provider,* Vol 358(1440), 1997–2009.

Rockström, J. *et al.,* (2007). Managing Water in Rainfed Agriculture, In *Water for Food, Water for Life* (in press).

Rosegrant, M., Cai, X., Cline, S., and Nakagawa, N. (2002.) *The Role of Rainfed Agriculture in the Future of Global Food Production.* EPTD Discussion paper no. 90. IFPRI, Washington, DC, USA. 105 pp.

Sanchez, P., Swaminatha, M.S., Dobie, P., and Yuksel, N. (2005). *Halving hunger: it can be done.* Summary version of the report of the Task Force on Hunger. The Earth Institute at Columbia University, New York. USA.

Sreedevi, T.K., Wani, S.P., Osman, M., Emmanuel D'Silva (2006). Improved Livelihoods and Environmental Protection through Biodiesel Plantations in Asia. Asian Biotechnology and Development Review. Vol. 8(2), 11–29.

SIWI (2001). Water harvesting for upgrading of rain-fed agriculture. Policy analysis and research needs. *SIWI Report II. Stockholm International Water Institute (SIWI), Stockholm, Sweden.* 104 pp.

UNEP (United Nations Environment Programme). (1997). *World Atlas of Desertification, Second Edition.*

Wani, S.P., Pathak, P., Tam, H.M., Ramakrishna, A., Singh, P., and Sreeedevi, T.K. (2002) "Integrated Watershed Management for Minimizing Land Degradation and Sustaining Productivity in Asia". Proceedings of a Joint UNU-CAS International Workshop, Beijing China – 8–13 September 2001. Paper published in Integrated Land Management in Dry Areas. pp. 207–230.

Wani, S.P., Maglinao, A.R., Ramakrishna, A., and Rego, T.J. (eds.) (2003a). Integrated watershed management for land and water conservation and sustainable agricultural production in Asia: Proceedings of the ADB-ICRISAT-IWMI Project Review and Planning Meeting, 10–14 December 2001, Hanoi, Vietnam, Patancheru 502 324, Andhra Pradesh, India : International Crops Research Institute for the Semi-Arid Tropics. 268 pp.

Wani, S.P., Pathak, P., Jangawad, L.S., Eswaran, H., and Singh, P. (2003b). Improved management of Vertisols in the semi-arid tropics for increased productivity and soil carbon sequestration. *Soil Use and Management*, **19**, 217–222.

Wani, S.P., Singh, H.P., Sreedevi, T.K., Pathak, P., Rego, T.J., Shiferaw, B., Shailaja Rama Iyer. (2003c). Farmer-participatory integrated watershed management: Adarsha watershed, Kothapally, India, An innovative and upscalable approach. A case study. In: R.R. Harwood and A.H. Kassam (Eds.). Research towards integrated natural resources management": Examples of research problems, approaches and partnerships in action in the CGIAR. Interim Science Council, Consultative Group on International Agricultural Research. Washington, DC, USA: pp. 123–147.

Wani, S.P., Ramakrishna, Y.S., Sreedevi, T.K., Long, T.D., Thawilkal Wangkahart, Shiferaw, B., Pathak, P. and Kesava Rao, A.V.R. (2006a). Issues, Concepts, Approaches and Practices in the Integrated Watershed Management: Experience and lessons from Asia in Integrated Management of Watershed for Agricultural Diversification and Sustainable Livelihoods in Eastern and Central Africa: Lessons and Experieces from Semi-Arid South Asia. Proceedings of the International Workshop held 6–7 December 2004 at Nairobi, Kenya. pp. 17–36.

Wani, S.P., Ramakrishna, Y.S., Sreedevi, T.K., Thawilkal Wangkahart, Thang, N.V., Somnath Roy, Zhong Li, Yin Dixin, Zhu Hong Ye, Chourasia, A.K., Shiferaw, B., Pathak, P., Piara Singh, Ranga Rao, G.V., Rosana P Mula, Smitha Sitaraman and Communication Office at ICRISAT (2006b). Greening Drylands and Improving Livelihoods. Brochure published by ICRISAT

Chapter 10

Development of drought and salinity tolerant crop varieties

Vincent Vadez

International Crops Research Institute for Semi-Arid Tropics (ICRISAT), Patancheru, Hyderabad, India

10.1   Introduction

Water deficit, often referred to as "drought", is the most prominent abiotic stress, severely limiting crop yields and opportunities to improve livelihoods of poor farmers in the semi-arid tropics. For example, it is estimated that annually, drought causes US$ 520 M losses in the case of groundnut alone. One of the major research objectives is then to exploit modern science techniques to develop crops that are more resilient to low and erratic rainfall patterns. Conventional breeding of early maturing varieties, capable of escaping drought and maturing before water deficit becomes too severe, has been a common approach. However, substantial variation for drought tolerance, beyond earliness has neither been identified nor utilized extensively. However, conventional, yield-based selection for drought tolerance is difficult due to the large genotype by environment (GxE) interactions for yield under stress. In view of this, the application of molecular biology technologies holds great promise not only to improve, but also speed up the breeding of drought tolerant genotypes (Ribaut *et al.*, 1996).

Soil salinity is also an important limiting factor for crop yield improvement. It affects 5–10% of arable lands, with extent variously estimated at 75–100 million ha worldwide (Munns *et al.*, 2002). In fact, salinity affected areas are increasing because of mismanaged irrigation. A proper irrigation management could curb this increase, but such approach requires an important initial investment that often contradicts with the immediate economic choices of concerned farmers. The genetic option of improving crop for salt tolerance appears then to be the most direct alternative. Yet, despite the importance of salinity on the crop production worldwide, and the abundance of knowledge gathered on genes/mechanisms involved in salinity tolerance, there has been surprisingly little effort to breed for improved salinity tolerance, except for wheat, rice, barley, or soybean (Flowers, 2004). Developing salt tolerant crop varieties is critical to maintain productivity at current level in areas jeopardized by increased salinization, and to further increase world crop production in unexploited saline areas.

In this paper, we report on the steps and approaches to develop drought and salt tolerant varieties, using modern molecular techniques, taking examples of on going research at ICRISAT.

10.2   Approach, method and measurements for salinity tolerance

There are broadly two types of "salinity": (i) an excess Na in soil where pH remains within an optimal range for crop growth, which refers to coastal or dry land salinity

(Munns *et al.*, 2002), and is called "salinity"; and (ii) an excess Na in soil where pH is above 8.5–9.0, which refers to transient salinity, and is called "sodicity", (Rengasamy, 2002). In this review, we will focus on salinity only. Although the genetic basis for salinity tolerance and evidence of genetic diversity between and within species is well established, the difficulty remains as how to assess salinity tolerance and whether to approach salt stress research from the field or from the lab. Controlled environment evaluation of salinity tolerance uses artificial conditions, often using much higher salt levels than in natural saline conditions, does not evaluate salt tolerance based on harvestable yield (grain or fodder), and exposes plants to light and vapor pressure deficit (VPD) levels usually lower than those in the field. A plethora of work has evaluated salinity tolerance at germination or very early stages. Although it is fast and easy, its applicability is questionable and studies have shown a poor relation between the performance at germination or early stages under salinity to that at later stages (e.g. Flowers, 2004). However, controlled conditions ensure that salt treatment is homogenous, and that a non-saline control is used. Field evaluation of salinity tolerance results in large experimental errors because of heterogeneity of salinity in the field requires large plots to get reliable estimates of yield under field conditions, which limits the number of genotypes that can be assessed and which often cannot use a control non-saline field. An obvious advantage of field evaluation is that it assesses directly what a farmer would eventually need: yield under salinity. A combination of both controlled environment and field-testing is then needed.

At ICRISAT, a facility that combines the advantages of the controlled conditions and also allows assessing yield under saline conditions has been set up. This facility is located outdoors, is equipped with moveable rainout shelters, and uses large pots filled with natural soil. Salt application is made on a per unit soil basis, dissolved in irrigation water to ensure uniform distribution, at a rate previously determined for each crop. Adequate watering maintains salinity levels constant throughout the experimental period. Using this facility, it has been found that yield evaluation showed little relation to vegetative stage evaluation for salinity tolerance, across crops. Seed yield under salinity was in many cases related to the yield potential under control conditions, so that salinity tolerance per se accounts only for the part of the variation in the yield under salinity that is not explained by the yield potential. Therefore, these data show the need to evaluate plants under salinity along with a non-saline control.

10.3 Physiological and biochemical mechanisms of salt tolerance

The mechanisms by which plants tolerate salt are well described in the literature. These mechanisms are: (i) minimizing the entry of sodium into the plant; (ii) increasing the Na efflux from the root; (iii) avoiding Na loading into the xylem; (iv) storage of Na in stem tissues; (v) storage in non-vital cell compartments (vacuoles, apoplast); (vi) storage in old leaf blades; and (vii) cellular protection of key enzymes by specialized metabolites. Mechanisms, (i), (ii), (v) and (vii) are those receiving the most attention from the scientific community, in particular for the molecular studies, with a number of cell membrane and tonoplast transporters involved in Na extrusion or compartmentation, and a number of metabolites involved in cell protection. We argue that the importance given to mechanisms (i) and (ii) may be related to the high salt treatment often used to assess plant materials, whereby avoiding Na "invasion" becomes a prerequisite. We also argue that one aspect, i.e. the tolerance of the reproductive process to saline stress, has largely been overlooked.

10.4 Exploitation of the diversity for salinity tolerance

One key to efficiently use molecular breeding for traits conferring abiotic stress tolerance is to develop segregating populations from parents having maximum contrast for these traits. We have used that approach at ICRISAT and explored the diversity for salinity tolerance in large set of genotypes, including mini-core collection (Upadhyaya and Ortiz, 2001), i.e. set of germplasm that represent most of the diversity available in the entire collection.

10.4.1 Pearl millet

Pearl millet [*Pennisetum glaucum (L.) R. Br.*] Is often grown in saline lands and is known to be relatively more tolerant to salinity than other crop plants, particularly maize and legumes (Ashraf and McNeilly 1987), although there are relatively few studies on the effect of salinity on pearl millet. Pearl millet responses to salinity have been evaluated at germination stages, although poor relation was found with later evaluation (Krishnamurthy *et al.*, 2007a). A range of improved hybrid parental lines, hybrids, germplasm lines, and parents of mapping populations have been screened under saline conditions in the facility described above. We have found that a salt treatment of 21.06 g/9 kg Alfisol was optimal to reliably screen salinity tolerance. Several pairs of parents of existing RIL (Recombinant Inbred Lines) populations have been identified with a large contrast for salinity tolerance based on repeated trials. The contrast is sufficiently large to justify the search of QTLs (Quantitative Trait Loci) for salinity tolerance using these RIL populations. In pearl millet also, we have found little relation between the yield and the relative reduction in yield under salinity and the biomass or the relative reduction biomass at booting stage, showing that genotypic differences for salinity tolerance are likely explained by differences in the sensitivity of the reproductive stages to salt stress. Future focus of work in pearl millet would be to evaluate a large and representative set of germplasm to fully explore the salinity tolerance variability in pearl millet, to map QTL for salinity tolerance using the populations that appear to be suitable for that purpose, and to develop few more populations to validate these putative QTLs.

10.4.2 Sorghum

Sorghum [*Sorghum bicolor (L.) Moench*] is known to be relatively more tolerant to salinity than other crops such as maize or legumes, and has the potential to replace maize in saline soils (Igartua *et al.*, 1995). Many studies in sorghum have assessed salinity tolerance at the germination stage, although little relation was found with tolerance assessed at later stages (Munns *et al.*, 2003; Krishnamurthy *et al.*, 2007b). At ICRISAT, a range of improved hybrid parental lines, hybrids, germplasm lines, and parents of mapping populations have been screened under saline conditions in our facility. We have found that a salt treatment of 21.06 g/9 kg Alfisol was optimal to reliably screen salinity tolerance and that large contrast for salinity tolerance as assessed by the seed yield under salinity or the ratio of seed yield salinity/seed yield control exists. Significant positive relation between the seed yields in saline and non-saline conditions (r2 > 0.50) was found. As in pearl millet, we found no relation between the seed yield or seed yield ratio and the biomass or biomass ratio at vegetative stages, showing that assessment for salinity tolerance needs to be made at maturity and that salt effects on reproductive stages were the major explanation for genotypic difference in seed yield under saline conditions. Forthcoming efforts on sorghum tolerance to salt stress will focus on screening a large representative set of 300 sorghum genotypes for salinity tolerance, investigating the reasons for the sensitivity of reproductive stages to salt stress, and map QTL for salinity tolerance, from existing and new mapping populations.

10.4.3 Chickpea

Saline soils are very common in West and Central Asia and Australia, where chickpea is widely grown. It has been earlier stated that chickpea was fairly susceptible to salinity, and that there was not enough genetic variation to warrant breeding for that trait (Saxena, 1984). However, fairly large variation in sodicity tolerance was found later, and several tolerant sources were identified in India (Dua and Sharma, 1995). A salinity tolerant desi chickpea variety (CSG 8962) was released in India for salt affected soils. Work at ICRISAT has re-assessed the previous statements by Saxena (1984) and explored a large range of genotypes, including the mini-core. We found that a treatment of 8.88 g/7.5 kg Vertisol was optimal to reliably screen salinity tolerance. A 5-6-fold range of variation for seed yield under saline conditions was found across chickpea germplasm, with desi types being more salt tolerant than kabuli (Vadez et al., 2007). There also, little relation was found between the yield or yield ratio and biomass or biomass ratio at vegetative stages under salinity. In fact, salinity tolerance was well correlated to the ratio of seed number, indicating that tolerant genotypes were those able to maintain a large number of viable reproductive structures and those genotypic differences in salinity tolerance were explained by differences in the sensitivity at the reproductive stages. Na accumulation in shoots was low (0.1–0.6%) and not related to salinity tolerance, indicating that Na accumulation in shoot was not linked to differences in salinity tolerance. Forthcoming work will concentrate on identifying QTL for salinity tolerance using existing RIL and new populations developed specifically for that purpose, and investigate why and what processes during reproduction are affected by salinity that later lead to genotypic differences in salinity tolerance.

10.4.4 Groundnut

Very little work has been done on the effect of salinity on groundnut or peanut. Interest is increasing in countries like India where there is need for increased groundnut production. Salt-affected areas are potential targets for such increase in the groundnut production area. However, there has been no clear trait identified related to salinity tolerance. We have set up a protocol where suitable salt stress could be applied and salinity tolerance assessed in a large range of genotypes, including the mini-core collection. We have found that a salt treatment of 10.53 g/9 kg Alfisol is optimal to reliably screen for salinity tolerance (Srivastava et al., 2006). A five fold range of variation was found for pod yield among the genotypes tested. No relation was found between seed yield under saline and non-saline conditions and a poor relation was found between the biomass at maturity and seed yield under saline condition. Research orientation at the moment is focused on confirming the contrast in salinity tolerance in a large set of genotypes, developing RIL populations for QTL mapping based on contrasting genotypes, and assessing putative traits using contrasting genotypes to better understand the mechanisms of tolerance in groundnut.

10.4.5 Pigeonpea

It is not clear under what experimental basis it has been asserted that pigeonpea was relatively tolerant to salinity. Rather, pigeonpea appears to be among the most sensitive legume species to salinity. There appears to be appreciable genotypic differences in tolerance to salinity, although these differences were considered insufficient to undertake a breeding program. It has been reported that wild relatives of pigeonpea, especially *C. plathycarpus, C. sericeus and C. albicans* have more tolerance than cultivated types. Although F1 hybrids between these wild and cultivated germplasm were indeed more tolerant than the cultivated pigeonpea (Subbarao et al., 1990), the work has not been taken any further. We have recently screened the mini-core pigeonpea collection, along with a set of genotypes from 10 species of wild relatives of pigeonpea (Srivastava et al., 2007). It appears that given the high sensitivity of pigeonpea to salinity, there is a high rate of mortality in the trials, although many accessions do produce high amounts of biomass. Therefore, we have considered in this particular case that a biomass evaluation at a sufficiently advanced stage would be sufficient to screen tolerant materials. We found that a salt treatment of 5.26 g/9 kg Alfisol was optimal to reliably screen salinity tolerance. A large range of variation for biomass and biomass ratio exists. A number of *C. plathycarpus, C. sericeus and C. scaraboides* accessions were tolerant, whereas the other species tested were mostly sensitive. Given the extremely high sensitivity of pigeonpea to salt stress and the relative tolerance found in the wild relative, we believe that wide hybridization is the most promising avenue. Further work is now focusing on confirming the contrast in materials screened so far, develop inter-specific segregating populations, and study the mechanisms of tolerance, in particular Na loading in the xylem.

Therefore, it appears across all the crops tested for salinity tolerance at ICRISAT that a large range of variation exists and is exploitable for breeding. Pearl millet and sorghum are the most tolerant among ICRISAT mandated crops, followed by groundnut and chickpea, pigeonpea being extremely sensitive. In all cases except pigeonpea, we have found that there was no relation between an evaluation of salinity tolerance made at vegetative stage and the evaluation based on yield, which clearly indicate that salt tolerance needs to be evaluated at maturity. We also found across the crops a lack of relation between the relative reduction in biomass under salt stress and the relative reduction in seed yield, which indicate that the reproductive processes are likely to be the most vulnerable to salt stress and that understanding the mechanisms of salt tolerance probably requires the understanding of tolerance during reproduction.

10.5 A trait-based approach based on the yield architecture under drought

Drought screening is difficult under field conditions because of large genotypes-by-environment interactions. These are in part due to a lack of a clear understanding of the key traits conferring "drought tolerance" in a given environment. This is also explained by the difficulty to apply similar drought treatment to genotypes that do vary a lot in their phenology, a key factor to consider while tackling drought stress. An alternative to this is to approach drought tolerance through its putative components. In this respect, research is on going to exploit the genetic differences in drought avoidance mechanisms, i.e. better water capture (T) or transpiration efficiency (TE). These are two components of a simple model defined by Passioura (1977) where yield (Y) is defined as $T \times TE \times HI$, where HI (harvest index) represents how biomass is converted into grain. Beside the rationale, these component traits may have a higher heritability than yield, easing the identification of QTL. The trait-based approach has indeed yielded good success for some abiotic stresses (Sinclair *et al.*, 2000).

10.5.1 *Roots*

Work on chickpea has spearheaded research activities on roots at ICRISAT during the past 15 years. Chickpea is exclusively grown during the post-rainy season, and almost exclusively depends on stored moisture, thereby, facing terminal drought conditions. It has been found that besides earliness, deep and more profuse rooting was a direct contributor to the seed yield under terminal drought. Breeding for root traits is therefore ongoing. A dramatic improvement in the methods used to assess root traits has been achieved (Kashiwagi *et al.*, 2005). Using these methods, large variation for root traits has been found. One QTL accounting for over 30% of the variation in root length density has been identified (Chandra *et al.*, 2004) from a population developed between variety Annigeri (shallow roots) and ICC4958 (deep and profuse roots). New segregating populations have been developed to validate these QTLs and their phenotyping-genotyping is on going. Next step toward MAS breeding in chickpea is to initiate the introgression of these QTLs into locally adapted varieties.

Roots may also play a potential part in terminal drought QTL of pearl millet and staygreen QTL in sorghum. However, work is needed to assess the range of variation for root traits in these two crops and to assess the relation between rooting and drought tolerance. In preliminary experiments we also confirmed previous statements that root have large range of variation in groundnut.

10.5.2 *Transpiration efficiency (TE)*

Groundnut has led the work on TE during the past 15 years at ICRISAT. This crop is usually grown under rain fed conditions, and is often exposed to erratic rainfall patterns, thereby exposing the crop to intermittent drought spells at every stage of crop development. Current research has focused on developing varieties capable of efficiently using erratic rainfalls, having high TE. Genetic variability for TE has been found in groundnut (Wright *et al.*, 1994). Although surrogate traits of TE have been identified and used in previous breeding program, recent work shows the limit of their use for breeding purpose. A breeding approach to increase yield under drought using surrogates has not been superior to a conventional approach, in part due to a negative relation between TE and high HI. Rather, we have optimized the use of a gravimetric measurement of TE, where we can assess large number of entries. We have used these methods to assess the range of variation for TE in a large and representative set of groundnut germplasm and found a 4-fold range of variation for TE. Segregating populations for TE will be developed to map QTLs for TE.

10.5.3 *Towards improving HI under drought*

From Passioura's model, $Y = T \times TE \times HI$ (1977), virtually no effort has been made to improve the HI under water deficit. It is well known that reproduction is extremely sensitive to any abiotic stress, in particular drought (see lifetime work by JS Boyer, e.g. Boyer and Westgate, 2004). In the end, the success of agriculture is tightly bound to the success of reproduction. Preliminary work indicates that genotypic variation exists for the sensitivity of reproduction to water deficit in groundnut and chickpea. A thorough assessment of that variation is really needed. Also, we need to investigate how roots and TE, the other two components of the model, interact with HI. For instance, we found a positive relation between HI and profuse rooting in deep layer under severe drought. Therefore, research is needed to address T, TE, and HI in a comprehensive way, and investigate how these traits interact with each other.

10.6 Successes in marker assisted breeding for terminal drought tolerance

10.6.1 *Marker development and technology*

Breeding of drought tolerance requires having the necessary tools to breed. This includes the development of sufficient number of markers and/or the development of efficient marker technologies, so that putative molecular marker-phenotypic trait associations can be efficiently pinpointed. Over the past several years we have been able to develop and map a small number (ca. 40) of EST (Expressed Sequence Tag) – SSR (Single Sequence Repeat) markers in pearl millet, and establish protocols for doing this on a larger scale in sorghum (where much larger EST resources are available). We have also initiated exploitation of markers detected by conserved intron-spanning primers (Feltus *et al.*, 2006), and are adding these gene-based markers to base maps for both sorghum and pearl millet. Although sorghum enjoys genome wide genetic and physical maps, and soon genome sequence data, other ICRISAT mandated crops lack a reasonable number of molecular markers or good intraspecific genetic maps. Therefore, efforts are being to develop the genomic resources for these crops.

10.7 Successes in marker assisted breeding

10.7.1 *Pearl millet*

Research on genetic variation in grain yield under post-flowering stress indicated that as much as half of the variation in yield of pearl millet under terminal drought is explained by two non-stress related parameters: yield potential and phenology. Therefore, a drought response index (DRI) was calculated to quantify the remaining part of the variation, associated with tolerance/susceptibility (Bidinger *et al.*, 1987) and to identify traits linked to tolerance. Panicle harvest index (PNHI) was found to be closely related to the DRI, and was evaluated as a phenotypic selection criterion and as a target trait for QTL identification. A major QTL, which accounts for significant variation in grain yield and PNHI under terminal drought on linkage group 2 was identified in two mapping populations (Yadav *et al.*, 2002, Bidinger *et al.*, 2007). MABC transfer of the tolerant allele at this QTL into the background of terminal drought sensitive H77/833-2 and 841B, produced introgression lines with superior terminal drought tolerance under the same environment in which the QTL was identified (Hash *et al.*, 2005). Recent work on selected QTL introgression lines suggested that more profuse rooting in deeper soil layer might be a major underlying factor to that QTL. Efforts are on going to further characterize the role of roots in that QTL.

10.7.2 Sorghum

The focus has been on reduced leaf senescence, the "stay-green" trait, which prolongs photosynthesis under declining soil moisture. We have used published information (Haussman *et al.*, 2002) to add QTLs stg1, stg2, stg3, stg4, stgA and stgB, from the donor parent B35 to a number of senescent lines in a MABC program. An evaluation of BC 1 and BC 2 lines confirmed: (i) the expression of the trait in post rainy season line R 16, (ii) an improvement of stover quality, (iii) a modest improvement in grain filling and grain yield under terminal stress. Evaluation of BC 3 and BC 4 backcross progenies from the same recurrent parent with several stay green QTL is on going. Initial data indicate that some of these are much better agronomically than the BC1 s and significantly less senescent than R 16. A number of markers have been added in the vicinity of QTL Stg3, which will help in background selection to more quickly recover the recurrent parent phenotype. More markers in the vicinity of QTLs that are tightly linked to undesirable characters would be highly needed to speed up and precise that process. We also hypothesized that the staygreen trait, i.e. the maintenance of functional green leaves late into the maturation cycle, would necessarily require drought avoidance mechanisms (increased water uptake, or water saving strategies). We found that under water deficit roots of staygreen materials are about 60 cm deeper than in senescent materials.

10.8 Transgenics, a form of "functional" diversity

This approach could be very useful to speed up the process of molecular introgression of putatively beneficial genes that offer enhanced tolerance to drought. Several gene transfer approaches have been attempted to improve drought tolerance in different crops. However, a "single-gene" transgenic approach may not be suitable for developing drought tolerance since abiotic stress tolerance is a multigenic trait. A wiser approach may be using transcription factor/s, i.e. major "switch" to trigger a cascade of genes in response to a given stress (Chinnusamy *et al.*, 2005). This approach has been undertaken at ICRISAT in groundnut and chickpea by using the drought responsive element (DRE) DREB1A from *Arabidopsis thaliana* driven by the stress-responsive promoter rd29A from *A. thaliana*. A large number of independently transformed events of groundnut and chickpea have been developed. These are undergoing evaluation under laboratory and greenhouse conditions. Fourteen transgenic events of groundnut showed significant differences in the kinetics of transpiration response to soil drying. Several T3 generation transgenics had higher TE than the wild type parent under well-watered conditions, and that one event had higher TE than the wild type parent across moisture conditions. These materials showing a large variation for TE are isogenics to JL24, and offer an ideal material to study the physiological basis for the differences in TE. We have also found that DREB1A events, grown in large and long PVC tubes had increased rooting when submitted to water deficit, and this correlated with a high water uptake than in wild type JL24.

10.9   Conclusions

Advances towards the use and application of MAS (Marker Assisted Selection) breeding to develop new varieties with drought and salinity tolerance are making significant progress, at different stages across crops. We have been successful at exploiting the diversity to reveal superior genotypes for different traits. In the case of groundnut, backcross populations from re-synthesized and cultivated groundnut are also likely to reveal interesting recombinants for drought. In case MAS breeding can readily take place, further saturation of the QTL location to speed up and ease the identification of suitable introgression lines, or to get rid of linkage drag is needed. Considerable effort is being dedicated to root and there is a need to further refine/simplify the phenotyping protocols for root traits, and achieve a quantum leap in root research, by focusing more on what roots "do" rather than what root "are", and then develop the phenotyping capacities for such alternative traits. These would be used to explain the role of root in sorghum and pearl millet QTLs. Finally, there is also a need to explore genotypic variation for HI under drought conditions, and to investigate the relations, synergies, trade-off between water uptake, TE and HI. In this, transgenic materials presented here will be very useful, in particular because preliminary data indicate that DREB1A also influences the root development, and the success of reproduction under water deficit.

References

Ashraf, M. & Mcneilly, T.M. (1987) Salinity effects on five cultivars/lines of pearl millet (*Pennisetum americanum [L] Leeke*). *Plant Soil*, 103: 13–19.

Bidinger, F.R., Mahalakshmi, V. & Rao, G.D.P. (1987) Assessment of drought resistance in pearl millet [*Pennisetum americanum (L.) Leeke*]. I. Factors affecting yield under stress. *Australian Journal of Agricultural Research*, 38: 37–48.

Boyer, J.S., & Westgate, M.E. (2004) Grain yields with limited water. *Journal of Experimental Botany*, Water Saving Agriculture Special Issue, 1–10.

Chandra, S., Buhariwalla, H.K., Kashiwagi, J., Harikrishna, S., Sridevi, R.K., Krishnamurthy, L., Serraj, R. & Crouch, J.H. (2004) Identifying QTL-linked Markers in marker-deficient Crops. In: Fischer, T. *et al.* (2004). *New directions for a diverse planet*: Proceedings for the 4th International Crop Science Congress, Brisbane, Australia.

Chinnusamy, V., Jagendorf, A. & Zhu, J.K. (2005) Understanding and Improving Salt Tolerance in Plants. *Crop Sci.*, 45: 437–448.

Dua, R.P. & Sharma, P.C. (1995) Salinity tolerance of Kabuli and Desi chickpea genotypes. *Int. Chickpea and Pigeonpea Newsletter*, 2, 19–22.

Feltus, F.A., Singh, H.P., Lohithaswa, H.C. Schulze, S.R. Silva, T.D. & Paterson A.H. (2006). A Comparative Genomics Strategy for Targeted Discovery of Single-Nucleotide Polymorphisms and Conserved-Noncoding Sequences in Orphan Crops. *Plant Physiol.*, 140: 1183–1191.

Flowers, T.J. (2004) Improving crop salt tolerance. *J. Exp. Bot.*, 55: 307–319.

Hash, C.T, Rizvi, S.M.H., Serraj, R., Bidinger, F.R., Vadez, V., Sharma, A., Howarth, C.J. & Yadav, R.S. (2005) Field Assessment of Backcross-derived Hybrids Validates a Major Pearl Millet Drought Tolerance QTL. Paper presented at the Crop Science Society of America Congress, Seattle.

Haussmann, B.I.G., Mahalakshmi, V., Reddy, B.V.S., Seetharama, N., Hash, C.T. & Geiger, H.H. (2002) QTL mapping of stay-green in two sorghum recombinant inbred populations. *Theor. Appl. Genet.*, 106: 133–142.

Igartua, E., Garcia, M.P., Lasa, J.M. (1995). Field responses of grain sorghum to a salinity gradient. *Field Crops Res.*, 42: 15–25.

Kashiwagi, J., Krishnamurthy, L., Serraj, R., Upadhyaya, H.D., Krishna, S.H., Chandra, S. & Vadez, V. (2005) Genetic variability of drought-avoidance root traits in the mini-core germplasm collection of chickpea (*Cicer arietinum L.*). *Euphytica*, 146, 213–222.

Krishnamurthy, L., Serraj, R., Rai, K.N., Hash, C.T., & Dakheel, A.J. (2007a) Identification of salt tolerant pearl millet (*Pennisetum glaucum (L.) R. Br.*]) breeding lines. *Euphytica* (in press).

Krishnamurthy, L., Serraj, R., Hash, C.T., Dakheel, A.J. & Reddy, B.V.S. (2007b) Screening sorghum genotypes for salinity tolerance biomass production. *Euphytica* (in press).

Munns, R., Husain, S., Rivelli, A.R., James, R.A., Condon, A.G., Lindsay, M.P., Lagudah, E.S., Schachtman, D.P., Hare, R.A. (2002) Avenues for increasing salt tolerance of crops, and the role of physiologically-based selection traits. *Plant and Soil*, 247, 93–105.

Passioura, J.B. (1977) Grain yield, harvest index and water use of wheat. *J. Aust. Inst Agric. Sci.*, 43: 21.

Rengasamy, P. (2002) Transient salinity and subsoil constraints to dryland farming in Australian sodic soils: an overview. *Aust. J. Exp. Agriculture*, 42, 351–361.

Ribaut, J.M., Hoisington, D.A., Deitech, J.A., Jiang, C. & Gonzalez-de-Leon, D. (1996) Identification of quantitative trait loci under drought conditions in tropical maize. 1. Flowering parameters and the anthesis-silking interval. *Theor. Appl. Genet.*, 92: 905–914.

Saxena, N.P. 1984. Chickpea. In: *The Physiology of Tropical Field Crops*, Eds Goldworthy and Fisher. pp. 419–452, John Wiley & Sons Ltd. New York.

Sinclair, T.R., Purcell, L.C., Vadez, V., Serraj, R., King, C.A., Nelson, R. (2000) Identification of soybean genotypes with N2 fixation tolerance to water deficits. *Crop Science*, 40 (6): 1803–1809.

Sinclair, T.R. et al. (2000) Identification of soybean genotypes with nitrogen fixation tolerance to water deficit. *Crop Sci.*, 40: 1803–1809.

Srivastava, N., Vadez, V., Krishnamurthy, L., Saxena, K.B., Nigam, S.N., & Aruna, R. (2007) Screening technique for salinity tolerance in groundnut (*Arachis hypogaea*) and pigeonpea (*Cajanus cajan*). *Indian Journal of Crop Science* (accepted).

Srivastava, N., Vadez, V., Upadhyaya, H.D., & Saxena, K.B. (2006) Screening for intra and inter specific variability for salinity tolerance in pigeonpea (*Cajanus cajan*) and its related wild species. *Journal of SAT Agriculture Research*, 2, <http://www.icrisat.org/journal/cropimprovement/v2i1/v2i1screeningfor.pdf>

Subbarao, G.V., Johansen, C., Jana, M.K., & Rao, J.V.D.K. (1990) Physiological-Basis of Differences in Salinity Tolerance of Pigeonpea and Its Related Wild-Species. *Journal of Plant Physiology*, 137: 64–71.

Upadhyaya, H.D. & Ortiz, R. (2001) A mini core subset for capturing diversity and promoting utilization of chickpea genetic resources in crop improvement. *Theo. Appl. Genet.*, 102: 1292–1298.

Vadez, V., Krishnamurthy, L., Gaur, P.M., Upadhyaya, H.D., Hoisington, D.A., Varshney, R.K., Turner, N.C. & Siddique K.M.H. (2007) Large variation in salinity tolerance is explained by differences in the sensitivity of reproductive stages in chickpea. *Field Crop Research* (Accepted).

Wright, G.C., Nageswara Rao, R.C. and Farquhar, G.D. (1994) Water-use efficiency and carbon isotope discrimination in peanut under water deficit conditions. *Crop Science*, 34: 92–97.

Yadav, R.S., Hash, C.T., Bidinger, F.R., Cavan, G.P. & Howarth, C.J. (2002) Quantitative trait loci associated with traits determining grain and stover yield in pearl millet under terminal drought stress conditions. *Theor Appl Genet.*, 104: 67.83.

Chapter 11

System of Rice Intensification (SRI) to enhance both food and water security

Norman Uphoff

CIIFAD, Cornell University, Ithaca, NY, USA

11.1 Introduction

SRI is a biologically-based technology that differs from what is currently understood as 'biotechnology' in that it involves no genetic manipulation or modification. Instead it capitalizes upon productive potentials that already exist within the genomes of rice (and other crops), inducing different, more productive phenotypes from available genotypes (Uphoff, 2007). While the reasons for methodology's effectiveness are still not fully understood, positive results from its alternative management practices have already been seen in >25 countries (Satyanarayana *et al.*, 2006; Uphoff, 2006). This means that to the important question are no longer whether SRI is effective, but why? Explanations for the increased productivity of resources used in irrigated rice production that can be attributed to SRI methods are mostly coming from a better understanding of soil biological factors and of plant-soil-microbial interactions (Randriamiharisoa *et al.*, 2006).

The central concern of this book – how to achieve more crop per drop? – is directly addressed by SRI in that its methods, discussed below, improve the capacity of plants to utilize soil water resources by enhancing their growth of larger, healthier root systems. Improved root functioning is a generalizable 'solution' to the problem of how to make cropping systems less vulnerable and more resilient in the face of climate stress. Yet surprisingly little attention has been paid to roots in rice science or in crop science generally (DeDatta, 1981; Matsuo *et al.*, 1997).

11.2 The System of Rice Intensification

SRI was developed over two decades by Henri de Laulanié, S.J., a Frenchman who made Madagascar his home for 34 years. He was trained in agriculture before entering a Jesuit seminary in 1941 so he had some scientific training before being sent to Madagascar in 1961. Concluding that enhancing smallholders' rice production could do more to reduce hunger and poverty in this country than any other innovation, he devoted himself to finding ways that would raise the productivity of the land, labor, water and capital devoted to rice cultivation without being dependent on external, purchased inputs, recognizing that most rural households had little access to such inputs, whether for logistical or financial reasons.

Over 20 years, Laulanié assembled a set of practices – some from observing unusual farmer practices, some from himself adapting recommended 'modern' practices, and one through serendipitous good luck (Laulanié, 1993) – that could increase rice yields not just incrementally but actually by multiples. Such increases were achieved by synergistic interaction among the practices, promoting larger root systems and canopies, each giving positive feedback to the other, and further by promoting soil fertility by enhancement of the abundance and activity of soil biota.

These practices were put together empirically, quite inductively, without any theory behind them, trying to determine what growing conditions were most favorable for rice plants. However, the various mechanisms and processes involved become better understood in the last 6–8 years (Randriamiharisoa *et al.*, 2006). Initially there was considerable skepticism and even resistance among rice scientists (Surridge, 2004). But as 'the SRI effect' is being reported from more and more countries, acceptance is increasing. Scientists in the three largest rice-producing countries, China, India and Indonesia, have validated SRI results (e.g., Zheng *et al.*, 2004; Zhu, 2006; Subbiah *et al.*, 2006; Gani *et al.*, 2002). The 25th and 26th countries where SRI benefits have been confirmed are Zambia and Iran, with Bhutan, Iraq and Burkina Faso expected to follow suit soon.

11.2.1   *SRI in operational terms*

SRI is not a technology, i.e., some package of practices or physical inputs that is fixed or static. Rather, SRI is based on a number of insights into how rice can be grown most successfully. In his only published article on SRI (1993), Laulanié wrote that he learned from the rice plant: 'the rice plant is my teacher (*mon maître*).' SRI as a set of insights and principles changes, often radically, the rice-growing practices that farmers have followed for generations. At the same time, rather than being simply adopted or implemented, the practices derived from these practices are to be *adapted* by farmers' to own conditions, encouraging further innovation. There is no need to go into the specific practices here as they can be accessed with detailed discussions from the SRI home page (http://ciifad.cornell.edu/sri/). When SRI practices are used together, the resulting rice plants are more productive and healthier *phenotypes* that are derived from whatever genotypic potential they start with. Examples are seen from Figures 11.1 and 11.2.

SRI can be presented as six basic ideas that constitute it operationally as follows:

(1) *Whenever establishing a rice crop by transplanting, use very young seedlings* – preferably 8–12 days old, but no more than 15 days old. It can be shown that older seedlings have less potential for profuse tillering and prolific root growth than do young seedlings. This can be explained in terms of *phyllochrons*, which are regular intervals of plant growth that apply for all gramineae species (Nemoto *et al.*, 1995; explained in Stoop *et al.*, 2002). This same effect has been demonstrated by researchers in India – and has been seen by farmers – with finger millet (*Eleusine coracana*). Early transplanting also enhances finger millet's root and shoot growth (see Figures 11.3 and 11.4). So we are finding that there can be broader application of SRI concepts and practices than was at first evident just with rice.

(2) *Rice plants should be planted singly* – rather than in clumps or hills of 3–6 plants. It is well known that when plants of other species are planted close together, this inhibits their growth of roots and canopy. Rice (*Oryza sativa*) is not really different from other plants. Rice seedlings should be placed in the soil very shallow and should be handled carefully, understanding that any trauma to tender roots will impair the plant's subsequent performance.

Plate 11.1 Indonesian farmer holding two rice plants of same variety and maturity, the one on left grown with SRI practices and the one on right grown with conventional practices. Picture courtesy of Shuichi Sato, Nippon Koei. (See *Colour plate 11.1*)

(3) *Rice plants benefit from wider spacing* – preferably in a square pattern rather than in rows so that the plants get optimum exposure to sunlight and air on all sides. The recommended spacing to start with is 25×25 cm (16 plants/m^2); but if the paddy soil is more fertile (or becomes more fertile), wider spacing will give still higher yields. Wider spacing, rather than crowding, helps achieve, throughout the entire field, what agronomists have long recognized as 'the edge effect' because it ensures that all the leaves on all the plants receive enough solar radiation for photosynthesis. No leaves need to be 'subsidized' by other leaves' photosynthesis because of shading. It is counterintuitive that a lower plant population/m^2 will give a higher yield; but conventional spacing within/between hills is too dense for best plant growth. Spacing is, of course, something to be optimized rather than maximized, because one wants to have the greatest number of fertile tillers/m^2 and the most grains/tiller, to get highest yield – not just per plant.

(4) *Paddy soils should be kept moist but not continuously saturated.* Contrary to what rice farmers have believed and rice scientists have written (DeDatta, 1981), rice is not an aquatic plant, nor does it not perform best when grown under submerged, hypoxic soil conditions. This should have been appreciated long ago because in any field that is not well-leveled, the rice plants which grow in the lower, waterlogged parts are usually stunted, compared to those that grow in the

Plate 11.2 Farmer in Dông Trù village, Hanoi province, Vietnam, holding up two rice plants, SRI (on left) and non-SRI (on right). Behind her are the fields from which the plants were taken, shown after a typhoon passed over the village. Note lodging of the non-SRI rice crop on right and resistance to lodging of the SRI crop on left. Picture courtesy of Elske van de Fliert, FAO IPM Program, Hanoi. (See *Colour plate 11.2*)

higher, better-drained areas, which thrive. With SRI, farmers are advised either to apply small amounts of water daily, not flooding their fields so that soil and plants are well exposed to the sun and air; or to practice alternate wetting and drying of their fields, flooding them for a few days but then leaving them unflooded for some number of days. Anaerobic soil conditions not only limit the oxygen available to plant roots but also change the composition and size of populations of soil organisms, reducing or even eliminating aerobic microorganisms and most meso- and macrofauna in the soil.

(5)   *As much organic matter as possible should be added to the soil.* SRI was developed with chemical fertilizer during the 1980s, but when government subsidies were removed and small farmers could no longer afford fertilizer, Laulanié found that compost – any decomposed biomass – gave even better results and involved only labor, no cash costs. SRI is pragmatically rather than ideologically 'organic.' While almost all agronomists acknowledge that organic nutrition is most beneficial for soil fertility, many argue that organic fertilization is not feasible or

Plate 11.3 Comparison of typical finger millet (*Eleusine coracana*) plants grown with different crop management methods. On left is modern variety (A404) grown with adaptation of SRI methods; in center, same variety with conventional methods; on right, local variety grown with conventional methods. Picture courtesy of Ashish Anand, PRADAN Khunti team, Jharkhand state, India. (See *Colour plate 11.3*)

economic, mixing scientific with practical considerations. However, this sidesteps the point that sufficient organic fertilization can, by itself, give the best results – although combining organic and inorganic sources of fertilization can also work effectively with other SRI methods. Compost should be evaluated not so much in terms of its own nutrient content, but more in terms of what it does to improve soil structure and biological diversity and activity (Uphoff 2006). The long-standing motto of organic farmers – don't feed the plant; rather, feed the soil, and the soil will feed the plant – is relevant here, for reasons discussed more below.

(6) *Aerate the soil actively, as much as possible.* When paddy fields are not kept continuously flooded, weeds become more a problem. They can be controlled by manual weeding or by herbicides, but these practices do not aerate the soil as does the use of a rotary weeder (or conoweeder). These simple and inexpensive implements churn up the soil and bury weeds while breaking up the top soil horizons, working oxygen and nitrogen into the soil. Doing more mechanical weedings than are needed just to control weeds can add several tonnes/hectare to rice yields simply as a result of soil aeration. Investments of labor in this kind of weed control become thus a benefit rather than just a cost.

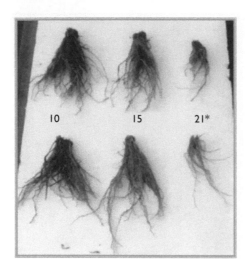

Plate 11.4 Roots of two varieties of finger millet (*Eleusine coracana*) at 60 days of age, having been transplanted at 10, 15 or 21 days after emergence. As with SRI, seedlings transplanted beyond 15 days demonstrate less growth potential. Picture courtesy of Dr. A. Satyanarayana, director of extension, Acharya N. G. Ranga Agricultural University, Hyderabad. (See *Colour plate 11.4*)

11.2.2 SRI results

What is the effect of these practices when used together? Table 11.1 summarizes the results of 11 evaluations done in eight countries by a variety of institutions – national or state agricultural universities in China and India, government agencies (district agricultural development office in Nepal, national IPM program in Vietnam), NGOs in Bangladesh and Cambodia, a donor agency (GTZ in Cambodia), an international research center (International Water Management Institute in India and Sri Lanka), a private company (Syngenta in Bangladesh), and a consulting firm (Nippon Koei in Indonesia).

We see that results vary considerably because of inter-country differences in growing conditions, but also because SRI practices were not (yet) all being used fully or always being used by all farmers as recommended. Because agriculture is a biologically-based/driven enterprise, the same practices are unlikely to have the same effects in all environments. It is a fact of life (of nature) that biological potentials and processes can differ considerably.

Even so, the average results are quite remarkable: an average yield increase of 52% vs. control or comparison yields, with a 44% average reduction in water use and 25% lower costs of production per hectare, which leads to more than doubling of farmers' net income per hectare. This represents unprecedented gains in productivity. SRI is the only innovation that I know of where increases in the productivity of land,

Table 11.1 Summary on SRI effects on yield, water saving, cost reduction, and net income

Country	Evaluation done by/for:	Yield increase	Water saving	Reduction in costs	Increase in net income	Data base
BANGLADESH	BRAC/SAFE/ POSD/BRRI/ Syngenta-Bangladesh Ltd.	24%	Not measured	7%	59% (32–82%)	On-farm evaluations (N = 1,073), funded by IRRI PETRRA
CAMBODIA National survey: covering 5 provinces	German Agency for Development Cooperation (GTZ)	41%	Flooding at TP from 96.3% ↓ to 2.5%	56%	74%	Random sample survey of 400 SRI users and 100 non-users (N = 500)
Long-term SRI users: 36 villages in 5 provinces	Center for Study and Dev. of Cambodian Agriculture (CEDAC)	105%	50%	44%	89%	Farmers who had used SRI for 3 years (N = 120)
CHINA Xinsheng village, Sichuan province	China Agricultural University (CAU)	29%	44%	7.4%	64%	SRI use had gone from 7 (2003) to 398 (2004) (N = 104)
INDIA Tamil Nadu: Tamiraparani basin	Tamil Nadu Agricultural University (TNAU)	28%	40–50%	11%	112%	Supervised on-farm (N = 100)
Andhra Pradesh: All 22 districts	AP Agricultural University (ANGRAU)	38%	40%	NA	NA	Supervised on-farm trials (N = 1,525)
West Bengal: Purulia district	International Water Management Institute (IWMI) – India Programme	32%	Rainfed version of SRI	35%	67%	SRI use in demo villages had gone from 4 farmers to 150 in 3 seasons
INDONESIA S. Suluwesi and Nusa Tenggara provinces	Nippon Koei – Decentralized Irrigation System Improvement Project	84%	40%	24%	412%	3 years of on-farm evaluation trials on 1,363 ha (N = 1,849)
NEPAL Morang district	District Agricultural Development Office, Biratnagar	82%	43%	2.2% Rotary does not yet widely available	163%	SRI users in the district went from 1 in 2003 to >1,400 in 2005 (for data: N = 412)
SRI LANKA Ratnapura and Kurunegala districts	International Water Management Institute (IWMI)	44%	24%	11.9–13.3%	90–117%	Survey of 60 SRI users and 60 non-users, randomly (N = 120)
VIETNAM Dông Trù village, Hanoi province	National IPM Program	21%	60%	24%	65%	Record-keeping on SRI results done by Farmer Field School alumni
AVERAGE		52%	44%	25%	128%	

From: Uphoff (2006), which gives references for data sources.

of labor, of water and of capital can be achieved all at the same time – without some tradeoff where the increased productivity of one factor is achieved with a reduction in another factor's productivity.

11.3 Costs and benefits

There are invariably some costs involved in any improvement. The most certain is that for best results, good water control needs to be maintained. Investment in infrastructure and/or organization for managing water more carefully may have to be made, to be able to supply smaller amounts of water on a regular, reliable basis. This cost can be offset by the value of water saving. But it is an important consideration.

Also, more labor is likely to be required initially per hectare or per season while farmers are learning the new methods and becoming skilled in them. This is more than paid for by higher labor productivity, more output of rice per hour or per day of labor. But some farmers may find the greater labor expenditure difficult to provide (Moser and Barrett, 2003). Interestingly, it is increasingly reported that – once SRI techniques have been mastered – farmers using SRI are able to *save labor* as well as to reduce their requirements for water, seed and cash. Once the benefit of labor saving becomes better known and is well documented, SRI should become even more attractive to farmers.

Additional benefits that are by now confirmed in more than two dozen countries include:

- *Greater resistance of the SRI crop to pest and disease losses,* making the use of chemical means of protection either not necessary or not economic. This may be explainable by the theory of trophobiosis proposed by Chaboussou (2004).
- *Greater resistance to abiotic stresses* – water stress and drought, storm damage, cold snaps, heat waves, etc. This is attributable to the larger and stronger root systems which enable SRI plants to resist lodging (falling over), even despite their larger and heavier panicles of grain. This was seen from Figure 11.2.
- *Higher outturn of milled rice* – in addition to the greater production of paddy (unmilled rice), when SRI paddy is milled, there is about 15% more of polished rice (kg/bushel of paddy). This is because with SRI, there is less chaff (fewer unfilled grains) and less shattering (fewer broken grains). This is a bonus with SRI methods, which produce grains that are heavier and denser but not necessarily larger.

Potential benefits that have not yet been adequately studied, but which we expect are associated with SRI on the basis of theory, observations and some fragmentary evidence are:

- *More nutritional value and grain quality* – for the same reasons that SRI paddy is heavier and resists shattering. SRI root systems are larger and go deeper into the soil, so uptake of micronutrients is likely to be greater. We know that chalkiness in SRI grain is 30–65% less.
- *Reduction in greenhouse gas emissions* – methane emissions from paddy fields will be reduced when rice is grown under aerobic soil conditions. Rice paddies are a major source of this greenhouse gas. However, no studies have been done to see whether there is more production of nitrous oxide when previously-flooded soils

are not kept flooded. If there are no applications of inorganic nitrogen fertilizer to drive this process, there could be little or no increase in nitrous oxide generation.

11.4   Biological dynamics contributing to SRI results

These beneficial outcomes from SRI practice are driven by two basic elements that form the foundation of SRI, as elaborated in Randriamiharisoa *et al.* (2006):

- *Root systems* that are larger and healthier, and
- *Populations of soil biota* in, on and around these roots that are more abundant, diverse and active, providing a host of services to plants.

These two elements are interactive and symbiotic. Root systems, especially ones with larger and photosynthetically more active canopies, produce more root exudates, i.e., carbohydrates, amino acids, organic acids, vitamins, phytohormones and other compounds, that are put into the soil immediately adjacent to the roots, known as the rhizosphere (Pinton *et al.*, 2001; Römheld and Neumann, 2006). These compounds benefit a huge array of soil organisms that constitute a dynamic food web (Thies and Grossman, 2006). Microorganisms that dwell on and around the roots are 'grazed' by protozoa and other soil organisms, which in turn are themselves food for still larger organisms in the soil.

Larger soil organisms improve soil structure by aggregation of small mineral particles and by making pores that permit air and water to penetrate and circulate in the soil. Of particular importance to crops, they make soil more absorptive and retentive of water. At the same time, smaller organisms attend to the nutritional needs of plants by fixing nitrogen, cycling N, solubilizing phosphorus and other minerals, producing phytohormones that stimulate plant roots whose growth in turn furnishes (through exudation) more energy supplies to soil organisms, inhibiting root pathogens, or even producing antibiotic substances that protect roots, and many other services (e.g., Uphoff *et al.*, 2006).

11.4.1   *Eliciting more productive phenotypes through biological activity*

A demonstration of the differences in phenotype that can result from the same genotype is seen in Figure 11.5, showing two rice plants of same age (52 days after planting) and same variety (VN 2084). Both were started in the same nursery, but the SRI plant on the right was removed at 9 days and put into an SRI growing environment with wide spacing, soil organic matter, and soil aeration. The plant on the left was taken out of the nursery at 52 days and before being transplanted was compared with an SRI plant randomly pulled up from the SRI plot. We have no laboratory data on soil organisms or phytohormones in the rhizosphere of these plants, but such profuse root growth needs to have some active stimulation and support from the environment, e.g., phytohormones such as are known to be secreted by aerobic soil bacteria and fungi, not being a matter just of nutrients being passively available in the rhizosphere.

Research has been done for several decades on the contributions that soil organisms make to plant growth and health, but there has been little impact on soil science, which has continued to analyze soils mostly in terms of their chemical and physical

Plate 11.5 Two rice plants of same age (52 days) and variety (VN 2084), grown by Luis Romero, San Antonio de los Baños, Cuba, as explained in text. Picture courtesy of Dr. Rena Perez. (See *Colour plate 11.5*)

properties. It has left the more complex, ambiguous and frustrating work on soil biology to soil microbiologists and ecologists, who are fewer in number. Recent work has been showing, however, how intimate and synergistic are plant-microbial interactions, which can be dated in the fossil record back more than 400 million years.

For example, rhizobacteria previously known to have beneficial effects on leguminous crops are also active in rice, either living as endophytes in rice plant roots where the contribute to higher yield (Dazzo and Yanni, 2006) or in the phyllosphere in and on leaves where they also enhance yield (Feng *et al.*, 2005). It gives a very different perspective on crop production and performance to know that soil bacteria live beneficially in plant roots to mutual advantage and that they also migrate from there up to plants' leaves, where they increase the level of chlorophyll and of photosynthetic activity, contributing thereby to higher crop productivity.

Crop research at the molecular level is now demonstrating hormonal signaling between root zone and canopy that affects gene expression in the cells of tissues in the plant leaves. To understand why tomatoes grown with 'modern' methods were yielding 30% less than the same variety under more 'organic' management, Mattoo and Abdul-Baki (2006) analyzed certain genes known to affect plant senescence and resistance to pests. They found that plants with higher doses of N fertilizer (200 kg/ha) and black plastic mulch had their gene for senescence switched on and another affecting chitinase production switched off; conversely, plants with leguminous mulch and reduced applications of N fertilizer (100 kg/ha) had the opposite genetic responses – senescence switched off and chitinase production on – related to hormonal signaling.

We have not had the research support needed to study such processes and mechanisms in SRI rice, but the repeated and evident superiority of its productivity and health supports explanations that focus on root-microbial interactions, affecting phytohormones, genetic expression, and physiological performance. The most visible difference between the 'modern' and 'organic' tomato plants evaluated by Matoo and Abdul-Baki was in the size and evident health of their respective root systems.

11.4.2 *Extension of SRI learning to other crops*

These dynamics are not restricted to rice, as seen from the extrapolations of SRI concepts and methods to other crops. Yields of finger millet (*Eleusine coracana*) and sugar cane (*Saccharum officinarum*) have been increased, with doubling and even tripling in some cases, without relying on inorganic fertilization. An NGO known as PRADAN (Professional Assistance for Development Action) has begun working with rainfed farmers on what it calls the System of Finger Millet Intensification. A comparison of the phenotypic differences between SF_mI plants and millet grown with conventional methods in Jharkhand state of India is shown in Figure 11.4. In Andhra Pradesh, a set of management practices is being promoted called the Sugarcane Renewed Intensification system (http://www.financialexpress.com/fe_full_story.php?content_id=147399) with good effect.

Farmers elsewhere in India have been encouraged by their experience with SRI to adapt its concepts and methods to their production of cotton, vegetables and other crops. Thus, the attention that SRI has focused on plant spacing, on the handling of seedlings, on soil aeration, and on greater use of organic matter to enhance soil fertility – all with the aim of benefiting soil organisms as well as plant roots – is shaking up centuries-old practices in order to capitalize on previously underutilized and even unrealized genetic potentials. Getting plants with larger and better-functioning root systems and nurturing more fertile soils based on the abundance, diversity and activity of soil biota represent promising elements of a strategy for sustaining Indian agriculture.

11.5 Broader relevance of SRI experience

The System of Rice Intensification has emerged on the world scene at an opportune time. The Green Revolution was perhaps the largest single poverty-reduction program ever implemented. However, in the last 10–15 years, there has been a stagnation in the yield increases achieved through genetic improvements coupled with intensified use of agricultural inputs (fertilizers, pesticides, herbicides, etc.), as reviewed in Uphoff (2006a). The Green Revolution has been losing momentum, and the question is now: what do we do for an encore? Doing 'more of the same' is no longer very promising.

Experience with SRI pushes everyone to think more about 'soil systems' than about 'soil' per se (Uphoff, 2006). The physical-chemical elements of soil that have preoccupied most agronomists are always best understood in relation to the functioning of biological elements (or the absence thereof) in soil systems, since soil productivity is a function of the abundance and activity, as well as diversity, of soil organisms; 'The microbial flora causes a large number of biochemical changes in the soil that *largely determine* the fertility of the soil' (DeDatta, 1981:60, emphasis added). Having and retaining sufficient water in the soil to support these populations is crucial for crop fertility.

It is encouraging that SRI concepts have been adapted to rainfed conditions in India, Myanmar and the Philippines with doubling and even tripling of yields that are not dependent on irrigation facilities (Gasparillo *et al.*, 2003; PRADAN, 2006; Kabir, 2006). The provision of organic matter to the soil, along with different plant and soil management practices can make substantial improvements in soil structure and water retention which in turn support the fertility-enhancing effects of soil organisms. There is much that is still not understood or scientifically demonstrated about such systems, but the results are real and objective. It is not a matter of discovering something new but of determining what are the mechanisms and limits of such fertility increases, as well as knowing better the sustainability of such systems and how to support their long-term productivity.

Part of the success of SRI hinges on the participatory approach which has been part of the SRI methodology from the beginning. The new rice system has been promoted in Madagascar by an NGO established by Fr. de Laulanié and Malagasy colleagues, *Association Tefy Saina*. SRI is regarded an entry point for human development, not just as a means to grow more rice. Farmers are encouraged to begin on a small scale, to master and gain confidence in these new techniques that change 'the ways of the ancestors,' which are revered in traditional culture and belief. Farmers' engagement in experimentation and evaluation is part of the dissemination strategy, so that they become not adopters but adapters – partners who will make further improvements in their agriculture and other aspects of their lives.

As stated above, SRI is not considered as a technology, and thus the common concept of 'transfer of technology' is quite inappropriate. SRI is not something to be transferred but rather something to be learned. One of the most gratifying aspects of working with SRI has been to see how many farmers have become inspired by this 'windfall' of benefit – getting more from less – so that they spend some of their own time and resources to spread the innovation to other farmers. Also, the progress made with SRI has been spurred by many farmer innovations and improvements beyond the original set of practices assembled by Laulanié. He would be the most pleased to see his system being further extended and modified by farmers (Laulanié, 2003).

Beyond SRI yield increases, greater water productivity and other benefits, we anticipate that this innovation can promote some rethinking of agricultural practices more generally. For achieving both food security and water security in the 21st century, SRI suggests that we pay more attention to plant root development and to promoting the abundance, diversity and activity of soil organisms as allies in an agroecological strategy to meet human needs in consonance with the interests and requirements of natural ecosystems.

References

Chaboussou, F. (2004). *Healthy Crops: A New Agricultural Revolution*. Jon Anderson, Charnley, UK.

Dazzo, F.B. and Y.G. Yanni (2006). The natural rhizobium-cereal association as an example of plant-bacteria interaction. In: N. Uphoff *et al.*, eds., *Biological Approaches to Sustainable Soil Systems*, CRC Press, Boca Raton, FL, 109–127.

DeDatta, S.K. (1981). *Principles and Practices of Rice Production.* J.W. Wiley, New York.

Feng, C., S-H Shen, H-P Cheng, Y-X Jing, Y. G. Yanni and F. B. Dazzo (2005). Ascending migration of endophytic rhizobia, from roots to leaves, inside rice plants and assessment of benefits to rice growth physiology. *Applied and Environmental Microbiology* 71: 7271–7278.

Gani, A., T.S. Kadir, A. Jatiharti, I.P. Wardhana and I. Lal (2002). The System of Rice Intensification in Indonesia. In: N. Uphoff et al., eds., *Assessments of the System of Rice Intensification*, 58–63. CIIFAD, Ithaca, NY (http://ciifad.cornell.edu/sri/proc1/sri_14.pdf).

Gasparillo, R., R. Naragdao, E. Judilla, J. Tana and M. Magsalin (2003). Growth and yield response of traditional upland rice on difference distance of planting using *Azucaena* variety. Broader Initiatives for Negros Development, Bacalod City, Philippines. (http://ciifad.cornell.edu/sri/countries/philippines/binuprst.pdf).

Laulanié, H. (1993). Le système de riziculture intensive malgache. *Tropicultura* 11, 110–114.

Laulanié, H. (2003). *Le Riz à Madagascar: Un dèveloppement en dialogue avec les paysans.* Editions Karthala, Paris.

Kabir, H. (2006). The adaptation and adoption of the System of Rice Intensification (SRI) using the Farmer Field School (FFS) approach. PhD thesis, University of Honolulu (http://ciifad.cornell.edu/sri/theses/kabirthesis.pdf).

Margulis, L. and D. Sagan (1997). *Microcosmos: Four Billion Years of Microbial Evolution.* University of California Press, Berkeley.

Matsuo, T., Y. Futsuhara, F. Kikuchi and H. Yamaguchi, eds. (1997). *Science of the Rice Plant*, three volumes. Food and Agriculture Policy Research Center, Tokyo.

Mattoo, A.K. and A. Abdul-Baki (2006). Crop genetic responses to management: Evidence of root-shoot communication. In: N. Uphoff et al., eds., *Biological Approaches to Sustainable Soil Systems*, CRC Press, Boca Raton, FL, 221–230.

McDonald, A., P. Hobbs and S.J. Riha (2006). Does the System of Rice Intensification (SRI) outperform conventional best management? A synopsis of the empirical record. *Field Crops Research* 96, 31–36.

Nemoto, K., S. Morita and T. Baba (1995). Shoot and root development in rice related to the phyllochron. *Crop Science* 35: 24–29.

Pinton, R., Z. Varanini and P. Nannipieri, eds. (2001). *The Rhizosphere: Biochemistry and Organic Substances at the Soil-Plant Interface.* Marcel Dekker, New York.

PRADAN (2006). Increased food grain production through rainfed SRI: Report from 2005 season by PRADAN team working in Purulia district, West Bengal, India, April (http://ciifad.cornell.edu/sri/countries/india/inpradan406.pdf).

PRADAN (2006a). Storing water onfarm and improving food security with 5 percent technology (http://www.iwmi.cgiar.org/smallholdersolutions/index.asp?nc=5&id=681&msid=132).

Randiramiharisoa, R., J. Barison and N. Uphoff (2006). Biological contributions to the System of Rice Intensification. In: N. Uphoff *et al.*, eds., *Biological Approaches to Sustainable Soil Systems*, CRC Press, Boca Raton, FL, 409–424.

Römheld, V. and G. Neumann (2006). The rhizosphere: Contributions of the soil-root interface in sustainable soil systems. In: N. Uphoff et al., eds., *Biological Approaches to Sustainable Soil Systems*, CRC Press, Boca Raton, FL, 91–107.

Satyanarayana, A., T.M. Thiyagarajan and N.Uphoff (2006). Opportunities for water-saving with higher yield from the system of rice intensification. *Irrigation Science*, 1432–1319 on line.

Stoop, W., N. Uphoff and A. Kassam (2002). A review of agricultural research issues raised by the System of Rice Intensification (SRI) from Madagascar: Opportunities for improving farming systems for resource-poor farmers. *Agricultural Systems* 71, 249–274.

Subbiah, S.V., K.M. Kumar and J.S. Bentur (2006). DRR's experience of SRI method of rice cultivation in India. Directorate of Rice Research, Hyderabad. Paper for International Dialogue on Rice and Water, International Rice Research Institute, Los Baños, March 7–8.

Surridge, C. (2004). Feast or famine? *Nature* **428**, 360–361.

Thies, J.E. and J.G. Grossman (2006). The soil habitat and soil ecology. In: N. Uphoff et al., eds., *Biological Approaches to Sustainable Soil Systems*, CRC Press, Boca Raton, FL, 59–78.

Uphoff, N. (2006). Increasing water savings while raising rice yields with the System of Rice Intensification (SRI). Paper for 2nd International Rice Congress, New Delhi, October 9–13.

Uphoff, N., A. Ball, E. Fernandes, H. Herren, O. Husson, M. Laing, C.A. Palm, J. Pretty, P.A. Sanchez, N. Sanginga and J.E. Thies, eds. (2006). *Biological Approaches to Sustainable Soil Systems*, CRC Press, Boca Raton, FL.

Uphoff, N. (2006a). Opportunities for overcoming productivity constraints with biologically-based approaches. In: N. Uphoff et al., eds., *Biological Approaches to Sustainable Soil Systems*, CRC Press, Boca Raton, FL, 693–713.

Uphoff, N. (2007). Agroecological alternatives: Capitalizing on existing genetic potentials. *Journal of Development Studies*, **43**:1.

Zheng J.G., X.J. Lu, X.L. Jiang and Y.L. Tang (2004). The System of Rice Intensification for super-high yields of rice in Sichuan basin. In: T. Fischer et al., eds., *New Directions for a Diverse Planet: Proceedings of the 4th International Crop Science Congress, Brisbane, Australia* (http://www.cropscience.org.au/icsc2004/poster/2/3/319_zhengjg.htm).

Zhu D. F., *et al.* (eds). (2006). *The Theory and Practice of SRI* (in Chinese). Chinese Publishing Company of Science and Technology, Beijing.

Chapter 12

Aerobic rice – An efficient water management strategy for rice production*

H.E. Shashidhar

Department of Genetics & Plant Breeding, College of Agriculture, UAS, GKVK, Bangalore, India

12.1 Introduction

Asia's food security depends largely on the irrigated rice fields, which produces three quarters of all rice harvested. But rice is a profligate user of water, consuming half of all developed fresh water resources. The increasing scarcity of water threatens the sustainability of the irrigated rice production system and hence the food security and livelihood of rice producers and consumers. In Asia, 17 million ha of irrigated rice areas may experience "physical water scarcity" and 22 million ha may have "economic water scarcity" by 2025 (Tuong and Bouman, 2001). Therefore, a more efficient use of water is needed in rice production. Several strategies are being pursued to reduce rice water requirements, such as saturated soil culture (Borrel *et al.*, 1997), alternate wetting and drying (Li 2001, Tabbal *et al.*, 2002), ground cover systems (Lin Shan *et al.*, 2002), system of rice intensification (SRI, Stoop *et al.*, 2002), aerobic rice (Bouman *et al.*, 2002), and raised beds (Singh *et al.*, 2002). It is reported that SRI and AWD systems have high water productivity with some amount of saving (approx. 20 per cent) without any compromise on productivity. However, water requirement of these production systems is also very high as land preparation consists of soaking, followed by wet ploughing or puddling of saturated soil. Further, when standing water is kept in the field (5–10 cm) during crop growth, large amount of water (about 10–15 per cent) is lost through seepage and percolation. Every drop of water received at the farmer's field by way of rainfall, surface irrigation or pumped from aquifers, is valuable and needs to be used effectively. Aerobic rice provides for effective use of rain that falls on the farmer's field, as there is no standing water and the farmer can skip irrigation if soil moisture status is sufficient for crop. This is not possible if water is already standing in the filed.

Irrigated rice has a very low water-use efficiency as it consumes 3000–5000 liters of water to produce 1 kg of rice. The traditional rice production system not only leads to wastage but also causes environmental degradation and reduces fertilizer use efficiency. Along with high water requirement, the traditional system of transplanted rice production in puddled soil on long run leads to destruction of soil aggregates and

*The following are the members of the Aerobic Rice team : G.S. Hemamalini, R. Latha, R. Venuprasad, Hima Bindu, Manjunath Janmatti, Erpin Sudheer, Adnan Kanbar, Vinod M. S, Naveen Sharma, M. Toorchi, H.R. Prabuddha, R. Dhananjay, R. Vasundhara, R.P. Veeresh Gowda, Kalmeshwer Gouda Patil, K. Manjunatha, T.M. Girish, K.K. Manohara, Patil Malagouda, B.T. Sridhar.

reduction in macro pore volumes, and to a large increase in micro pore space which subsequently reduce the yields of post rice crops, ex. wheat.

Added to this, irrigated rice fields will cut off the oxygen supply from the atmosphere resulting in the anaerobic fermentation, of soil organic matter. Methane, a major end product of anaerobic fermentation, is released from the submerged soil to the atmosphere through roots and stems of rice plants. Its concentration in the atmosphere has more than doubled during the last 200 years. Its current atmospheric concentration of 1.7 ppm by volume, up from 0.7 ppm in the pre industrial times, is much lower than the 360 ppm of carbon dioxide, up from 275 ppm. The global annual emission of methane is estimated to be 500 Tg (1 Tg = 1 million tonnes; Wahlen *et al.*, 1989) with an apparent net flux of 40 Tg/yr (Cicerone and Oremland, 1988). The current burden of methane in the atmosphere is approximately 4700 Tg. But one molecule of methane traps approximately 30 times as much heat as does a molecule of carbon dioxide. The heating effect of the atmospheric methane increase is approximately half that of the carbon dioxide increase (Dickinson and Cicerone, 1986, Ramanathan *et al.*, 1985). Continued increase in atmospheric methane concentrations at the current rate of approximately 1 per cent per year is likely to contribute more to future climatic change than any other gas except carbon dioxide (Cicerone and Oremland, 1988). Aerobic rice cultivation will curb methane production and saves water without affecting the productivity. It is the time to save water from the irrigated system of rice cultivation by adopting the aerobic rice cultivation. Varieties or hybrids with enhanced productivity for aerobic cultivation must be bred to address the water scarcity and pollution.

12.2 Water requirement for paddy

Water requirement of low land rice varies from 1,650 to 3000 mm (Table 12.1). Aerobic rice production system eliminates continuous seepage and percolation losses, greatly reduces evaporation as no standing water is present at any time during the cropping season, and effectively uses the rainfall and thus helps in enhancing water productivity, concomitant loss of soil sediments, silt and fertility from the soil. A comparison of water requirement of lowland flooded rice and aerobic rice system clearly shows that aerobic rice system can save about 45 per cent of water (Table 12.1).

Table 12.1 Comparison of seasonal water requirement between lowland flooded rice and aerobic rice.

	Seasonal water requirement (mm)	
	Lowland flooded rice	Aerobic rice
Land preparation	150–300	100
Evaporation	200	100
Transpiration	400	400
Seepage and percolation	500–1.500*	335
Application loss (at 60% efficiency)		335
Total seasonal water requirement	1650–3000	935

Source: Tuorg and Bouman, 2003; Lampayan and Bouman, 2005.

Soil with seepage loss of 5–15 mm/day.

Some of the challenges faced by the paddy farmers are:

Lack of adequate water through-out the season, over-pumping of ground water causing aquifer resources to decline and the high 'cost' of water when available.
Continuous cropping and yield decline in irrigated habitats;
Lack of remunerative prices for the produce;
Non-availability of adequate labor at the right time, at affordable prices and in required numbers;
Contamination of surface and ground water due to leached nitrogenous fertilizers and pesticides:
Contamination of atmosphere due to methane emissions from paddy plants (ebblusion), paddy fields, and nitrous oxide leached out of fields;
Salinization of prime (those nearest the reservoirs) paddy lands.
Over-irrigation, over-fertilization, over-protection due to intensive cropping & Depletion of Silicon an essential element for paddy, in paddy soils.

Many varieties developed for the rainfed ecosystem have not been adapted by the poor and marginal farmers since they failed to meet the local requirements. In India over 700 varieties are officially released for cultivation. In spite of this, only a few varieties are in actual cultivation. Grain quality characteristics, cooking quality and sensory evaluations along with physical characteristic need to be considered seriously before release of varieties.

12.3 Development of drought tolerant quality rice varieties

> **Morphological and Anatomical**
> Grain Yield; Maximum Root length, Root Volume, Root Dry Weight, Root Thickness; Biomass; Harvest index; Leaf drying; Leaf tip firing; Delay in flowering.
> **Phenological**
> Earliness; Delay in Flowering; Anthesis-Silking Interval; Seedling vigor; Weed competitiveness; Photosensitivity; perenniality.
> **Physiological & Biochemical traits**
> Osmotic Adjustment; Carbon Isotope Discrimination; Stomatal conductance; Remobilization of stem reserves; Specific leaf weight; ABA; Electrolyte leakage; Oxygen scavenging; Heat shock proteins; Stay green; Cell Wall proteins; Leaf water potential; Water use efficiency; Aquaporins; Nitrogen use efficiency; Epicuticular wax; Dehydrins; feed forward response to stress

Sugarcane and paddy are among the first options to be explored if we have to effect substantial savings in water used in agriculture, as these are considered as the most water intensive among all crops. Breeding for drought tolerance has and is being attempted by several scientists and groups across the globe by spending considerable amount of money, time, and effort on this activity. Several traits have been found to contribute to enhanced drought tolerance in crop plants and each scientist seems to have confidence in a particular trait or a set of traits. Effects on enhanced drought tolerance have also been demonstrated. It is at the point of establishing the link between these individual or group of traits and grain yield that many scientists have failed. There are just a few instances where the utility of a trait has been shown to have a demonstrated impact on enhanced yield.

Water deficit for a crop could occur at a particular stage or at different stages during crop growth. While breeding for drought tolerance, it is very tricky to estimate the exact intensity, timing or severity of stress that a crop could encounter. Thus the ability to withstand stress at any stage of crop growth would be an invaluable asset to a plant that has to contend with the unpredictability of drought.

The concept of deficit irrigation (DI) was proposed by English et al. (1990). DI advocates applying less than optimum levels of water to the crop so that there is no wastage due to several other factors. The concept of DI has been researched in several crops across the world. DI is known to enhance the quality of the produce in fruit crops. It is expected to save considerable amount of water drawn from reservoirs and effect savings in electricity, investment of public funds in water storage devices, etc.

By applying the concept of DI to rice we perceived that the whole cropping season can be subjected to certain levels of sub-optimum water regimes which would effect in savings in water. It is proposed that farmer could not irrigate the crop, even though there could be water in his reservoir or in the canals.

Aerobic rice is both a concept of growing rice and appropriate genotypes suited for such a growth. While it is similar to upland rice which is topographically high altitude rice, given a toposequence, aerobic rice is one which grows any rice field that is never flooded right through the cropping season. These could be slightly sloping lands, may be sandy and have higher degrees of percolation unlike lowlands. The difference between aerobic and conventional irrigated rice is given in the Table 12.2. The overall stages of Aerobic rice development is presented below in different heads.

12.3.1 Selection for roots

Budda, an indigenous drought tolerant low yielding cultivar with deep root system and good combining ability for the root and shoot traits (Dhananjay, 2001) has been crossed with IR64, a lowland high yielding variety which manifests good osmotic adjustments (Ingram et al., 1994). The F1 generation was grown in the Main Research Station, Bangalore and the F2 generation was studied during the dry season 2003. The genotypes were grown in light-gray polyvinyl chloride (PVC) cylinders (100 cm long and 18 cm diameter) filled with clayey soil and farmyard manure (FYM), one healthy plant was maintained in each cylinder. Hundred and forty four F2 plants of Budda/IR64 were evaluated for their root morphology under well watered condition and to estimate root parameters. Sampling in all experiments was done at 70 DAS. At the time of root sampling, cylinders with soil and plant were thoroughly soaked in water over-night and sampled the next day. Sampling was done as described by Hemamalini et al. (2000) with care taken to retain roots, root hairs, and root branches. The following observations were recorded at the time of root sampling viz., maximum root length (cm), total root number, root number at 15, 30 and 60 cm depth, root volume (cc), root dry weight (g), plant height (cm), number of tillers, leaf area (cm), shoot dry weight (g), relative water content (%), leaf rolling, total dry weight (g), root

Table 12.2 Differences between aerobic and irrigated rice.

Eco-Friendly Aerobic Rice	Conventional Irrigated Rice
No need of land levelling	Land should be leveled
Direct seeding	Nursery raising is needed
Reduced seed rate	Higher seed rate (25 kg per ac)
40–50% water saving; no constant maintenance of water level	Constant maintenance of water level is necessary
Weeding can be mechanized (bullock pair)	Not possible (Rotary weeder)
No need for trimming bunds and plugging holes	Requires constant attention by way of trimming of bunds and plugging holes
Labour requirement is less	More
Intercropping of any other arable crop is possible	Not possible
Crop rotation can be practiced (with pulses- for balanced nutrition)	Not common
Aerobic condition in soil	Anaerobic condition prevails
Soil structure is maintained	Destroyed. Subsurface hard pan is made by repeated plowing
Faster organic matter decomposition	Slower
Oxygenated rhizosphere is found	Not found
Nitrogen use efficiency is more	Less
Nitrous oxide is not produced	Produced
Better water use efficiency	Low
Very low mosquito population	Severe mosquito incidence
Efficient utilization of rain water	Less efficient utilization of rain water
No occurrence of methanogenesis	Methanogenesis occurs
Better mineral nutrient dissolubility (Lerman, 1979)	Less dissolubility
Adsorption of NO_2/NH_4 to soil particles	Not so
Production of toxins like ethanol and lactate is absent (Kirk, 2004)	Toxins are produced
Energy efficiency is more – efficient glycolysis (Kirk, 2004)	Less
Reduced humidity in microclimate; healthy crop	High humidity
Incidence of diseases and pests is significantly low	High
Cost of cultivation is significantly low	High

to shoot ratio, root dry weight/tiller (g), specific root length (cmg^{-1}), total growth rate ($cmday^{-1}$), leaf area index and specific leaf area ($cm^2 g^{-1}$).

12.3.2 *Participatory plant breeding (PPB)*

Selected fifty F3 families were raised in the farmer's field at Shettigere, a village in the outskirts of Bangalore in the wet season 2003 with aerobic method of cultivation. At maturity forty farmers were invited to the field day from about five villages from Bangalore North district to select the superior genotypes. Each invitee was asked to select five superior plants in the field based on his requirements, and give his impressions about the choice of the plant. The informations were pooled to find out the general preferences of the farmers of that locality and a target-oriented selection was practiced to have a variety of farmer's choice, which will have no problem in adoption, because they are getting the variety of their choice. About seventeen plants from the farmer's selections were randomly selected to screen for the drought stress. Similar PPB was adopted in the F4, F5 and F6 generations during wet season 2004 and dry season 2005 in the farmer field.

Selection of segregants/families by farmers was made using index cards. Each farmer was provided with five slips bearing his registration number. He was taken round the plot and asked to vote for the five best lines. They were asked questions to justify their choice.

12.3.3 Screening for drought tolerance

Seventeen farmers' selections and seventeen breeders' selections based F3 PPB and plant type respectively were evaluated under two moisture regimes of 0.8 IW/CPE (Well-watered) and 0.6 IW/CPE (Low moisture stress) in the same farmers' fileds. These ratios are decided by calculating the water holding capacity of the soil. The measured quantity of water (40 mm) was given through Parshall Flume and the cumulative pan evaporation was considered based on the USWB CLASS-A-open pan evaporimeter. Depending upon the daily evaporation in USWB CLASS-A open Panevaporimeter, irrigations scheduled at 0.8 IW/CPE ratio received irrigations once in five to six days whereas irrigations scheduled at 0.6 IW/CPE ratio received delayed irrigations at 10 to 12 days in the vegetative phase. With the advancement of the reproductive development, the stress level was brought to 1 IW/CPE ratio respectively. Irrigation was similar to any aerable crop like maize, wheat or sorghum. Plots could be bunded or unbunded, to reduce the conveyance losses.

12.3.4 Advanced yield trial

Six stabilized aerobic rice lines are in All India trials across six locations, since 2005. Kharif. Lines for the trials are nominated by breeders of the drought network and scientists from IRRI, Philippines. In each location the nominated lines are evaluated in three replications each in three hydrological situations, namely aerobic, drained puddles and irrigated conditions. All the six lines performed significantly superior over national and international checks at all locations (from two year data).

Crop establishment was excellent, with no symptoms of deficiencies or toxicities of Iron, Zinc or any other micronutrient considering the fact that the plants were not being grown under irrigated condition with standing water as is usually done. Because planting densities are lower, crops are relatively healthier. Irrigation is by furrow, as with maize or sorghum, and can be limited to once in every 5 or 6 days. Yield averages 4 to 5 tonnes per hectare with good grain quality parameters (Table 12.3).

12.4 Innovations about the technology development package

Key features of this production technology package include:-

Direct seeding;
Deficit irrigation;
Avoidance of destruction of soil structure through puddling
Deep and extensive rooting;
Decreased seed rate and density of plants in field;
Distance between plants increased to provide adequate sunlight, aeration, ventilation and also facilitate mechanical weeding;
Designer fertilizers specific to needs of direct seeded paddy which would include silicon.

Table 12.3 Performance of newly bred "Eco-friendly Aerobic rice" – accessions and checks under different seasons, hydrological condition, years & locations.

Season/Year	Summer, 2005	Summer, 2005	Kharif, 2005	Kharif, 2005	Kharif, 2005	Kharif, 2005	Kharif, 2005	Kharif, 2005	Kharif, 2005	Kharif, 2005	Kharif, 2006	Kharif, 2006	Kharif, 2006	Kharif, 2006	Kharif, 2006	Grand total	Grain type
State	KAR	KAR	KAR	KAR	CHG	CHG	CHG	JRK	JRK	JRK	KAR	CHG	CHG	CHG	CHG		
Location	DJL	Mandya	DJL	Mandya	Raipur	Raipur	Raipur	HZB	HZB	HZB	DJL	Raipur	Raipur	Raipur	Raipur		
Situation	Aerobic	Aerobic	Aerobic	Aerobic	Rainfed	Lowland	Control	Rainfed	Lowland	Control	Aerobic	Aerobic	Irrigated	TSD MS	TSD LS		
Genotype × Rep	11 × 3	6 × 3	11 × 3	6 × 3	63 × 3	63 × 3	63 × 3	65 × 3	65 × 3	65 × 3	10 × 3	100 × 3	100 × 3	100 × 3	100 × 3		
Aerobic Rice Genotypes																	
BI33 (ARB6)	3979 a	4840 a	4270	5140 a	1272 a	4970 a	3987 ab	2604 a	3875 a	4938 a	5055 a	4951 a	3590	4770	3686	4128.5	MS
BI34 (ARB3)	4270 a	4390 b	5062	4510 b	1439 a	5202 a	4000 ab	2875 a	4000 a	5021 a	4855 b	4768 a	3296	3200	3483	4024.7	MS
BI43 (ARB4)	3312 a	4560 b	4187	4540 b	1574 a	3411 b	3676 b	1500 b	2438 b	4813 ab	4711 b	–	–	–	–	3520.2	MS
BI48 (ARB5)	3333 a	–	5500	–	1113 a	3581 b	3766 ab	1813 b	3396 ab	5313 a	4654 b	5347 a	3657	6300	2733	3885.1	MS
Check																	
Rasi (IET1444)	4625 a	3550 b	4854	3850 b	–	–	–	–	–	–	3900 b	–	–	–	–	4155.8	SB
MTU1010	–	–	–	–	490 b	2809 b	4544 a	959 c	2709 b	4584 ab	–	2867 b	4389	4310	3876	3449.6	–
IR64	–	–	–	–	–	–	–	1959 b	2438 b	4021 b	–	–	–	–	–	2806	–
CD 5%	2601	1120	2869	1108	510	790	846	547	790	851	1050	800	–	–	–	–	–

KAR = Karnataka; CHG = Chhattisgarh; JRK = Jharkhand; DJL = Doddajala; HZB = Hazaribagh; TSD MS = Terminal Stage Drought – More Severe, TSD LS = Terminal Stage Drought – Less Severe, ARB = Aerobic Lines Bangalore, CD = Critical Difference, MS = Medium Slender, SB = Short Bold.

Diversified cropping systems to accommodate principles of crop rotation or intercropping with pulses, legumes and large seeded cereals.

References

Borell, A., Garside, A., and Shu, F. K. (1997). Improving efficiency of water for irrigated rice in a semi-arid tropical environment. *Field Crops Res.*, 52:231–248.

Bouman, B.A.M., Wang Hua Qi, Yang Xiao Guang, Zhao Jun Fang, and Wang Chang Gui. (2002). Aerobic rice (Han Dao): a new way of growing rice in water-short areas. In: Proceedings of the 12th International Soil Conservation Organization Conference, 26–31 May 2002, Beijing, China. Tsinghua University Press, p. 175–181.

Cicerone, R. J., and Oremland, R. S. (1988). Biogeochemical aspects of atmospheric methane. *Global Biochem. Cycles*, 2: 299–327.

Dhananjaya, M.V. (2001). Genetic analysis of root and shoot morphological characters related to drought avoidance in rice. Ph. D thesis submitted for University of Agricultural Sciences, Bangalore (unpublished)

Dickinson, R. E., and Cicerone, R. J. (1986). Future Global Warming from Atmospheric Trace Gases, *Nature*, 319: 109–115.

English, M. J., Musick, J. T., and Murty, V.V. (1990). Deficit irrigation. In: *Management of farm irrigation systems* (eds. Hoffman, G. J. Towell T. A. and Solomon, K. H., St. Joseph, Michigan, United States of America, ASAE.

Hemamalini, G. S., Shashidhar, H. E., and Hittalmani, S. (2000) Molecular marker assisted tagging of morphological and physiological traits under two contrasting moisture regimes at peak vegetative stage in rice (*Oryza sativa* L.). *Euphytica*, 112: 69–78.

Ingram, K. T., Bueno, F. D., Namuco, O. S., Yambao, E. B., and Beyrouty, C. A. (1994). Rice root traits for drought resistance and their genetic variation. In: Kirk, G. J. D. (Ed.) *Rice roots: nutrient and water use.* (ed. Kirk, G. J. D.) International rice research Institute, Los, Banos, Laguna, Philippines, p. 63–77.

Lampayan, R. M., and Bouman, B. A. M. (2005). Management strategies for saving water and increase its productivity in lowland rice-based ecosystems. In: *proceedings of the First Asia-Europe Workshop on Sustainable Resource Management and Policy Options for Rice Ecosystems (SUMAPOL)*, 11–14 May 2005, Hangzhou, Zhejiang Province, P.R. China. On CDROM, Altera, Wageningen, Netherlands.

Li, Y. H. (2001). Research and practice of water-saving irrigation for rice in China. In: Barker, R. Li, Y. and Tuong T. P., *Water-saving irrigation for rice.* Proceedings of an International Workshop, 23–25 Mar 2001, Wuhan, China. Colombo (Sri Lanka): International Water Management Institute. p. 135–144.

Lin Shan, Dittert K, and Sattelmacher B. (2002). The ground cover rice production system (GCRPS)—a successful new approach to save water and increase nitrogen fertilizer efficiency? In: *Water-Wise Rice Production.* (eds. Bouman, Hengsdijk, Hardy, Bindraban, Tuong, Lafitte and Ladha). Los Baños (Philippines): IRRI, (in press).

Ramanathan, V., Cicerone, R. J. Singh, H. B., and Kiehl, J. T. (1985). Trace gas trends and their potential role in climate change. *J. Geophys. Res.*, 90: 5547–5566.

Singh, A. K., Choudury, B. U., and Bouman, B. A. M. (2002). The effect of rice establishment techniques and water management on crop-water relations. In: *Water-Wise Rice Production* (eds. Bouman, Hengsdijk, Hardy, Bindraban, Tuong, Lafitte and Ladha) Los Baños (Philippines): IRRI, (in press).

Stoop, W., Uphoff, N., and Kassam, A. (2002). A review of agricultural research issues raised by the system of rice intensification (SRI) from Madagascar: opportunities for improving farming systems for resource-poor farmers. *Agric. Syst.*, 71:249–274.

Tabbal, D. F., Bouman, B. A. M., Bhuiyan, S. I. Sibayan, E. B., and Sattar, M. A. (2002). On-farm strategies for reducing water input in irrigated rice; case studies in the Philippines. *Agric. Water Manage*, 56(2): 93–112.

Tuong, T. P., and Bouman, B. A. M. (2001). *Rice production in water-scarce environments.* Paper Presented at the Water Productivity Workshop, 12–14 Nov 2001, Colombo, Sri Lanka.

Tuong, T. P., and Bouman, B. A. M. (2003). Rice production in water-scarce environments. In *Water Productivity in Agriculture: Limits and Opportunities for Improvement* (Eds. Kijne, J.W., Barker, R., Molden D.), p. 53–67. CABI Publishing, UK.

Wahlen, M., Tanaka, N., Henry, R., Deck, B., Zeglen, J., Vogel, J. S., Southon, J., Shemesh, A., Fairbanks, R., 5 and Broecker, W. (1989). Carbon-14 in methane sources and in atmospheric methane: The contribution from fossil carbon, *Science*, 245, 286–290, 1989.

Chapter 13

Towards developing low water-need and drought-tolerant super rice in China

Hanwei Mei & Lijun Luo

Shanghai Agrobiological Gene Center, Shanghai, China

13.1 Introduction

China is facing two major challenges: food security and water crisis, particularly in the rainfed areas of China where a significant portion of population still live in absolute poverty. Rice is the staple food for most Chinese people and is regarded as a strategic commodity and one of the key components of traditional culture in China. However, rice production consumes 70% of the fresh water used in agriculture, which consumes 80% of the total fresh water resources in China. In other words, water consumption in traditional rice production accounts for near a half of the total fresh water (Padolina 1996). The renewable fresh water endowment per capita in China is about 2,200 m^3, which is only one fourth of the world average. Half of the area in China receives less than 400 mm rainfall per year. Thus, there is a serious water shortage in many parts of China.

China's agriculture has been seriously affected by drought. About 26 million ha of crops per year suffered from drought from 1991 to 2005. About 35 million tonnes of grains were lost per year for about five years (Zhang 2006). It was estimated that the annual economic loss from drought alone reaches more than 25 billion dollars (Deng 1999). Drought has been a major problem in two major rice ecosystems. The first is the hilly areas of Central and Southwestern China, where rice is largely grown under the rainfed conditions. Rice yield is about 1–3 tonnes/ha, much less than the national average (6.5 tonnes/ha). The second system includes a significant portion of the traditional rice growing areas where irrigation systems are insufficient or the water resources are diminishing. As over 70% rice lands are easily affected by drought, the water shortage has become increasingly serious. To achieve long-term food security and sustainable development, the drought tolerance or water saving rice cultivars and its application technologies are urgently needed.

In the past 30 years, many researches have been conducted to investigate the genetic basis and physiological mechanism of the crop's drought tolerance (DT) together with the great efforts on developing the DT rice varieties. Unfortunately, because of the extreme complexity of plant's drought tolerance and the lack of effective methodology, the progress on these researches has been very limited and slow. In 1999, we initiated a research program supported by the national and local governments, and the private foundations (e.g. the Rockefeller Foundation). The major objectives are: (1) to establish the scientific-based field DT screen facilities and evaluation methodology; (2) to reevaluate the drought tolerant rice germplasm and construct the DT core

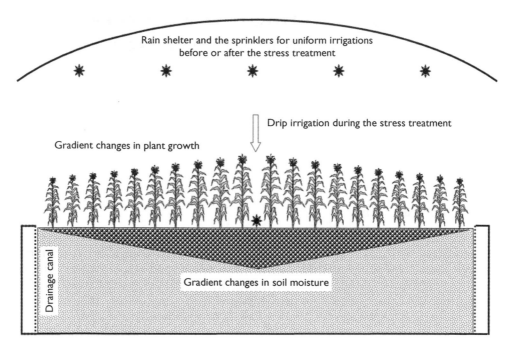

Plate 13.1 Illustration of the facility for drought tolerance screening showing the gradient changes in both soil moisture and the growth of the plants. (See Colour plate 13.1)

collection for the further researches and breeding activities; (3) to identify the drought tolerant genes/QTLs and to understand the genetic basis of drought tolerance in rice; (4) to develop low water-need or drought tolerant super rice cultivars by integrating the conventional breeding and the molecular technologies.

13.2 Field-based drought tolerance screening system targeting the final grain yield

13.2.1 DT screening in rain-off facilities and in field during dry season

Following the idea of the line source system (LSS) irrigation like the system used by Lanceras et al. (2004), special screening facility was designed to produce gradient changes in soil moisture in a compact space under the shelf of green house (Plate 13.1; Liu et al., 2006). Instead of the sprinklers in LSS, drip irrigation was made use of to provide water into the soil between two columns of plots. In this case, soil water gradient was developed and maintained from irrigated end of each row of 15–20 rice plants to another end near the drainage canal where the rice plants suffer from water stress. The research practices in Shanghai Agrobiological Gene Center (Shanghai, China) for three years showed the effective control of soil moisture and the gradient drought stress, according to the changes in soil water content and the phenomenon in rice plants. This screening facility is suitable for drought tolerance screening for both breeding materials and a whole set of separating population with about 200 lines. For example, Zou *et al.* (2005) reported a reliable and stable phenotyping of drought tolerance in a RI population according to the ANOVA (Analysis of Variance) and the QTL (Quantitative Trait Loci) mapping results showing low interactions with years as an environmental factor. Another large scale field facility was built for the purpose of screening of drought tolerance of rice by Huazhong Agricultural University in Wuhan, China (Yue *et al.*, 2005). In this facility, the whole roof above the screening area is mobile so that the rice plants are covered by the rain shelter only during rain fall. In this case, the plants are grown up in environment similar to the normal rice field as much as possible. But in a coastal region like Shanghai, the design with movable roof is prone to serious damage from typhoons. So the green houses with large space are suitable to serve as the rain shelter if only as much covering area as possible can be open during the non-raining time in order to enhance the ventilation.

Such rain-off systems are very useful for the screening of drought tolerance in the regions without typical dry seasons or during the rainy seasons. More and more screening facilities have been set up in agricultural research institutes in China recently (e.g. in Guangdong Academy of Agricultural Sciences), or will be built in the near future to meet the sharply increased research activities on crop drought resistance.

For generations, rice breeders in China usually grow breeding lines in Hainan, which is a tropical island, in the winter season every year. From November to April in the succeeding year is the dry season in this region. The weather with hardly any rain, strong solar radiation and high temperature in March and early April is very suitable to screen drought tolerance of rice plants on a large scale during the reproductive stage in the open field, if only the spatial variances in the field is acceptable and strong irrigation and drainage system is available. In our breeding practice, either the drought tolerance under severe water deficit or the ability of restored growth from drought stress can be investigated undoubtedly in the experimental field (Figure 13.1).

13.3 Evaluating the importance of drought tolerance related traits based on grain yield as the primary target

The grain yield under drought stress is the primary parameter for selection in screening rice germplasm and breeding materials. But measuring the grain yield is not reliable for breeding lines at their early generations. The measurement of each components of grain yield is also labor and time consuming for a large number of accessions. In this case, a series of drought tolerance related characteristics of rice plants were recognized as secondary traits, by which the drought tolerance can be evaluated quickly to screen large populations, or to select the individuals to make cross during heading stage, or to have an insight into the physiological mechanism under the adaptation of plants to water stress. A series of researches show high correlations between grain yield and secondary traits like leaf rolling, leaf death, delay in head date, leaf water potential, relative water content, canopy temperature, and so on. Among them, canopy temperature (CT) seems to be a reliable and most easily detected characteristic. It is practical to make the primary selection for breeding lines in early generation based on the optimal observation on stress phenomenon like leaf rolling and desiccation, delay

Figure 13.1 Rice plants under severe water deficit and their restored growth after full irrigation in the open field grown in dry season in Hainan, China.

Table 13.1 Phenotypic correlations and path analysis between grain yield and its components in either well-watered or water stress conditions (Adapted from Zou *et al.* 2005).

Trait	Well-watered vs Stress	GY under well-watered			GY under water stress		
		Correlation coefficient	Direct effect	Indirect effect	Correlation coefficient	Direct effect	Indirect effect
GY	0.56**						
PN	0.69**	0.19**	0.27	−0.08	0.43**	0.49	−0.06
SN	0.84**	0.65**	0.41	0.24	0.27**	0.54	−0.27
SF	0.69**	0.51**	0.34	0.17	0.66**	0.60	0.06
TGW	0.88**	0.17*	0.14	0.03	0.28**	0.28	0.00

GY: Grain yield; PN: Panicle number; SN: Spikelet number per panicle; SF: spikelet fertility; TGW: 1000 grain weight; *P < 0.05; **P < 0.01.

in heading date and decreased spikelet fertility, together with quick scanning of canopy temperature.

The reproductive stage of rice plants is one of the most sensitive periods to drought stress. Water deficit at this stage causes severe spikelet sterility and final yield loss. Spikelet fertility can be visually estimated under field conditions and has been used as an important secondary characteristic (Garrity & O'Toole, 1994; Fukai *et al.*, 1999). The relationship between grain yield and four yield components was investigated under both well-watered and water stress conditions by the authors (Table 13.1). It was found that spikelet number per panicle (i.e. panicle size) was the most important factor under well-watered condition, followed by spikelet fertility as the second important one. But spikelet fertility became the most important contributor to the grain yield under drought stress, while the panicle number per plant became the second important one. Large panicles still had large direct effect on grain yield in path analysis, but binding with large negative indirect effect probably coming from more sterile spikelets. Yue *et al.* (2006) also reported that relative spikelet fertility or fertility under drought stress conditions is not only a highly informative indicator for severity of drought stress, but also the most important determinant of yield under drought stress conditions.

13.4 Development and demonstration of conventional and hybrid drought tolerant rice cultivars

13.4.1 *Developing conventional drought tolerant rice cultivars*

During the initiation of the breeding program for drought tolerant rice about 10 years ago, the author's research group made great efforts to collect upland rice germplasm around the world. After evaluating the drought tolerance, local adaptation and integrated performance of near 1000 accessions by several years, a set of rice varieties were picked out as the core collection of upland germplasm for the breeding program.

The authors developed drought tolerant rice variety *Zhonghan 3* by pure line selection from *CNA6187-3* and released it nationwide in 2003. This line has such a good performance in both drought tolerance and grain yield that it was used as the check line in the Multiple Location Trial in China replacing *IAPAR9* from Brazil as the former check line.

A series of drought tolerant rice cultivars were developed by hybrid breeding with crosses between elite paddy rice and upland rice varieties. Three drought tolerant varieties developed by the author's group, including *Huhan 3*, *Huhan 7* and *Huhan 15*, were released by the Chinese Ministry of Agriculture or the government of Shanghai City from 2004 to 2006.

There were a total of 24 drought tolerant rice varieties or combinations released by Chinese Ministry of Agriculture from 2003 to 2006 (Table 13.2). Among them, 18 varieties belong to *japonica* type (including 3 glutinous ones) while only 6 varieties belong to *indica* type. Compared to the great input and distinguished progress in paddy rice breeding during several decades last century, the breeding practice of drought tolerant rice in China is at its early stage. The yield potential of the released drought tolerant rice varieties is still much lower than that of modern paddy rice varieties or hybrid combinations in China. On the other hand, there is a great development space for the breeding program of drought tolerant rice, not only in increasing the yield potential, but also in enhancing the performance on grain quality, insect and disease resistance, depending on the abundant knowledge and materials created in paddy rice breeding and researches of modern life sciences.

13.5 Screening the upland CMS maintainer and developing the upland CMS line

There are great achievements in hybrid rice development in China. However, these hybrid combinations were created based on well watered conditions. For the purpose of developing 3-line hybrid drought tolerant rice, a total of 784 upland rice varieties were screened for the ability to maintain the male sterility of cytoplasmic male sterile (CMS) line. *Zhenshan 97A*, one of the most widely used CMS lines in China, was used as the testing parent. Only one line (H3034) was screened out as full CMS maintainer which is a pure line selected from the field population of *EAIC139-55-1-23*.

Following the backcross breeding procedure for CMS lines, a drought tolerant CMS line, *Huhan 1A*, was developed and released in 2004. This line had strong drought tolerance, good grain quality and high combining ability (Yu *et al.*, 2005). It was not only used in the breeding program in the author's group, but was also released to many provincial agricultural academies to screen local hybrid combinations as an activity of the Research Network of Drought Tolerant Rice in China.

13.6 Pyramiding drought tolerance with other characteristics among introgression lines (ILs)

A protocol of developing multi-parental introgression lines was employed in the International Rice Molecular Breeding Program (Li *et al.*, 2005). A huge set of introgression lines (ILs) was developed following this strategy by a research network in China (Luo *et al.*, 2005). Then the bulked seed samples of BC_2F_2 or the populations of more advanced BC lines were evaluated for different target traits. For example, more than 2000 ILs were developed by the author's group using a CMS restorer, *Zhong 413*, as the recurrent parent and about 200 donor parents with diverse origin. These lines were screened for grain appearance, drought tolerance (Mei *et al.*, 2005) and low nitrogen

Table 13.2 Drought tolerant rice varieties released by the Chinese Ministry of Agriculture from 2003 to 2006 (Translated from the National Rice Data Center in China, http://www.ricedata.cn/variety/shengding/index.htm)

Year	Names	Parental lines	Types	Breeding institutes
2003	Liaoyou 14	Liao 30A/C4115	Japonica, 3-line hybrid	Rice Research Institute, Liaoning Academy of Agricultural Sciences
	Xiahan 51	Qingxuan 1/Jingxi 1	Japonica, conventional	Hebei University, etc
	Handao 65	Akihikari/Sanbangqishiluo	Japonica, conventional	College of Agriculture and Biotechnology, Chinese Agricultural University
	Handao 9	Akihikari/Banli 1	Japonica, conventional	College of Agriculture and Biotechnology, Chinese Agricultural University
	Liaojing 27	Xiuyanbufujing (pure line)	Japonica, conventional, glutinous	Rice Research Institute, Liaoning Academy of Agricultural Sciences
2004	Liaohan 109	HJ29 (pure line)	Japonica, conventional	Panjing North Agricultural Technology Developing Limited Company
	Hanfeng 8	Shennong 129/Han 72	Japonica, conventional	Shenyang Agricultural University
	Zhonghan 209	Xuan 21/IR55419-04-1	Indica, conventional	Chinese Rice Research Institute
	Huhan 3	Mawannuo//IRAT109/P77	Japonica, conventional	Shanghai Agrobiological Gene Center
	Jingganghandao 1	IAPAR9 by irradiation	Indica, conventional	Rice Research Institute, Jiangxi Academy of Agricultural Sciences
	Huanhanyou 1	N422S/R8272	Japonica, 2-line hybrid	Rice Research Institute, Anhui Academy of Agricultural Sciences, etc
	Danhannuo 3	Danjing 5 (pure line)	Japonica, conventional, glutinous	Dandong Academy of Agricultural Sciences of Liaoning Province
	Liaoyou 6	Liao 30A/C272	Japonica, 3-line hybrid	Rice Research Institute, Liaoning Academy of Agricultural Sciences
	Tianjing 4	H701/D7-38	Japonica, conventional	Rice Research Institute, Jilin Academy of Agricultural Sciences
	Tianjing 5	Songjing 3/Jiyujing	Japonica, conventional	Rice Research Institute, Jilin Academy of Agricultural Sciences
	Danhandao 2	Danjing 5 (pure line)	Japonica, conventional	Dandong Academy of Agricultural Sciences of Liaoning Province
	Zhongzuo 59	C57-2/Zaofeng	Japonica, conventional	Crop Science Institute, Chinese Academy of Agriculture Sciences
2005	Luhan 1	Line 6527 by mutation	Indica, conventional	Natural Nourishing Food Engineering Institute, Anhui Academy of Agricultural Sciences
	Gannonghandao 1	IAPAR9 (pure line)	Indica, conventional	College of Agriculture, Jiangxi Agricultural University
	Zhenhan 6	Zhengzhouzaojing/Zhengdao92-44	Japonica, conventional	Food Crop Research Institute, Henan Academy of Agricultural Sciences
2006	Handao271	Han 2/Khaomon	Japonica, conventional	Chinese Agricultural University
	Liaohan 403	S299/S2026	Japonica, conventional	Rice Research Institute, Liaoning Academy of Agricultural Sciences
	Jingyou 13	Zhongzuo 59A/Luhui 3	Japonica, 3-line hybrid	Beijing Agro-Biotechnology Research Center
	Danhandao 4	Danjing 5by irradiation	Japonica, conventional	Dandong Academy of Agricultural Sciences of Liaoning Province
	Zhonghan 221	Shuangtounonghu/IAPAR9	Indica, conventional	Chinese Rice Research Institute
	Handao 15	Qixiuzhan/Zhonghan 3	Indica, conventional	Shanghai Agrobiological Gene Center
	Luodao 998	Handao277(pure line)	Japonica, conventional	Luoyang Institute of Agricultural Sciences of Henan Province
	Hannuo 303	Xiuzinuo/Xiaobairen	Japonica, conventional, glutinous	Rice Research Institute, Liaoning Academy of Agricultural Sciences
	Liaoyou 2015	Liao 20A/C4115	Japonica, 3-line hybrid	Rice Research Institute, Liaoning Academy of Agricultural Sciences

tolerance (unpublished data). The IL populations showed great potential in both direct applications as promise lines in the breeding program and the pyramiding of multiple elite characteristics in new varieties. For example, drought tolerant ILs were crossed with ILs with low nitrogen tolerance. The segregating population (like F_2 or F_3 families) had similar growth period and general performance in the field because such lines had similar genetic background from the same recurrent parent. After screening under drought and low nitrogen conditions separately, derived lines with both drought and low nitrogen tolerance were selected and confirmed in F_4–F_5 generations. In this case, the drought tolerance was able to be integrated with resistance to other abiotic stress, good grain quality, and high yielding characteristics into a single line or hybrid parents.

13.7   Development and demonstration of drought tolerant hybrid rice

A series of hybrid combinations were evaluated for drought tolerance and yield potential with *Huhan 1A* as female parent and either restorers in paddy rice or newly developed upland rice restorers as the male parent. Two high yielding, good quality and drought tolerant hybrid rice combinations, *Hanyou 2* and *Hanyou 3*, were developed in author's group and released by the city government of Shanghai in 2006.

Both *Hanyou 2* and *Hanyou 3* were demonstrated in several provinces in China like Shanghai, Zhejiang, Guangxi, and Yunnan. The grain yields varied from about 6 tonnes/ha to 7.5 tonnes/ha with much less water applications, gaining 20%–100% yield increase compared to the local dominant hybrid rice combinations grown under water stress conditions. For example, *Huhanyou 3* was grown in three rainfed conditions in Anji, Zhejiang in 2006. The amount of water consumption including irrigation and water fall was estimated to be about 3,720 m^3/ha in Treatment I with irrigation at seedling stage followed by rainfed. The water consumptions were about 2,430 m^3/ha and 3,225 m^3/ha in Treatment II and Treatment III respectively, based on rainfed for whole growing stage. The surface flow of rain fall was not prevented in the field of Treatment II while the rainfall was kept in the field of Treatment III by ridge of the field. Compared to the normal rice production, about 58%, 73% and 64% of water consumption were saved in such rainfed conditions respectively, while about 100% grain yield increases were achieved in proportion to the yield of paddy hybrid rice combinations as the control. In this experiment, the application of pesticide was decreased from 5–6 times in paddy hybrid rice field to 2 times as the humidity in the canopy was much lower than the irrigated field. As the irrigated water applied was much less, and thoroughly blocked in field, it was observed that the growth of drought tolerant rice in water saving strategy is propitious to controlling the non-spot pollutions (Mei Daoliang *et al.*, personal communication).

13.8   Long-term researching and breeding on the DT Super Rice and the Green Super Rice

To face the challenges of indiscriminate application of pesticides, overuse of fertilizers, and increasingly frequent occurrence of drought in rice production, the Green Super Rice will be developed by pyramiding the resistance to multiple insects and diseases, high nutrient efficiency and drought tolerance using the genomic-based strategies (Zhang, 2005).

Integrated research approaches are necessary for us to understand the genetic basis for drought resistance. QTL mapping was widely employed to identify the loci influencing a series of traits about the response of rice plant to drought stress. The authors and their colleagues found some QTLs with large effect on multiple traits related to the drought tolerance (Liu *et al.*, 2005; Zou *et al.*, 2005). Based on the observation in traits measuring fitness, grain yield and the root system using PVC pipe protocol, Yue *et al.* (2005, 2006) not only mapped QTLs for a series of characters about drought tolerance, but also successfully separated the drought tolerance and drought avoidance as two mechanism of drought resistance. It was indicated that drought tolerance and drought avoidance have distinct genetic basis according to the correlation analysis and QTL mapping results. There are a large amount of drought tolerant genes/QTLs mapped or isolated by the researchers around the world. Maker assisted selection and gene transformation are widely used to introduce or pyramid drought tolerant genes not only mapped or isolated from rice itself, but also from other organisms.

Based on the rapid progress in the researches on drought tolerance, breeding of super paddy rice, the wide application of DT Super Rice or the Green Super Rice in China is expected to be achieved in 5–10 years, depending on the long-term cooperation of rice research and production community.

References

Deng, N. (1999) Formulating the outline of a national program of agricultural science and technology development, and enhancing the revolution of agricultural science and technology in China. *Review of China Agricultural Science and Technology*, 2: 3–8 (in Chinese).

Fukai, S., Pantuwan, G., Jongdee, B., Cooper, M. (1999) Screening for drought resistance in rainfed lowland rice. *Field Crops Res.*, 64, 61–74.

Garrity, D. P. & O'Toole, J. C. (1994) Screening rice for drought resistance at the reproductive phase. *Field Crops Res.*, 39, 99–110.

Lanceras, J. C., Pantuwan, G., Jongdee, B., Toojinda, T. (2004) Quantitative trait loci associated with drought tolerance at reproductive stage in rice. *Plant Physiol.*, 135, 384–99.

Li, Z. K., Fu, B. Y., Gao, Y. M., Xu, J. L., Ali, J., Lafitte, H. R., Jiang, Y. Z., Rey, J. D., Vijayakumar, C. H., Maghirang, R., Zheng, T. Q., Zhu, L. H. (2005) Genome-wide introgression lines and their use in genetic and molecular dissection of complex phenotypes in rice (*Oryza sativa* L.). *Plant Mol. Biol.*, 59, 33–52.

Liu, H.Y., Mei, H. W., Yu, X. Q., Zou, G. H., Liu, G. L., Luo, L. J. (2006) Towards improving the drought tolerance of rice in China. *Plant Genetic Resources*, 4: 47–53.

Liu, H. Y, Zou, G. H., Liu, G. L., Hu, S. P., Li, M. S., Mei, H. W., Yu, X. Q., Luo, L. J. (2005) Correlation analysis and QTL identification for canopy temperature, leaf water potential and spikelet fertility in rice under contrasting moisture regimes. *Chinese Science Bulletin*, 50, 130–139.

Luo, L. J. (2005) Development of near isogenic introgression lines and molecular breeding on rice. *Molecular Plant Breeding*, 3, 609–612 . (in Chinese with English abstract).

Mei, H. W., Luo, L. J., Xu, X. Y., Yu, X. Q., Tong, H. H., Wang, Y. P., Guo, L. B., Ying, C. S., Wu, J. H., Chen, H. W., Yang, H., Li, M. S. (2005) Development and screening of introgressive line population based on the genetic background from high yielding restorer Zhong 413. *Molecular Plant Breeding*, 3, 649–652 (in Chinese with English abstract).

Padolina, W. G. (1996) Efficient water use: key to sustainable water resources management. In:Gurrero RD and Calpe AT (ed.) *Water Resources Management-A Global Priority*. Manila: Rainbow Graphic Corp., Philippines, p.3–7.

Yu, X. Q., Mei, H. W., Li, M. S, Liang, W. B., Xu, X. Y., Zhang, J. F., Luo, L. J. (2005) Perspective and breeding strategy of drought tolerant hybrid rice. *Molecular Plant Breeding*, 3, 637–641 (in Chinese with English abstract).

Yue, B., Xiong, L., Xue, W., Xing, Y., Luo, L., Xu, C. (2005) Genetic analysis for drought resistance of rice at reproductive stage in field with different types of soil. *Theor. Appl Genet*, 111, 1127–1136.

Yue, B., Xue, W., Xiong, L., Yu, X., Luo, L., Cui, K., Jin, D., Xing, Y., Zhang, Q. (2006) Genetic basis of drought resistance at reproductive stage in rice: separation of drought tolerance from drought avoidance. *Genetics*, 172, 1213–1228.

Zhang, Q. F. (2005) Genomics-based strategies for the development of "Green Super Rice". Rice Genetics V, Proceedings of the Fifth International Rice Genetics Symposium, 19–23 November 2005, Manila, Philippines. (http://www.irri.org/ science/abstracts/029.asp).

Zhang, Z. B. (2006) *Researches and developments on high WUE agricultural of dry area in China*. Science Press.

Zou, G. H., Mei, H. W., Liu, H. Y., Liu, G. L., Hu, S. P., Yu, X. Q., Li, M. S., Wu, J. H., Luo, L. J. (2005). Grain yield responses to moisture regimes in a rice population: association among traits and genetic markers. *Theor. Appl. Genet.*, 112: 106–113.

Chapter 14

Overview and integration

U. Aswathanarayana (Editor)

Section I: Biophysical dimensions of food security

This Section covers four inter-connected themes: Remote Sensing tools (chaps. 1–4), Soil (chap. 5), Water (chaps. 6–8) and Agriculture (chaps. 9–13). In the interests of comprehensive coverage of issues which are not covered by the chapters, short notes are added in respect of Soil Health Kits (under Soil), and Wastewater Reuse (under water).

Remote Sensing: Because of the advantage of the repetitive coverage and capability for synoptic overview, satellite remote sensing has emerged as a powerful and cost-effective tool covering all aspects of water, soil, crop, and ecosystem management. Remote sensing methodologies are sought to be made commercially viable through the launching of dedicated satellite systems, develop new retrieval algorithms for remote-sensing data and formatting them into GIS packages.

In Chapter 1, Nemani *et al.* described how remote sensing methodologies could be used in ecosystem management. Forecasting provides decision-makers with insight into the future status of ecosystems and allows for the evaluation of the *status quo* as well as alternatives or preparatory actions that could be taken in anticipation of future conditions. The chapter describes an integrated system called the Terrestrial Observation and Prediction System (TOPS). TOPS is a data and modeling software system designed to seamlessly integrate data from satellite, aircraft, and ground sensors with weather/climate and application models to expeditiously produce operational now-casts and forecasts of ecological conditions. TOPS has been operating at a variety of spatial scales, ranging from individual vineyard blocks in California, and predicting weekly irrigation requirements, to global scale producing regular monthly assessments of global vegetation net primary production. TOPS can be adapted for various agroclimatic situations to help agriculture administrators and farmers in crop management. For instance, Leaf Area Index (LAI) derived from satellite data is used in ecosystem process models to compute water use and irrigation requirements to maintain crops at the optimal levels of water requirement. By integrating leaf area, soils data, daily weather, and weekly weather forecasts, TOPS can estimate spatially varying water requirements within the farm so that managers can adjust water delivery from irrigation systems.

In Chapter 2, Dwivedi explains how remote sensing technologies could be made use of to design strategies of arresting soil degradation. Food security could be enhanced, among other means, through arresting the loss of the biological potential of the soils caused by soil degradation. Remote sensing has emerged as a powerful and cost-effective tool to monitor and quantify various manifestations of soil degradation, and assist in decision-making about strategies of reclamation of degraded soils and utilizing them (e.g. biodiesel plantations). Spceborne spectral measurements have been used to design methods of grouping of ravines for purposes of reclamation. Quantity and mineralogy of salts together with soil moisture, colour and roughness are the major factors affecting reflectance of salt-affected soils. Better delineation of salt-affected soils could be achieved by digitally merging the data from different sensors. Multispectral video systems which record spectral response pattern of vegetation in the visible and near infrared regions of the spectrum have been used for mapping crop variations due to soil salinity. Multispectral imaging (DMSI) systems have been used for assessment of vegetation degradation due to various factors including land degradation. Waterlogging manifests itself in the form of dramatic decrease in the albedo. Soil compaction and crusting take the form of lithospheric and biogenic crusts. The effects of a lithospheric crust are more pronounced in saline soils. Crusted soil is characterized by higher spectral reflectance relative to the same soil with the crust broken.

In Chapter 3, Sesha Sai *et al.* describe how remote sensing methodologies could be used for crop management. Crop management is an integral part of the Sustainable Agriculture (SA). For the intensification of agriculture, remote sensing data on existing land use/land cover pattern is integrated with information on soils, irrigation facilities and socio-economic data to arrive at the most appropriate crop to be grown on a sustainable basis. For instance, post-*kharif* rice fallows offer considerable scope for achieving sustainable production by the introduction of short duration leguminous crops. IRS-LISS III data was used for deriving spatial distribution of cotton crop, while soil maps prepared by the National Bureau of Soil Survey and Land Use Planning, India were used to derive the soil suitability regimes for cotton cultivation. This information is useful for agricultural extension workers in advising the cotton farmers. The resources are optimized using the high resolution satellite data, and through data assimilation techniques, with locale-specific information being made available to the farmers through Village Knowledge Centres.

In Chapter 4, Varshneya *et al.* describe the methods of preparation of agroclimate calendar. In any economy, forecasting of rainfall and renewable water supplies for the upcoming water year is critically important in water management process for crop production, industrial use, hydropower generation, drinking water as well as sustenance of aquatic species. This is all the more so for India as about two-thirds of the agricultural land in India is rainfed. The chapter deals with various approaches for the preparation of crop calendar, based on the predicted temporal and spatial distribution of rainfall. The India Meteorological Department makes use of ten parameters to prepare long-range forecasts. The National Centre for Medium range Weather Forecasting (NCMRWF) predicts location-specific rainfall for four days twice a week, based on Numerical Weather Prediction (NWP) technique. The average accuracy of rainfall forecast on Yes/No basis for Pune was 70% and for Anand in Gujarat it was 68–76% in monsoon season. Based on traditional approaches, Anand Agricultural University prepared a Nakshtra-Charan wise forecast for eight agroclimatic zones of Gujarat state in India. The validation of this forecast on Yes/No basis indicated that accuracy varied from 42% to 73% for various zones. On the basis of the Sea Surface Temperatures, and other atmospheric phenomena, and statistical analysis of the past climate data, the Meteorological Department of South Africa has started providing as a service customized high-resolution hydroclimatic calendars for large farms, resorts, ranches, wildlife parks, etc. This approach holds great promise.

Soil: Soil Health Cards: A soil may have inherent fertility arising out of its mineralogy, humus content, and ability to hold moisture. Fertility may be induced in the soil by the addition of suitable fertilizers. The use of manure for fertilizing the soil is probably as old as the agriculture itself (the Romans even had a particular god, Sterculius, to preside over the protection of the fertility of the soil!). The increase in harvest due to the use of fertlizers varies greatly depending upon the nature of the soil, agroclimatic conditions, and crop management. It is now generally agreed that manure and fertilizer are complementary but not competitive. For purposes of management of soil nutrients and conditioning, it is necessary to keep track of the soil health in a farm. The health of a soil is characterized by the following soil properties: (i) Texture and structure, (ii) Water-soluble salts (electrical conductity), Keen box data (density, porosity, Maximum water holding capacity, field capacity), soil pH, CEC (Cation Exchange Capacity) of soil/roots, Organic carbon, and Assay of biological processes. Farmers are provided with the Soil Health Cards for their farms, on the basis of which they are advised on fertilization, depending upon the crop that is proposed to be grown, and agroclimatic calendar. The common field-scale deficiencies are those of micronutrients, such Zn, Mn, Fe, S, etc. (vide *Curr.Sci.*, v.82, no.7, p. 797–807, 2002).

In Chapter 5, Johri *et al.* deal with ways of using soil microorganisms to optimize soil productivity. The maintenance of the biological basis of land-soil productivity is an integral part of the strategy for sustainable agriculture. The Below Ground Bio Diversity (BGBD) of the soil microorganisms has a significant bearing on soil productivity because of the involvement of the microorganisms in nutrient cycling, sequestration of atmospheric carbon, modification of soil physical structure and water regimes, enhancement of plant health by interacting with pathogens and pests, predators and parasites, and enhancement of plant defense through induced systemic resistance and other mechanisms. The erosion of the biological component of the soils led to the general decline in the soil fertility of the Indo-Gangetic Plains (IGP) of India which produces about 50% of food grains of the country. It is possible to reverse this trend through the management of biodiversity of soils, which needs to be implemented at three different scales: (i) Keystone biota level management – biological nitrogen fixation, mycorrhizas, earthworms etc., (ii) Soil level management – organic matter inputs, balanced mineral fertilizers and amendments, tillage and irrigation, and (iii) Cropping system level management – cropping system design in space and time, choice crops and varieties, genetic manipulation, microsymbioses, rhizosphere microbial dynamics, etc.

Water: In Chapter 6, Chandrasekharam traces the pathways of arsenic from water to food in West Bengal, India. Apart from drinking water, food is now known to be a source of arseniasis in West Bengal (India). Bioaccumulation of arsenic in food crops is strongly influenced by the arsenic content of the irrigated water, and chemical, physical, and microbial characteristics of the soil. The arsenic content of vegetables and cereals grown using arsenious groundwater (85 to 108 µg/L) in West Bengal is 300% greater compared to the mean concentration generally reported in vegetables and cereals elsewhere in the world. For instance, people consuming 100 gm of "*arum*", a leafy vegetable, that could contain 0.22 mg/kg of arsenic, will reach the maximum daily allowable limit of arsenic by eating this leafy vegetable alone. Even food cooked with arsenic contaminated groundwater showed high values (0.12–1.45 mg/kg) that fell well above the limit prescribed by WHO. Arsenic accumulation in rice roots is high where tube wells are used for irrigation. When the roots of the rice plants are ploughed back into the soil after harvesting, the arsenic in rice roots gets mobilized and infiltrates in to the shallow aquifers. Thus arsenic from groundwater, after entering the food chain, gets concentrated in the roots and enters the shallow aquifer thus establishing an "*arsenic flow cycle*" in the rice fields. This is a catastrophe waiting to happen.

This can only be prevented by using canal water for irrigation, and rooftop rainwater harvesting for drinking water.

In Chapter 7, Aswathanarayana deals with ways of making do with less water. There is little doubt that almost any water use could be accomplished with less water than being currently used. Rice is the staple food in many parts of Asia – Pacific region. Irrigated rice is a highly water-intensive crop (3000–5000 L/kg of rice). In countries such as India and China, rice crop alone uses more than 50% of all the total water consumed in the country. Any attempt to reduce the quantity of water used in agriculture should address the problem of ways and means of reducing the water needs of rice. Two approaches are being attempted in China and India: (i) developing low water-need and drought-tolerant rice paddy strains, (ii) agronomic methods, such as SRI, which need less water. There should be economic incentives/ discentives such that a farmer would voluntarily use the least amount of water in his own economic interest. This is best accomplished by metering/ rationing of water, on the lines of electricity. *All this can be achieved without imposing any major economic burden on the farmer.* Industries are using more and more of recycled water. Also, they either use less water per unit of production or switch over to dry technologies, for the simple reason that the less the quantity of water used in processing, the less would be the effluents that need to be disposed of. Singapore which imports all its water needs encourages its citizens to use less water for bathing, washing clothes, etc.

In Chapter 8, Murthy gives an account of the role of irrigated agriculture in food security. About 18% of world's arable land which is under irrigation accounts for about 40% of the crop output. As the demand for water for domestic and industrial needs is increasing rapidly, it becomes imperative to develop new technologies and practices to produce more food with less water. For maximizing crop production water has to be supplied to the crops at different growth stages depending on its physiological nature as well as soil moisture depletion allowed in the root zone. The most important aspect of the irrigation system management is the delivery of water from the storage or diversion to the farmer's fields. There is considerable scope for the improvement of irrigation efficiency through the reduction in conveyance losses. Drainage should be built in any irrigation system. Improper irrigation leads to water logging and soil salinization. In some states in India, such as Andhra Pradesh, Water Users Associations have been instituted under participatory irrigation management, to improve equity and efficiency in the use of irrigation water. Some of the technical advances, such as low water-need paddy strains, drip and sprinkler irrigation, computerized system of delivering irrigation water on the lines of electricity supply, etc. have great potential to improve the efficiency of irrigated agriculture.

With increasing urbanization, municipalities generate ever increasing quantities of wastewater. The acute water shortage in northern China necessitated the use of domestic and industrial wastewater for irrigation. For instance, Tianjin city supports 153,000 ha of sewage irrigated farmland. China has developed comprehensive recycling of waste water: Domestic sewage → Preliminary treatment → Facultative ponds → storage ponds or lagoons (for fish farming) → irrigation of farmland → Effluent recharging groundwater. Fish production in the bioponds proved very lucrative (4500–6000 kg of fish/ha /year).

Applying desalinization technology to agriculture is generally not cost-effective. There are, however, some salt-tolerant tree species, such as, jojoba (*Simmondsia chinensis*, whose seeds yield valuable industrial oil), and Kharijal (*Salvadora persica*, whose seeds can be used in soap making) which can be irrigated with salt water.

Agriculture: Chapter 9 constitutes the pith of Section 1. In it Wani *et al.* explain at great length how food security could be achieved through the unlocking of the potential of the rainfed agriculture. Eighty per cent of the world's agricultural land area is rainfed and generates 58% of the world's staple foods. The key window for improvement of green water productivity entails shifting non-productive evaporation to productive transpiration, with no downstream water trade-off. Improved land and water management practices along with integrated nutrient management (INM) comprising of applications of inorganic fertilizers and organic amendments such as crop residues, vermicompost, farm manures, *Gliricidia* loppings as well as crop diversification with legumes, not only enhanced productivity but also improved soil quality. Normalized difference vegetation index (NDVI) estimation from the satellite images showed that within four years, vegetation cover could increase by 35% in Kothapally watershed. Reduced runoff and erosion reduced the risk of downstream flooding and siltation of water bodies and thus directly improved the environmental quality in the watersheds. The Self-Help Groups (SHGs) such as village seedbanks, not only provide timely and quality seeds, but also create a venue for receiving technical support and building the capacity of members like women for the management of conservation and livelihood development activities. Incorporating knowledge-based entry point in the approach led to the facilitation of rapport and at the same time enabled the community to take rational decisions for their own development. Investments in rainfed areas produce multiple benefits such as reducing poverty, developing social capital, community-empowerment, building institutions, protecting environment, reducing land degradation, conserving biodiversity, sequestering carbon and provide environmental services.

In Chapter 10, Vadez gives an account of the drought- and salinity-tolerant crop varieties. Among the ICRISAT-mandated crops, pearl millet and sorghum are the most salinity-tolerant, followed by groundnut and chickpea, with pigeonpea being extremely sensitive. The reproductive processes are likely to be the most vulnerable to salt stress. ICRISAT has successfully exploited the diversity to reveal superior genotypes for different traits. Considerable effort is being dedicated to root, by focusing more on what roots "do" rather than what roots "are", and then develop the phenotyping capacities for such alternative traits. These would be used to explain the role of roots in sorghum and pearl millet (Quantitative Trait Loci) Atlas. Current research has focused on developing varieties capable of efficiently using erratic rainfalls, and having high transpiration efficiency (TE). ICRISAT is exploring genotypic variation for HI under drought conditions, and to investigate the relations, synergies, trade-off between water uptake, TE and HI. In this, transgenic materials presented in the chapter will be very useful, in particular because preliminary data indicate that DREB1A also influences the root development, and the success of reproduction under water deficit.

In Chapter 11, Norman Uphoff holds that SRI has the potential to enhance both food and water security. The methodology known as the System of Rice Intensification (SRI), developed by Fr. Laulanié in Madagascar, capitalizes upon production potentials that already exist within the genomes of rice (and other crops) when more ideal growing conditions are provided above and especially below ground. SRI involves the individual transplanting of very young seedlings, with wide spacing, into paddy soils that are kept well aerated, moist but not continuously saturated. SRI has potential to enhance both food and water security in that it needs less water (25–50% less), gives larger yields (50% or more), and can double the income of farmers per hectare. There are other derived benefits, such as greater resistance to pest and diseases damage and to abiotic stresses, higher outturn of milled rice, and possibly more nutritional value and reduced production of methane. Experience is now showing that other crops can benefit by the application of SRI concepts and practices.

Shashidhar (Chap.12) describes the newly-developed aerobic rice. Irrigated rice is a highly water-intensive crop (3000–5000 L/kg of rice). Besides, methane, a greenhouse gas, is released from the submerged soil to the atmosphere through roots and stems of rice plants. Aerobic rice is produced by crossing *Budda*, an indigenous drought tolerant low yielding cultivar with deep root system and good combining ability for the root and shoots traits, with *IR64*, a lowland high yielding variety which manifests good osmotic adjustments. Compared to the water requirement of lowland flooded rice, the aerobic rice system can save about 45 per cent of water. Aerobic rice is both a concept of growing rice and appropriate genotypes suited for such a growth. The project is developed in four stages: (i) Selection of roots, (ii) Participatory Plant Breeding, (iii) Screening for drought tolerance, (iv) Advanced field trials. Some of the benefits of the aerobic rice are: (i) Direct seeding; (ii) Deficit irrigation; (iii) Avoidance of destruction of soil structure because of puddling; (iv) Designer fertilizers specific to needs of direct seeded paddy which would include Silicon; and (v) Diversified cropping systems to accommodate principles of crop rotation or intercropping with pulses, legumes and large seeded cereals.

In Chapter 13, Mei and Luo give an account of low water-need, drought-tolerant "super rice" in China. Rice is not only China's staple food but is also an integral part of its culture. Rice production, however, consumes 70% of the fresh water used in agriculture, and there in lies the problem. To achieve long-term food security and sustainable development, China needs the development of low water-need and drought-tolerant rice cultivars and its application technologies. A series of drought tolerant rice cultivars were developed by hybrid breeding with crosses between elite paddy rice and upland rice varieties (e.g. *Huhan 3*, *Huhan 7* and *Huhan 15*), and were released by the Chinese Ministry of Agriculture during 2004 to 2006. The reproductive stage of rice plants is one of the most sensitive periods to drought stress. Water deficit at this stage causes severe spikelet sterility and final yield loss. Spikelet fertility became the most important contributor to the grain yield under drought stress, while the panicle number per plant became the second important one. A series of hybrid combinations were evaluated for drought tolerance and yield potential with *Huhan 1A* as female parent and either restorers in paddy rice or newly developed upland rice restorers as the male parent. Two high yielding, good quality and drought tolerant hybrid rice combinations, *Hanyou 2* and *Hanyou 3*, were developed in author's group and released by the city government of Shanghai in 2006. The grain yields varied from about 6 tonnes/ha to 7.5 tonnes/ha with much less water applications, gaining 20%–100% yield increase compared to the local dominant hybrid rice combinations grown under water stress conditions.

Socioeconomic Dimensions of Food Security

Chapter 15

Fermented foods as a tool to combat malnutrition

J.B. Prajapati

*Department of Dairy Microbiology, SMC College of Dairy Science,
Anand Agricultural University, Anand, Gujarat, India*

15.1 Introduction

The problems of food security, malnutrition and poor food hygiene among the people in developing countries are summarised in Table 15.1. An important way to address these issues is to develop systematic and cost-effective food processing technology that can give safe, nutritious and healthy food to the people.

Our ancient people have developed a technology of fermentation, which is ultimately a process of bioconversion of organic substances by microorganisms and/or enzymes of microbial, plant or animal origin, that gives multi-fold benefits. Fermentation is an oldest form of preservation, which is applied globally (Prajapati and Nair, 2003).

Table 15.1 Fact sheets on population, food & nutrition in developing countries.

FAO reports	• World Population increases from 6 billion to 9 billion in 2050.
	• Food Shortage envisaged – 60% more food will be required. Out of this 85% demand will be in developing countries.
	• Problem of Malnutrition by 2030, around 440 million persons will be chronically undernourished.
	• To reach the World Food Summit goal of reducing hunger by half by 2015, the number of hungry people needs to fall by 22 million a year. Currently it is falling by only 6 million a year.
	• One in every five people (a total of 777 million individuals) in the developing world is chronically undernourished.
	• Fifty-five percent of the 12 million child deaths each year are related to malnutrition.
	• More than 2000 million people suffer from micronutrient deficiencies.
WHO reports	• 30% of world population suffer from malnutrition.
	• 12 million children die due to malnutrition, diarrhoea and other health problems every year.
	• More than 50% of the disease burden of the world is attributed to hidden hunger, unbalanced protein/energy intake or vitamin or mineral deficiencies.
	• WHO has reported that 2.1 million people in world died from enteric diseases in 2000, with diarrhea being a major cause of malnutrition in infants and young children.
The Indian situation	• 8% of the Indians do not get two square meals a day.
	• Every third child born is under weight.
	• 50% of pre-school children suffer from under-nutrition.
	• More than half the women and children are anemic.

Several kinds of indigenous fermented foods have come-up from various parts of the world since centuries and are strongly linked to culture and tradition, especially in rural households and village communities. It is estimated that fermented foods contribute to about one-third of the diet worldwide (FAO, 2004). Vedamuthu (2003) indicated that the suitable process/product for rural areas should have (i) Simple technology, (ii) Simple and easily maintainable equipment, (iii) Improved stability, shelf-life and safety, and (iv) Increased nutritional value with least cost. At the same time they should improve/maintain health of children and vulnerable sections of the society and also be an engine for economic empowerment of rural population. Fermentation satisfies all these criteria, and hence should be exploited for the benefit of rural people. An exhaustive review by Farnworth (2004) has indicated about various kinds of fermented foods known around the world.

The reasons, why we propose fermentation technology as a means of tackling problem of food security and under-nutrition are given below.

Strong cultural foundation

Fermentation is an ancient tradition of food preservation. Dahi (yoghurt) and butter milk were popular since the time of Lord Krishna. Mention was made of the therapeutic value of dahi in Vedas and Ayurvedic treatises in India. Existence of *idli and dosa* was documented 2000 years back.

Excellent exposition of biodiversity

Can be applied to a wide variety of raw materials to produce a variety of different finished food products . Fermented foods are manufactured and consumed in practically every part of the world by every type of communities regardless of class, creed, sex, religion, age or ethnicity. Cereals, pulses, root crops, vegetables, fruits, milk, meat and fish are preserved by one or the other methods of fermentation.

Unique process with great potential

It is environment friendly. Consume less energy. Produce less waste. Usable by both men and women. Does not need costly equipment. Easy to manage under house hold conditions of low income communities as well as on industrial scale for urbanised welfare societies.

Applicable for many purposes

Enhancing shelf-life of food. Protection and preservation of foods. Producing desirable taste, flavour, texture. Improvement of nutritional value. Producing required physico-chemical properties. Improvement of food safety and food security.

They offer immense opportunity for production of products, which can be classified as functional foods, organic foods, natural foods, health foods, convenience foods, ethnic foods, neutraceuticals, foods for clinical nutrition. probiotics, prebiotics, synbiotics.

15.2 Nutritional benefits of fermented foods

The nutritive value of any food depends on the quality and quantity of nutrients present in that food. The nutritive value of fermented foods changes during the process of making them either due to addition/deletion of nutrients or due to fermentation, which is explained schematically in Figure 15.1.

During the process of fermentation, microbial growth and metabolism results in the production of diversity of metabolites. These metabolites include enzymes which are capable of breaking complex carbohydrates, proteins and lipids present within the substrate and/or fermentation medium; vitamins, antimicrobial compounds (e.g. bacteriocins and lysozymes), texture forming agents (e.g. xthan gums); amino acids, organic acids and flavour compounds.

Carbohydrates

Carbohydrates are the principle substrates for fermentative microorganism. Different acids are produced from sugars which bring down the pH of the fermenting medium and help in absorption of proteins and minerals. During the fermentation of milk for yoghurt or dahi production, about 20–30% of the lactose is utilized resulting mainly in lactic acid production. Lactic acid and other organic acids like acetate, butyrate, propionate, etc produced during fermentation act as biopreservatives by reducing pH, which inhibits the growth of potential spoilage and harmful bacteria. Lactic acid also influences physical properties of the food. Lactic acid, the major metabolite of several fermented foods is reported to improve the absorption of calcium and phosphorous besides forming a ready source of energy to the body (Laxminarayana, 1984).

The partial hydrolysis of lactose in milk based fermented foods make them suitable for lactose-intolerant people.

Proteins

The native milk proteins, which are known to form hard curd in the stomach, are converted into a soft curd containing finely dispersed casein particles due to bacterial action in fermented products. As a result, fermented milk proteins are particularly useful to children, old people and persons suffering from stomach ulcers.

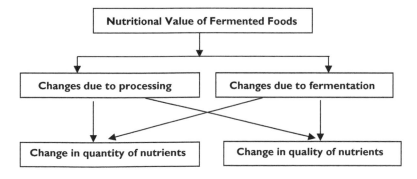

Figure 15.1 Nutritional value of fermented foods.

Any proteins, after fermentation, in general become more easily digestible due to enzymatic actions of microbes. The microbes do part of the work required to be done by the body to digest the proteins and hence, they are easier to digest. During fermentation, the proteolytic action of microbes release some peptides and amino acids. The soluble nitrogen content is higher in fermented foods.

The fermented food is also enriched by presence of microbial cell proteins. Most of the living cells would be digested in stomach releasing highly digestible proteins, rich in all essential amino acids (Laxminarayana, 1984). Overall, in a study by Laxminarayana, it was found that protein quality of dahi samples were higher by 3 to 30% than that of the milk used for preparing the dahi.

The content and quality of cereal proteins may be improved by fermentation. Hamad and Fields (1979) reported that natural fermentation of cereals increases their relative nutritive value and available lysine. The values shown in Figure 15.2 is an example of what fermentation can do for the improvement of essential amino acids in fermented foods.

In fermented foods made from mixtures of raw materials, the proteins from various sources become complementary to each other. For example, *kishk* is a fermented wheat and milk based product popular in Arabian countries, in which the proteins from milk

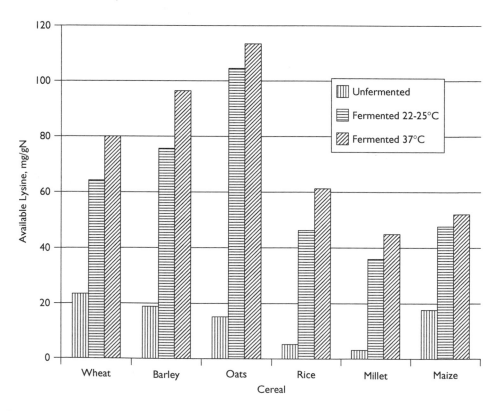

Figure 15.2 Influence of natural fermentation of cereals on available lysine. Data from Hamad and Fields (1979).

and wheat make good the deficiencies of each other and provide a better quality protein required by the body (Prajapati *et al.*, 2005).

Minerals

There are no quantitative changes in mineral content of food as a result of fermentation. However, fermentation improves the availability of minerals for the body as they are partly modified and are consumed with lactic acid. Fermented milk is an excellent source of calcium and phosphorus which are essential for the bones. Further, hydrolysis of chelating agents such as phytic acid during fermentation, improves the bioavailability of minerals. Because of these reasons, fermented foods are recommended to reduce the problem of mineral deficiencies.

Vitamins

During the process of fermentation, the microorganisms need certain vitamins for their own growth. Hence, it is reported that some vitamins of B-complex, orotic acid, etc., get reduced in fermented foods. However, several bacteria, while growing, synthesize many other vitamins too. There are several reports indicating the increase in the vitamin contents in fermented foods. However, the quantity and quality depends upon the culture and the method of manufacture of the product.

In general, most lactic acid bacteria are known to synthesize riboflavin, niacin, thiamine, and folic acid. These can have direct effects on health. For example, fermented foods are very good for the people subsisting largely on maize whose niacin or nicotinic acid contents are limited and pellagra is incipient. For people subsisting on polished rice, which contains limited amounts of thiamine, beri-beri is incipient (Radhakrishna, 2002).

Other nutritional benefits

Several raw materials, especially cereals and legumes, contain significant amounts of antinutritional and toxic substances like protease inhibitors, amylase inhibitors, metal chelates, flatus factors, hemagglutinins, saponins, cyanogens, lathrogens, tannins, allergens, acetylenic furan, isoflavonoids phytoalexins, phytates, enzyme inhibitors, etc. These substances reduce the nutritional value of foods by interfering with mineral bioavailability and digestibility of proteins and carbohydrates. The process of fermentation has been reported to reduce all such anti-nutritional compounds and liberate the nutrients and make them available to the consumers (Reddy and Pierson, 1994; FAO, 1999). *Idli* (fermented rice cake) has been found to increase the hemoglobin status in working women.

The process of fermentation makes the food more nutrient dense than the original raw material. A classical study involving rats as animal model showed that feeding yoghurt resulted in increased weight gain and increased feed efficiency compared to the milk from which it was manufactured (McDonough *et al.*, 1982). Several such studies on growth as well as other nutritional benefits have been complied in IDF bulletin (IDF, 1991). Nutritional composition of some of the Indian fermented milk products is given in Table 15.2.

Table 15.2 Nutritional composition of Indian fermented milk products.

Product Serving size	Dahi 200 g	Shrikhand 100 g	Lassi 1 cup (227 g)
Calories (kcal)	113	235	155
Calories from fat	103	33	NA
Fat, g	11	3	0.0
Saturated fat, g	7	2	NA
Sodium, mg	90	29	NA
Potassium, mg	300	57	NA
Total carbohydrate, g	11	44	17.2
Sugars (sucrose), g	0	40	NA
Protein, g	7	9	12.8
Vitamin A, IU	20	NA	NA
Calcium, mg	298	65	447
Iron, mg	0.5	2	NA
Phosphorus, mg	186	95	NA

15.3 Therapeutic benefits of fermented foods

Use of dahi and butter milk for getting relief from diarrhoea and other minor intestinal ailments has been recommended in Indian treatises and has been the *grandma's* therapy since centuries. In recent times, the number of clinical trials have shown several therapeutic benefits from fermented foods, such as, (1) Antibacterial property against common intestinal pathogens, (2) Used as therapeutic agents in gastro-intestinal disorders, (3) Can be digested by lactose intolerant people, (4) May have hypocholesterolemic activity, (5) Useful in preventing hepatic encephalopathy, (6) Anti-tumor or anticarcinogenic activities, (7) Immunostimulatory properties, (8) Anti-mutagenic activities, (9) Can be used after heavy antibiotic doses to replenish the intestinal flora, (10) Have been used in other diverse diseases.

Role of intestinal bacteria in human health is well known. Some of the bacteria employed in fermentation are native to the intestinal tract of man and animals and are the right microflora as probiotic organisms. The maximally utilized bacteria as probiotic organisms belong to *Lactobacillus acidophilus* and *Bifidobacterium bifidus*. The other lactic acid bacteria which are involved in fermentation too have regulatory role in the intestinal tract. Hence, their role in intestinal therapy is proved beyond doubt. Especially, fermented milks have been used for control of various types of diarrhoea as well as constipation, colitis, dysentery and other GI tract disturbances (IDF, 1991; Shah, 2005).

Lactic acid bacteria produce a number of antimicrobial compounds, like lactic acid, acetic acids, hydrogen peroxide, diacetyl and several types of bacteriocins and antibiotic like substances. These substances along with lower pH and Eh helps in control of growth of intestinal pathogens and thus can help in controlling intestinal diseases.

Several fermented probiotic products are available in the market with a specific culture of lactic acid bacteria which have been clinically proven to reduce traveller's diarrhea, antibiotic associated diarrhea and several other problems (Shah, 2003). In Tanzania, children given fermented porridge are less prone to diarrhoea relative to the children who are given unfermented porridge. Porridge is often contaminated with bacteria that cause diarrhoea because of impure water or poor hygiene. Fermentation helps to reduce contamination because harmful bacteria cannot multiply as easily in fermented food (Ashworth, 1997).

The other most convincingly proved attribute is the suitability of fermented milks for lactose intolerant people. This is because the lactose in milk is partly digested by the enzyme lactase produced by the cultures and its continued production in the intestine of the consumers too (Shah, 2005).

Some lactic acid bacteria have demonstrated in vitro and in vivo anti-tumor activities. As a result the use of fermented milks made using specific probiotic bacteria has been reported as supportive therapy for colon cancer (Patidar & Prajapati, 1997). Apparently the beneficial bacteria in the intestine reduces the activity of putrefactive bacteria and thereby impedes the chances of conversion of pro-carcinogens to carcinogens in the intestine.

Selected bacteria used in fermentation check translocation of undesirable bacteria through mucosal barriers. They increase the circulating and intestinal antibodies, produce specific antibodies and are also involved in increasing macrophage activity. They increase T-cell and B-cell activities and are also involved in the production of gamma-interferon. These observations demonstrate the potential of these bacteria for enhancing the immunity in the consumers (Patidar and Prajapati, 1997).

The fermented foods, especially milk based foods, can be beneficial to ageing people too as they slow down the age related changes. Calcium availability is higher particularly in fermented milks. Vitamin D is important for uptake of dietary calcium and in calcium metabolism. Protein content is high in fermented milk which is shown to slow down bone loss. It also slows down other age-related diseases, such as impaired lactose digestion, weakened immune defense system, less dehydration etc.

The Masai tribesmen in Africa possess low levels of serum cholesterol and a low incidence of clinical coronary disease despite a diet rich in milk and meat. While investigating this paradox, it was serendipitously discovered that consumption of large quantities of fermented milk by Masai was actually responsible for lowered serum cholesterol which overwhelmed the hypercholesterolemic action of a surfactant added to milk. There are numerous reports indicating that consumption of fermented milk products may favourably change the lipid metabolism. However, this property is strain specific and it is difficult to generalize. The cholesterol can be reduced directly or indirectly in fermented foods, but it will definitely have a beneficial effect, especially for the hypertensive people (Ashar and Prajapati, 2001).

Fermentation reduces the toxin (cyanide) that is naturally present in cassava, particularly in bitter varieties. The traditional way of making *gari* and *farinha* by grating cassava and then letting it soak in water to ferment cleverly allows the acid to release the toxin (Ashworth, 1997).

During the process of fermentation, especially from milk proteins, several bioactive peptides are found to be released by lactic acid bacteria. These peptides have antimicrobial, opiates, antithrombotic, antihypertensive and mineral utilizing properties. Several commercially available cultures showing strong anti-ACE activity have been reported (Barrett *et al.*, 2004).

A number of studies have reported that the levels of Conjugated linoleic acid are higher in fermented dairy products than in unprocessed milk (Barrett *et al.*, 2004).

15.4   Strategy for the promotion of fermented foods

The propagation and promotion fermented foods as a means to tackle the problem of mal-nutrition needs sincere efforts from several sides. The first requirement is to educate the common man about the health benefits of fermented foods and promote traditional fermented foods to the fullest extent. Media as well as social and health workers can play an important role in this campaign.

To manufacture fermented foods with improved technology and selected probiotic bacteria requires systematic research and development. This needs sincere efforts by academic institutions and the food industry. The following R & D areas need to be pursued to promote greater utilisation of fermented food: (1) Identification of new raw materials that can be fermented (e.g. ways of making bread from coarse grains, like sorghum and pearl millet), (2) Understanding and optimizing the biochemical and physicochemical mechanisms of fermentation, (3) Preparation of library of microbes, and customizing groups of them for fermentation of specific substances for specific fermentation protocols, (5) Assessment of nutritional improvements, (6) Clinical trials to evaluate therapeutic properties, and (6) Technology for preservation, packaging and distribution of fermented foods.

R & D is considered to be a costly item in the economy of any food company. However research and development could be effective and less expensive when it is carried out as collaborative network projects. Development of new foods and testing its special health benefits need input from many specialized areas of technology and science including agriculture, food processing, microbiology, biochemistry, nutrition, medicine, sociology, extension, etc. The SASNET-Fermented Foods has taken the initiative in this direction to build a network of all stakeholders who can promote R & D in fermented foods and propagate the use of fermented foods globally (Nair *et al.*, 2006).

Further, for reaching up to the last person in the society, there is need to evolve a special system. Alternatively some of the existing systems can be utilized. For example, the mid day meal program for school going children is running successfully in many States in India. This programme can conveniently include suitably packed curd-rice or butter milk or any such probiotic product which is designed to take care of nutritional enrichment and health promotion. There are several self-help groups (SHG's) working in India, which have contributed to the economic betterment and empowerment of rural and poor women like SEWA in Ahmedabad. The network of such groups can also be utilized for this purpose. Some of the fermented foods that can be easily put in the system are shown in Table 15.3.

15.5   Conclusions

Fermented foods offer a low-cost solution to the problem of malnutrition and ill-heath in the developing world. It has become a matter of great interest to scientists, medical practitioners, food companies, and marketing agencies as the demand for such foods is enormous and it is growing at a rapid rate. There is lot of scope to enhance use of traditional fermented foods and manufacture novel foods with health bacteria on an industrial scale. Thus the propagation of fermented foods on a wide scale can help in getting closer to the UN millennium goal of halving the under-nutrition by 2015.

Table 15.3 Some fermented foods that can be easily popularised.

Product and its description	Region/ country	Microflora associated	Fermentation conditions	Nutritive/ Therapeutic value
(1) Dahi (Set sour curd resembling yoghurt)	India	*Lactococci, Leuconostocs spp. Lactobacillus acidophilus, Lb. bulgaricus, Lb. casei, Str. thermophilus Some yeasts.*	12–14 h at ambient temp. Depends upon personal liking of sourness.	All nutrients of milk in more digestible form. Increased digestibility of protein and absorption of calcium.
(2) Lassi & Butter milkr (Stirred curd added with sugar or spices/ diluted with water to make a drink)	India	*Lactococci, Lactobacillus spp. Leuconostocs spp. Str. thermophilus Some yeasts.*	12–14 h at ambient temp. Depends upon personal liking of sourness.	All nutrients of milk in more digestive form. Energy value depends upon the level of dilution.
(3) Shrikhand (A concentrated dahi by partial draining of water and added with sugar and other condiments)	India	*Lactococci, Lactobacillus* spp. *Str. thermophilus Some yeasts.*	12–16 h fermentation of milk and then hanging the curd for draining water for 4–8 h.	Protein – 6% Fat – 6–8% Sugar – 20–40% More free amino acids and some B complex vitamins.
(4) Kishk (Dried balls made from fermented milk-wheat mixture)	Arab Countries	Lactobacilli Yeasts	Multiple steps of 24 h at 30–40°C	Rich source of lysine and other essential amino acids. Higher PER and NPU.
(5) Curd-rice	India	Lactobacilli Yeasts	12–14 h at ambient temp. Depends upon personal liking of sourness.	All nutrients of milk and rice in more digestive form.

References

Ashar, M.N., and Prajapati, J.B. (2001). Role of probiotic cultures and fermented milks in combating blood cholesterol – A Review. *Indian J. Microbiology.* **41**(1): 75–86.

Ashworth A (1997). *Fermentation.* Footsteps – Quarterly Paper No. 32, September 1997 edited by I. Carter, Yorkshire, UK.

Barrett E, Stanton C, Fitzgerald G and Paul Ross R (2004). Biotechnology of food cultures for the nutritional enhancement of milk and dairy products. In *"Handbook of functional dairy products"* ed. Shortt C and O'Brien J., CRC Press, Boca Raton, pp. 255–273.

FAO (2004). Biotechnology applications in food processing: Can developing countries benefit? *Electronic forum on Biotechnology in Food and Agriculture*: Conference 11.

FAO (1999). *Fermented cereals – A global perspective.* FAO Agril. Service Bulletin No. 138., FAO, Rome.

Farnworth ER (2004). The beneficial health effects of fermented foods – Potential probiotics around the world. *J. Neutraceuticals, Functional and Medical Foods*, 4(3/4):93–117.

Hamad AM and Fields ML (1979). Evaluation of protein quality and available lysine of germinated and fermented cereals. *J. Food Sci.* 44:456–460.

IDF (International Dairy Federation) (1991). Cultured dairy products in human nutrition. *Bulletin of international Dairy Federation*, No. 255, Brussels, pp. 1–24.

Laxminarayana (1984). Nutritive and therapeutic properties of fermented milk. *Indian Dairyman*, 36:329–336.

McDonough FE, Hitchins AD and Wong NP (1982). *J.Food Sci.* 47:1463–5 cited from FAO(2004).

Nair BM, Prajapati JB and Varshneya MC (2006). SASNET-Fermented foods: An initiative for meeting the challenges of poverty and hunger. *Indian Food Industry*, 25(5):13–15.

Patidar SK and Prajapati JB (1997). Methods for Assessing the Immunostimulating properties of dietary lactobacilli-A critical appraisal. *J. Food Sci. & Technol.*, 34:181–194.

Prajapati JB and Nair BM (2003). The history of fermented foods. In *"Fermented Functional foods"* edited by Edward R. Farnworth, CRC Press, Boca Raton, New York, London, Washington DC, pp. 1–25.

Prajapati Neha, Rema Subhash and Prajapati JB (2005). Manufacturer of Indian Kishk and a study of its nutritional properties. In Souvenir of Second International Conference on Fermented Foods, Health Status and Social Well-being, AAU, Anand, India, pp. 82–84.

Radhakrishna R (2002). "Food and Nutrition Security" in Kirit S. Parikh and R. Radhakrishna (eds.) *India Development Report*, 2002, Oxford University Press.

Reddy NR and Pierson MD (1994). Reduction in antinutritional and toxic components in plant foods by fermentation. *Food Research International.* 27:281–285.

Shah, NP (2005). Fermented functional foods- an overview. In Souvenir of Second International Conference on Fermented Foods, Health Status and Social Well-being, AAU, Anand, India, pp. 1–6.

Vedamuthu ER (2003). Fermented foods in the Indian context: Benefits and challenges. In Souvenir released at International Seminar on "Fermented foods, health status and social well-being", SMC College of Dairy Science, GAU, Anand, pp. 38–41.

Chapter 16

Changing patterns of food consumption in India and their dietary implications

Swarna S. Vepa

M.S. Swaminathan Research Foundation, Chennai, Madras, India

16.1 Introduction

Food consumption patterns depend upon a variety of factors, such as changes in tastes and preferences, affordability and availability. Food basket also differs from region to region and also across the income groups. This paper examines the broad food consumption patterns of the Indian population in rural and urban areas, over a period of time and at the across the income/expenditure deciles and across the regions at the average level. The aim is to see if the diets are adequate in calories protein and other essential micronutrients. The paper also tries to provide explanations for the major changes occurring in the dietary patterns and the consequences of such changes.

The study is based on the National Sample Survey data. The calculations are based mostly on the tabulated data provided by the survey reports, unless otherwise stated. Study based on the unit level data and the tabulated data differ from each other, in magnitude, though the direction of the trend remains the same.

The paper is organised into three sections. The first section deals with the cereal consumption levels and energy deficiency. The second section examines the consumption patterns of various food items at the average level. The third section examines the dietary deficiencies of micronutrients.

16.2 Cereal consumption

In the diets of the low-income population cereal is the main contributor the calories. Monthly per capita average consumption of cereals has declined between 1993–94 and 2004–05. It had fallen from 13.4 to 12.12 Kg/capita in rural India and from 10.60 to 9.9 kg/capita in Urban India. The decline is spread over all the expenditure classes and in all major states. Intake of coarse cereals and millets constituted only 1.28 kg out of the 12.12 kg of cereal consumption of rural areas. The reduced consumption of cereals has important consequences for essential nutrients such as calories and protein as well as micronutrient deprivation for various income groups.

Share of food expenditure as a percentage of total expenditure at the average level declined by 3.6% in rural areas and by 5.6% in urban areas between 1999–2000 and 2004–05; indicating an improvement in the incomes of the people. The diversification of food basket obviously reduces the cereal intake. The average cereal consumption of rural areas fell only marginally by about 600 grams/month. Cereal consumption has been declining across all deciles in all states (Table 16.1).

Table 16.1 Average Monthly Consumption – All India.

	Urban	Rural
Cereal	All class	
Rice	4.853	6.549
Wheat	4.646	4.293
Jowar	0.225	0.430
Bajra	0.113	0.389
Maize	0.025	0.307
Barley	0.001	0.006
Small millets	0.001	0.009
Ragi	0.076	0.131
Other cereals	0.002	0.005
Total	**9.942**	**12.118**
Pulses & Products	0.783	0.674
Cereal substitutes	0.033	0.043

Source: NSS 61st Round 2004–05.

There are two significant developments.

1. Cereal consumption at the average level has declined over years in all the expenditure deciles across the poor as well as the rich. Obviously, at the average level and especially in urban areas the income elasticity of cereal consumption turns negative and it declines sharply with affluence. This is a clear indication of increasing affordability of other foods by the urban middle class and the rich in both rural and urban areas. Food basket is definitely changing for the country as a whole. However it is not clear whether the quality of the food basket is for the better or worse. The 61st round data for the National Sample survey for 2004–05 on all the other items of the food has not yet been released.
2. Only in the lowest percentile there is a slight improvement in the cereal consumption over a period in the rural areas but substantial improvement in the urban areas. It appears that the income elasticity of cereal consumption is still very high for the very poor both in the urban and the rural areas. The income elasticity of cereal consumption is higher in rural areas as the cereal consumption increases across the expenditure deciles from the lowest to the highest. As the income increases cereal consumption goes up. The more backward and agricultural dependent the region is, the larger would be the cereal consumption.

Cereal consumption pattern of the states across the country also bring out some interesting facts about the factors contributing to higher cereal consumption. Poorer States, States that are less urbanised such as Bihar, Orissa, Madhya Pradesh and Uttar Pradesh, and where local availability of cereals is good the consumption of cereals appears to be high (Table 16.2).

There are reasons to believe that consumption out of homegrown produce, contributes to poverty reduction and higher cereal intake. The NSS report of 2003 as given in Table 16.3, clearly shows that cereal consumption is high in the farm household

Table 16.2 Changes in average per capita cereals consumption in quantity terms over the last decade in different percentile classes of population (ranked by MPCE), All India.

S.No.	Population percentile class	Monthly per capita cereal consumption (kg) in					
		Rural			Urban		
		1993–94	1999–2000	2004–05	1993–94	1999–2000	2004–05
1	0–5	9.68	9.78	9.88	8.91	8.99	9.25
2	5–10	11.29	11.15	10.87	10.11	10.15	10.04
3	10–20	12.03	11.64	11.33	10.61	10.25	10.09
4	20–30	12.63	12.27	11.70	10.75	10.75	10.24
5	30–40	13.19	12.56	11.98	10.89	10.61	10.12
6	40–50	13.33	12.89	12.16	10.99	10.80	10.25
7	50–60	13.72	13.03	12.37	10.91	10.69	10.08
8	60–70	14.07	13.36	12.61	10.95	10.66	10.09
9	70–80	14.41	13.45	12.77	10.73	10.50	9.97
10	80–90	14.59	13.67	12.72	10.68	10.52	9.63
11	90–95	14.98	13.73	12.77	10.19	9.94	9.50
12	95–100	15.78	14.19	13.50	10.29	9.72	9.10

Source: NSS 61st Round, Report No. 508, "Level and Pattern of Consumer Expenditure". 2004–2005.

Table 16.3 Land possession and consumption.

S.No.	State	Percentage of population without any land in their possession 1999–2000	Percentage population possessing less than 0.40 hectare 1999–2000	Avg. Qty. (kg/30 days) per capita consumption total cereals by farmer H.holds 2003	Avg. Value (Rs.) per capita consumption total cereals by farmer H.holds 2003	Implicit price per kilogram 2003
1	Andhra Pradesh	5.90	54.10	12.88	123.78	9.61
2	Assam	3.20	50.80	12.73	133.38	10.48
3	Bihar	11.10	56.50	14.31	111.79	7.81
4	Gujarat	10.00	47.00	10.38	80.00	7.71
5	Haryana	4.40	51.10	10.84	73.99	6.83
6	Himachal Pradesh	4.10	54.60	12.14	99.85	8.22
7	Jammu & Kashmir	0.90	37.20	13.06	126.63	9.70
8	Karnataka	3.60	44.20	11.43	97.28	8.51
9	Kerala	1.70	81.10	9.64	118.08	12.25
10	Madhya Pradesh	8.20	28.40	12.14	81.73	6.73
11	Maharashtra	14.30	39.40	11.1	86.77	7.82
12	Orissa	1.40	56.40	14.43	104.79	7.26
13	Punjab	9.80	56.00	10.48	73.46	7.01
14	Rajasthan	2.10	28.40	13.33	86.6	6.50
15	Tamil Nadu	10.70	65.10	10.63	103.04	9.69
16	Uttar Pradesh	4.20	50.00	13.29	90.38	6.80
17	West Bengal	4.90	72.10	14.02	132.4	9.44
	All India	7.20	51.00	12.71	100.87	7.94

NSS 55th Round. "Employment and Unemployment Situation in India, Report No. 458. '99–'00 .
NSS 59th Round" Consumption Expenditure of Farmer Households' Report no. 495, 2003.

Table 16.4 Changes in average per capita cereals consumption in physical terms over the last decade in the major states.

| S.No. | State | Monthly per capita cereal consumption (kg) in | | | | | |
| | | Rural | | | Urban | | |
		1993–94	1999–2000	2004–05	1993–94	1999–2000	2004–05
1	Andhra Pradesh	13.30	12.65	12.07	11.30	10.94	10.51
2	Assam	13.20	12.63	13.04	12.10	12.26	11.92
3	Bihar*	14.30	13.75	13.08	12.80	12.70	12.21
4	Gujarat	10.70	10.19	10.07	9.00	8.49	8.29
5	Haryana	12.90	11.37	10.66	10.50	9.36	9.15
6	Karnataka	13.20	11.53	10.73	10.90	10.21	9.71
7	Kerala	10.10	9.89	9.53	9.50	9.25	8.83
8	Madhya Pradesh #	14.20	12.94	12.16	11.30	11.09	10.63
9	Maharashtra	11.40	11.32	10.50	9.40	9.35	8.39
10	Orissa	15.90	15.09	13.98	13.40	14.51	13.11
11	Punjab	10.80	10.58	9.92	9.00	9.21	9.01
12	Rajasthan	14.90	14.19	12.68	11.50	11.56	10.84
13	Tamil Nadu	11.70	10.66	10.89	10.10	9.65	9.48
14	Uttar Pradesh ^	13.90	13.62	12.87	11.10	10.79	10.94
15	West Bengal	15.00	13.59	13.18	11.60	11.17	10.39
	All India	13.40	12.72	12.12	10.60	10.42	9.94

Source: NSS 61st Round, Report No. 508, "Level and Pattern of Consumer Expenditure". 2004–2005,
Note: *: Includes Jharkhand, #: Includes Chhattisgarh, ^: Includes Uttaranchal.

possessing land especially in less commercialised agriculture (IARI-FAO/RAP 2001).[1] The declining consumption of rice and increasing consumption of wheat in 2004–05 compared to 1999–2000 can be seen from the Tables 16.1 and 16.4. The relative increase in the price of rice compared to wheat could be one of the reasons for the shift. Higher output subsidy going to wheat than rice in production has made wheat more profitable to grow than rice, thereby increasing its availability and reducing prices (CACP Report 2006). As the profitability of rice production declined the area under rice also declined. The public distribution system provides wheat cheaper than rice in many states.

16.3 Adequacy of cereal consumption

Indian council of Medical Research norms prescribe a cereal intake of 12.6 kilograms per month per capita. It was seen that in 2004–05 most of the states recorded less than average intake of cereal. Nine out of the major 15 states had a cereal intake that was close to or far below the norm in the rural areas.[2] In urban India all the states with the exception of Orissa had below normal level of cereal consumption. Cereal intake was

1 The study is based on 50th NSS Round data (1993–94) and it reveals that the consumption out of home-grown produce reduces poverty.
2 The North Eastern states and Jammu and Kashmir being strategic to India's defence interests, they receive generous subsidised cereal supply through the public distribution system.

above the norm both in rural and urban areas of Orissa. Given this scenario, the most worrying factor is the suspicion that the lower deciles in certain states have such low levels of cereal consumption as well as calorie consumption due to non-affordability and extreme poverty rather than a change in tastes. The implications of declining cereal consumption could vary across the income deciles based on the dietary changes on the micronutrient intake are understood by looking at the micronutrient content of these foods. Cereals are highly energy and protein rich foods and rice protein is of very high quality. Some of the declining consumption could lead to dietary imbalances. The consumption of cereals with pulses further improves the protein quality (Gopalan *et al.*, 1989). Rice and pulses combination in the rice eating states leads to a better protein amino acid combination and easier digestibility of superior rice protein. A decline in the cereal diet with out a corresponding substitute may lead to some imbalances. Much more research is needed on the reasons for the changes in the food basket and the implications.

16.4 Calorie intake

Declining cereal consumption is of no consequence, if the food basket is much more diversified and balanced to achieve the calorie adequacy. A close look at the overall calorie intake shows that at the average level despite a low level of cereal consumption calorie intake was not very low. The calorie intake of almost all the states was above 2200 kilo calories per consumer unit per day. Particularly, the low cereal consumption of prosperous states such as Punjab and Haryana was compensated by other foods such as fats and sugars keeping in the calorie levels fairly high. However, the problem appears to exist with the lower ten percent population. Cereal contributes to large percentage of total calorie intakes of the poor. While the average intake of both urban and rural calorie intake for the country as whole was above the norm prescribed at 2400 Kcal for rural and 2100 Kcal for urban areas, The lowest deciles in the rural areas consumed only 1883 and 1889 kilo calories per consumer unit (MSSRF, December 2002), (Table 16.5).

Hence, it appears that the decline in the cereal consumption of the lower deciles has the impact of reducing the total calorie value of the diet. It would definitely affect the intake of other macronutrients such as protein and some micronutrients derived from cereals. There is an urgent need to ensure the availability and affordability of cereals for the lower income groups to ensure adequate consumption of cereal.

If we take into consideration the percentage of population consuming less than 1890 kilocalories/day/capita, the data reveals that at the average level of the expenditure classes, about 15% of the rural population and 16% of the urban population consume diets lower than 1890 kilo calories per day per consumer unit (NSS Report 471 2001).[3] 1890 kilocalories per consumer unit per day (consumer unit is adjusted for age and sex composition of the population) may be considered as a reasonable cut off point below which the population may be considered as undernourished. FAO has

---

3 The calculations are based on the tabulated data given in the National Sample Survey Report for 1999–2000. Calculation with unit level data would put the figure at a much higher level with persons consuming inadequate diets at about 30% across all classes. However, what is more relevant is non affordability and consequent low level of consumption in the low income classes and not in all classes. FAO calculations of Malnourished are not based on survey data.

Table 16.5 Per consumer unit/day calorie intake of lowest deciles and all classes.

S.No.	State	Rural		Urban	
		Per consumer unit calorie intake lowest ten % (Kcal)	Per consumer unit calorie intake all classes (Kcal)	Per consumer unit calorie intake lowest ten % (Kcal)	Per consumer unit calorie intake all classes (Kcal)
1	Andhra Pradesh	1749.53	2503.00	1841.57	2508.00
2	Assam	1650.02	2342.00	1876.11	2630.00
3	Bihar	1857.04	2638.00	1813.00	2645.00
4	Gujarat	1758.31	2460.00	1828.77	2518.00
5	Haryana	2027.32	3029.00	1691.90	2665.00
6	Himachal Pradesh	2495.01	3056.00	2222.28	3218.00
7	Jammu & Kashmir	2466.25	3266.00	2356.61	5955.00
8	Karnataka	1723.13	2508.00	1776.14	2494.00
9	Kerala	1631.83	2489.00	1580.95	2498.00
10	Madhya Pradesh	1827.00	2560.00	1867.18	2904.00
11	Maharashtra	1833.57	2500.00	1866.51	2484.00
12	Orissa	1967.00	2635.00	2100.00	2802.00
13	Punjab	2045.84	2955.00	1978.75	2667.00
14	Rajasthan	2182.46	3029.00	2071.24	2869.00
15	Tamil Nadu	1515.83	2265.00	1675.70	2509.00
16	Uttar Pradesh	2072.03	2918.00	1765.00	2610.00
17	West Bengal	1812.98	2573.00	1900.37	2597.00
18	Chandigarh	1687.30	2645.00	1802.70	2741.00
19	Delhi	1970.88	2128.00	1942.88	2623.00
20	Pondicherry	1686.20	2489.00	1664.74	2441.00
	All India	1883.64	2668.00	1889.96	2637.00

Source: NSS 55th Round, Report no-471 "Nutritional Intake in India" 1999–2000.

taken 1810 kilocalories per capita as the adequacy norm for India, after adjusting for sex and age for the calculations of as undernourished (FAO 2001).

16.5 Changes in the average consumption pattern of rural India

The National Sample survey has yet to release information on all the food items for the 61st round (2004–05). Hence the data for the previous years have been examined. Comparing the dietary patterns and the average per capita intake over the years will highlight the direction of change of the rural and urban diets. Average per capita intakes as estimated for the NSS 38th round (January–December 1983), NSS 50th round (1993–94) and NSS 55th round (July 1999 to June 2000) are taken for this purpose. It was seen that between the 38th and the 55th rounds, there was an increased diversification in diets with higher shares of food expenditure being devoted to non-cereal foods and an increase in milk consumption on average. It was observed that the per capita per month consumption of rice in rural India was declining steadily over the three NSS rounds. The consumption of pulses remained more or less steady over the period, though consumption of whole gram did fall drastically over the fifteen-year period, the percentage of households that reported consumption of vegetables like potatoes and onions, fruits and milk went up substantially. The per capita quantities consumed of these vegetables also increased over the 20-year period. Despite this unambiguous dietary diversification, it was noted that a fairly high percentage of households (about one-third) reported no consumption of fluid milk or fruits (Rekha Sharma and J.V. Meenakshi 2004)

In the case of quantity of milk intake, consumption increased between the 43rd and 50th NSS round but fell again in the 55th round, though it still remained above the 43rd round level. Thus, looking at the level of head count deprivation, there were still a large number of people who did not consume the "higher quality" foods. Consumption of edible oils has also remained more or less constant, as was also the case for consumption of meats. Egg consumption has increased steadily over the three NSS rounds. In the 55th round, all India egg consumption was at 1.09 kilograms, whereas in the 43rd round it was only 0.39 kilograms. However, this is still below the recommended daily Allowances in both rural and urban India. The consumption of dry chillies and tamarind has been falling over the three survey periods whereas turmeric, garlic and ginger consumption has been increasing. Decreasing consumption of gur has been compensated for by an increasing consumption of sugar.

16.6 Changes in the average consumption pattern of urban India

The cereal consumption has been declining steadily. Between NSS 38th round and NSS 55th round, rice consumption fell by almost 3%. There has only been a marginal increase in wheat consumption. The consumption of pulses such as moong, masur and grams and the consumption of edible oils like ghee, margarine has remained more or less the same. Milk consumption went up considerably over the 20-year period. The consumption of meats went up especially between the 50th round and the 55th round. Consumption of vegetables such as potatoes, tomatoes, onions and leafy vegetables had increased over the course of 1988 and 2000. Consumption of mangoes, coconuts and lemons has increased whereas in the case of the other fruits like apple and banana, there was a fall in consumption between the 50th and the 55th round. Consumption of sugar has risen steadily (1.27 kilograms per month per capita in 2000) and this was accompanied by a fall in the consumption of gur. Intake of dry chillies and tamarind fell whereas garlic, ginger and turmeric intake increased.

Urban India needs special attention, as processed foods are more important in the urban diets than in rural diets. Data on a large variety of foods has been made available in urban India as a whole from 1987–1988 to 1999–2000. Consumption has changed in favour of some prepared foods. Consumption of rice and wheat has decreased marginally. Milk and egg consumption records an increase. Substantial increases over the various time periods are seen in the consumption of tea, biscuits, salted snacks, prepared sweets, edible oils, and sugar. There is no evidence of a substantial shift of average urban diets towards nutritive and protective foods. The phenomenal growth in biscuit consumption and tea is more a result of prosperity and changing preferences (Vepa, 2003).

The changing dietary patterns, with preference for fat and sugar have resulted in obesity in the urban middle class. A study undertaken by Nutrition Foundation of India in 1999, in the urban areas of the country reveals that obesity prevails mostly

Table 16.6 Change in the pattern of consumption of selected food
items of the urban population (kg/month/capita).

S.No	Food items	1987–88	1993–94	1999–2000
1	Rice	5.26	5.13	5.10
2	Wheat/flour	4.37	4.44	4.45
3	Pulses	0.87	0.77	0.85
4	Liquid milk (litres)	4.26	4.89	5.10
5	Eggs (number)	1.43	1.48	2.06
6	Milk fat	0.04	0.05	0.07
7	Edible oils	0.41	0.46	0.74
8	Flesh foods	0.39	0.40	0.46
9	Vegetables	3.94	3.09	3.00
10	Leafy vegetables	0.40	0.15	0.17
11	Mangoes	n.a.	0.12	0.16
12	Bananas (number)	5.10	4.48	5.00
13	Lemons (number)	n.a.	1.23	1.39
14	Sugar/jaggary	0.97	0.97	1.32
15	Tea leaf (g)	60.43	63.93	70.44
16	Biscuits	0.07	n.a.	2.06
17	Salted refreshments	0.04	n.a.	1.36
18	Prepared sweets	0.11	n.a.	0.40

Source: National Sample Survey.

among the middle class than slum dwellers. One percent of the men and four percent of women in the urban slums are found to be obese. Among the middle classes 32.2 percent of men and 50 percent of women are obese (NFI, 1999).

16.7   Adequacy of consumption of various food items

16.7.1   *Rural Diets*

Pulses and edible oils were consumed far below the required level in most of the states. This was especially so in the case of West Bengal. Other than Himachal Pradesh, all the other states fell below the norm in their average per capita pulse intake. However, the shortfall was not much in most of these states. Edible oil consumption was below the prescribed level in 20 out of the 28 states. Assam, Orissa were in the worst situation. Punjab, had very high levels of edible oil intake.

Milk consumption was well above the required level in the northern and western states of Haryana, Punjab, Rajasthan, Jammu & Kashmir and Chandigarh. The northeastern states of Assam did not have the sufficient level of milk consumption of 4.5 kilograms per month. Nineteen out of the 28 states did not meet the required level of milk intake. It shows that rural people cannot afford to consume sufficient quantities of milk in many states.

In the case of the consumption of eggs, it was startling to note that all the states fell drastically below the prescribed norm level of 1.35 kilograms per month. The level of egg consumption was abysmal especially in Haryana (0.03 kilograms per month), Gujarat (0.05 kilograms per month) and Madhya Pradesh (0.04 kilograms per month).

Kerala had a fish and prawn intake above the prescribed the norm of 0.75 kilograms per month. All the other states fell below the prescribed norm for fish consumption. Interestingly, rural West Bengal had a deficit in fish intake unlike its urban counter part. Madhya Pradesh, Maharashtra and Chandigarh and Sikkim had a fish consumption that was less than 0.1 kilograms per month.

Meat consumption was below the norm for all the Indian states except for Nagaland. The other north-eastern states of Arunachal Pradesh and Meghalaya had moderate levels of consumption. The southern states of Kerala, Goa and Tamil Nadu were next in the levels of meat intake. Here the average meat consumption was in the range of 0.20 to 0.26 kilograms per month. Gujarat and Haryana had the lowest levels of meat consumption. The consumption of meat and fish did not seem to be related with the poverty levels in the states.

The prescribed level of vegetable consumption is 3.75 kilograms per month. Chandigarh had adequate levels of consumption. The remaining States consumed below the required level. In the case of Punjab and Nagaland the deviation was not much. Kerala had the lowest level of vegetable intake of only 1.81 kilograms per month. Assam was next with only 1.99 kilograms per month. Fruit consumption was also below the norm in almost all the states. Only Chandigarh and Delhi had consumption that was slightly above the norm. The deviation from the norm was greatest in the case of West Bengal and Bihar.

Twelve out of the 28 states consumed sugar at a level above the prescribed level of 0.9 kilograms per month. Kerala, Karnataka and Madhya Pradesh satisfied more than 90 percent of the requirement. The other states satisfied up to 70 to 50 percent of the requirement. The lowest recorded sugar consumption was in Orissa at 0.39 kilograms per month and Bihar at 0.44 kilograms per month.

For certain foods rural areas of some states have shown adequate consumption, irrespective of the poverty levels. Food habits seem to dominate the scene rather than the affluence. Cereal consumption was close to the norm in many states and higher in northeast states, Jammu and Kashmir and Orissa. For all the other items fewer states had met adequacy norms (Vepa, 2005), (Table 16.7).

16.7.2 Urban diets

In the urban diets the calorie decline due to less than recommended levels of cereal consumption was made up for by the higher levels of consumption of other energy foods such as sugar, edible oils and milk fats. In 11 out of the 20 states sugar consumption was above the ICMR norm. In the remaining states, sugar consumption varied between 98 and 74 percent of the requirement. Edible oil consumption was above the norm in 8 out of the 20 states. Barring Orissa, where the edible oil consumption was only 65 percent of the requirement, the per capita average consumption in the rest of the states was 79 percent of the requirement. An important component of the diet that provides both protein as well as calories is pulses. Adequate consumption of pulses was seen only in urban Himachal Pradesh and Chandigarh. For the other states, the shortfall in consumption was not much and ranged between 75 and 80 percent of the requirement.

Consumption of milk was adequate as per the requirement of 4.5 kilograms per month in the case of 10 out of 20 states. In fact, northern states of Punjab, Haryana,

Table 16.7 Per capita consumption (kg/month) of food items in rural areas.

S. No	State	Cereals	Pulses & Products	Milk milk*(0.9)	Edible oil	Eggs egg*(0.125)	Fish, Prawn	Meats (1 + 2 + 3)	b Kg/month	Fruits Kg/month	Sugar
	ICMR norms	12.6	1.2	4.5	0.66	1.35	0.75	0.75	3.75	1.5	0.9
1	Andhra pradesh	12.65	0.73	2.58	0.46	0.26	0.18	0.25	2.77	0.56	0.49
2	Assam	12.63	0.51	1.00	0.35	0.19	0.52	0.14	1.99	0.40	0.46
3	Bihar	13.75	0.78	2.17	0.40	0.06	0.15	0.06	2.60	0.36	0.44
4	Gujarat	10.19	0.92	4.88	0.82	0.05	0.03	0.04	2.83	0.56	1.16
5	Haryana	11.37	1.01	12.49	0.39	0.03	0.00	0.03	2.60	0.79	2.01
6	Jammu and Kashmir	14.68	1.11	8.58	0.70	0.35	0.00	0.33	3.15	0.73	0.87
7	Himachal pradesh	12.86	1.35	7.08	0.61	0.09	0.00	0.13	2.31	0.55	1.21
8	Karnataka	11.53	1.01	3.11	0.44	0.18	0.13	0.20	2.64	0.75	0.87
9	Kerala	9.89	0.55	2.67	0.42	0.32	1.75	0.26	1.81	0.77	0.85
10	Madhya Pradesh	12.94	0.87	2.44	0.43	0.04	0.05	0.05	2.41	0.31	0.82
11	Maharashtra	11.32	1.00	2.39	0.59	0.14	0.10	0.12	2.38	0.68	1.07
12	Orissa	15.09	0.46	0.58	0.26	0.06	0.31	0.06	2.65	0.31	0.39
13	Punjab	10.58	1.06	10.50	5.60	0.12	0.00	0.05	3.49	0.69	2.03
14	Rajasthan	14.19	0.67	8.66	0.43	0.00	0.00	0.08	2.07	0.29	1.17
15	Tamil Nadu	10.66	0.83	2.15	0.43	0.27	0.17	0.24	2.54	0.61	0.53
16	Uttar Pradesh	13.62	1.07	4.07	0.50	0.06	0.06	0.15	2.27	0.50	0.95
17	West Bengal	13.59	0.46	1.18	0.43	0.37	0.59	0.16	2.94	0.33	0.47
18	Chandigarh	9.37	1.21	8.93	0.69	0.19	0.04	0.12	5.16	1.79	1.69
19	Delhi	7.86	0.92	5.71	0.53	0.17	0.00	0.07	2.56	1.79	1.00
20	Pondicherry	11.04	0.82	2.39	0.53	0.47	0.44	0.21	3.22	0.79	0.50
	All India	12.72	0.84	3.41	0.50	0.14	0.21	0.15	2.55	0.50	0.84

Source: NSS, Report No.461 – "Consumption of Same Important Commodities in India-55 th Round" - 1999–2000.

Table 16.8 Per capita consumption (kg/month) of food items in urban areas.

S.No	State	1 Cereals	2 Pulses & Pulse products	3 Milk*	4 Edible oil	5 Eggs*	6 Fish & Prawn	7 Meats	8 Vegetables	9 Fruits	10 Sugar
1	Andhra Pradesh	10.94	0.87	3.96	0.60	0.32	0.08	0.29	2.93	0.97	0.67
2	Assam	12.26	0.75	1.93	0.55	0.34	0.79	0.23	2.25	0.80	0.64
3	Bihar	12.70	0.93	3.06	0.52	0.16	0.16	0.20	2.76	0.80	0.67
4	Gujarat	8.49	1.03	5.92	1.05	0.11	0.03	0.12	3.31	1.01	1.16
5	Haryana	9.36	1.05	8.13	0.63	0.12	0.00	0.07	2.81	1.21	1.49
6	Himachal Pradesh	10.33	1.40	9.07	0.73	0.31	0.01	0.18	3.06	1.23	1.16
7	Jammu & Kashmir	12.84	0.93	7.22	0.77	0.27	0.01	0.56	3.69	1.29	0.72
8	Karnataka	10.21	1.04	4.56	0.59	0.31	0.12	0.33	2.85	1.12	0.97
9	Kerala	9.25	0.69	3.14	0.46	0.40	1.88	0.33	1.66	1.02	0.88
10	Madhya Pradesh	11.09	1.00	3.90	0.64	0.14	0.05	0.12	3.22	0.82	1.02
11	Maharashtra	9.35	1.02	4.31	0.84	0.24	0.15	0.29	3.17	1.16	1.15
12	Orissa	14.51	0.74	1.77	0.43	0.21	0.35	0.19	3.41	0.78	0.66
13	Punjab	9.21	1.17	8.76	0.67	0.18	0.00	0.08	3.65	1.17	1.62
14	Rajasthan	11.56	0.96	6.95	0.62	0.07	0.00	0.11	2.97	0.76	1.16
15	Tamil Nadu	9.65	1.02	4.29	0.58	0.47	0.18	0.35	3.16	1.17	0.75
16	Uttar Pradesh	10.79	0.98	4.74	0.60	0.14	0.03	0.24	2.64	1.13	1.06
17	West Bengal	11.17	0.60	2.37	0.68	0.58	0.86	0.25	2.88	0.93	0.68
18	Delhi	8.61	1.17	7.86	0.74	0.24	0.12	0.26	3.29	1.94	1.07
19	Chandigarh	8.74	1.39	9.48	0.76	0.31	0.00	0.13	3.65	1.83	1.35
20	Pondicherry	9.62	1.00	4.18	0.64	0.56	0.43	0.26	3.33	0.93	0.67
	All India	10.42	1.00	4.59	0.72	0.26	0.22	0.24	3.02	1.06	1.00

Source: NSS 55 th Round Report No. 461
* One litre of milk is taken as 900 grams; one egg is taken as 125 grams

Gujarat, Delhi and Chandigarh had very high levels of milk intake. Consumption of fruits was very high in Delhi and Chandigarh, moderate in Punjab and Haryana, but low in the others. Rajasthan, Bihar, Orissa and Assam consumed well below the norm.

Vegetables were not consumed in adequate portions in any of the states. The norm requirement is 3.75 kilograms per month. All the states consumed below this level. Kerala had the lowest level of vegetable consumption of only 1.66 kilograms in a month. Uttar Pradesh, Assam and Bihar also had a relatively low intake. The better-off states were Jammu & Kashmir, Punjab and Chandigarh. The prescribed per capita level of fish consumption is 0.75 kilograms per month. Fish consumption was adequate only in Assam, Kerala and West Bengal. Haryana, Punjab, Chandigarh. Rajasthan showed negligible levels of fish consumption. In Gujarat, Himachal Pradesh, Uttar Pradesh and Jammu & Kashmir, consumption was abysmally low, almost nil. In all the other states, fish consumption ranges between 0.12 and 0.43 kilograms per month.

The consumption of eggs and meat was far below the requirement in all the states. The highest egg consumption was in West Bengal (0.58 kg/month) and Pondicherry (0.56 kg/month) but even this was well below the ICMR norm of 1.35 kilograms per month. All the states had very low levels of egg consumption especially so in the case of Gujarat and Rajasthan.

Only Jammu & Kashmir met the required meat consumption of 0.75 kg/month. All the other states had very low levels of meat consumption. Haryana (0.07 kg/month) and Punjab (0.08) had the next highest meat consumption. Southern states of Tamil Nadu, Karnataka and Kerala had relatively higher levels of consumption. The consumption of pulses was not high enough to compensate for the deficient consumption of eggs, meat and fish.

Th analysis reveals that on the whole the average diets are deficient in the consumption of various food items that provide micronutrients such as minerals and vitamins both in urban and rural diets and particularly in the average rural diets.

16.8 Dietary intake of micronutrients

The consequence of unbalanced diets deficient in certain food items results in deficiency of micronutrients, even though the protein calorie malnourishment is not high. The micronutrient content of the food consumed is only an approximation to the amount consumed per capita. The micronutrient content that is consumed depends upon the cooking methods and the quality and the combination in which it is consumed. The micronutrient content absorbed into the body also differs. Consumption by itself does not ensure healthy individuals. There is also considerable debate regarding the recommended daily allowance. The RDA figures differ across age groups and between sexes and between adult women, adult pregnant women and lactating women. The RDA levels recommended by the Indian Council of Medical Research differ from that of the World health organization figures. More over there was no single figure of RDA adjusted for age, sex and maternity, which is agreeable to all. Hence various studies adopt different norms and results differ.

Using data and results from the National Nutrition Monitoring Bureau (1999), India Nutrition Profile (1998) and the Institute for Applied Statistics and Development Studies (2002), the iron intake data for 18 states for the rural population was calculated and compared with the Recommended Daily Allowance in the diets of adults above the age of 18. It was found that the states of Bihar, Jharkhand, Orissa, Andhra Pradesh, Tamil Nadu, Kerala, Goa, Himachal Pradesh and all the eastern states except for Manipur and Tripura, consumed iron to an extent that was less than 50% of the Recommended Daily Allowance (RDA). Among the children (both 1–3 years and 4–6 years) these same states (except for Arunachal Pradesh), Karnataka, Chandigarh, Punjab and Delhi also recorded an iron intake less than 50% of the RDA requirement.

In the case of vitamin A, 12 out of 22 states studied by India Nutritional Profiles, (1998) and NNMB (1999) consumed less than half the prescribed level. This included all the southern states, Gujarat, Uttaranchal, Uttar Pradesh, Bihar, Jharkhand, Assam and Meghalaya. Average Vitamin C intake vis-à-vis the RDA is fairly good for India. Almost all the population consumed more than 90% of the RDA recommendation. Karnataka had the lowest intake levels of between 50–75% of the RDA recommended level. Isolated studies found more 50% of pregnant women and children to be deficient in folic acid. In the case of zinc, only few of the studies found children to be consuming less than the RDA requirement (Toteja & Padam Singh, 2004).

16.8.1 Micronutrients content of the average diets in the lowest and highest deciles

To examine the micronutrient content of the diets in the lowest deciles, National sample survey data has been used. The dietary content of the micronutrient estimated is only an approximate estimation of the micronutrient content in the diets as given by the Nutritive values of Indian Foods published by National Nutrition Monitoring Bureau and the average quantity of food items consumed as per the tabulated published data of the National Sample Survey on various food items at the average level. The RDA considered for defining adequacy was un-weighted simple average of the various levels prescribed for various groups and hence indicate the average requirement, which in effect means that adolescents, pregnant women and lactating mothers require higher levels. All the same, the computed figures of dietary nutrient content tell us how even the average levels of nutrients are not found in the diets, leaving alone the underestimation of deficiencies arising out a host of other factors. Further the computed dietary nutrient content gives the scope for to apply them to a different norm to derive own conclusions (Vepa, 2005).

The micronutrient intake of the all India and average levels of intake for two of the lowest per capita monthly expenditure classes of the National Sample Survey 55th round (1999–2000) has been estimated both for urban and rural India. The average intake of micronutrients was also estimated separately for urban and rural areas twenty states (Tables 16.9 and 16.10).

16.8.2 Micronutrient content of average rural diets across deciles

The consumption of micronutrients in the rural areas was found to be inadequate for all the classes, especially in the case of Calcium, Iron, Beta-carotene, Riboflavin and Zinc. For all the other nutrients namely Thiamine, Nicotinic Acid, Folic Acid, Vitamin C and Phosphorous, only the consumption of the high-income class exceeds the Recommended Dietary Allowance. The short falls are quite high for Calcium, Iron, Zinc and Carotene. While the RDA of Calcium is 600 milligrams per day, the consumption by the lowest class was only 170.40 milligram per day, the middle-income

Table 16.9 Micronutrient intake of rural India.

Nutrients	Calcium	Iron	Beta Carotene (Vitamin A)	Thiamine	Riboflavin	Niacin (Nicotinic Acid)	Ascorbic Acid (Vitamin C)	Folic Acid	Phosphorous	Zinc
RDA Average(mg/d)	600.00	29.54	2.34	1.03	1.22	13.42	42.78	0.11	1000.00	15.00
Exp. Class RDC										
All Classes	388.99	26.66	1.35	1.39	0.93	17.01	55.58	0.16	1293.00	9.71
	(64.83)	(90.26)	(57.78)	(134.92)	(76.15)	(126.77)	(129.93)	(149.79)	(129.30)	(64.73)
225–255	192.86	16.70	0.99	1.01	0.57	13.05	35.69	0.11	956.95	7.88
	(32.14)	(56.54)	(42.39)	(98.034)	(46.67)	(97.26)	(83.43)	(106.72)	(95.70)	(52.53)
000–225	170.40	14.10	0.94	0.88	0.49	11.06	28.63	0.10	841.96	7.02
	(28.40)	(47.73)	(39.96)	(85.42)	(40.12)	(82.43)	(66.93)	(89.31)	(84.20)	(46.80)

Note:Mg/d refers to milligram per day. Figures in the brackets refer to the percentage compared to the RDA adopted.

Table 16.10 Micronutrient intake of urban India.

Nutrients	Calcium	Iron	Beta Carotene (Vitamin A)	Thiamine	Riboflavin	Niacin (Nicotinic Acid)	Ascorbic Acid (Vitamin C)	Folic Acid	Phosphorous	Zinc
RDA Average(mg/d)	600	29.538	2.340	1.030	1.221	13.418	42.778	0.107	1000	15.00
Exp. Class RDC										
All Classes	435.17	24.480	1.400	1.360	0.990	16.260	65.890	0.169	1237.99	8.38
	(72.52)	(82.86)	(59.83)	(132.01)	(81.06)	(121.18)	(154.03)	(158.17)	(123.80)	(55.87)
300–350	226.8	19.380	0.970	1.150	0.660	15.030	43.570	0.130	1001.2	7.96
	(37.80)	(65.61)	(41.45)	(111.62)	(54.04)	(112.01)	(101.85)	(121.38)	(100.12)	(53.07)
000–300	180.05	17.980	0.750	0.960	0.540	12.290	31.760	0.108	852.86	6.83
	(30.01)	(60.87)	(32.05)	(93.18)	(44.22)	(91.59)	(74.24)	(100.98)	(85.29)	(45.53)

Note: Mg/d refers to milligram per day. Figures in the brackets refer to the percentage compared to the RDA adopted.

group consumes around 192.86 milligram per day and the highest income group consumes around 388.99 milligrams per day. In the case of Iron, the highest income group consumes 26.66 milligrams per day, the middle-income group consumes 16.70 milligrams per day and the lowest income group consumes 14.10 milligram per day as against the adequacy norm of 29.54 milligrams of iron per day. As against an adequacy norm of 2.34 milligrams of Beta-Carotene, highest group consumes about 1.35 milligrams of per day, the middle-income group consumes 0.992 milligrams per day while the lowest income group consumes only about 0.935 milligrams per day. Similarly, in the case of Zinc too, as against the adequacy norm of about 15 milligrams per day, the highest income group in the rural area consume only about 9.71 milligrams of Zinc per day. The middle-income group and the lower income group consume approximately about 7.88 milligrams and 7.02 milligrams per day.

On the other hand, in the case of micronutrients such as Phosphorous, the high-income group consumes about 1293 milligrams per day that is more than the Recommended Dietary Allowance of 1000 milligrams per day. The middle-income group consumes about 956.95, milligrams per day. The lowest income group consumes about 841.96 milligrams per day. On an average, it can be seen that the consumption of the micronutrients by the rural population is alarmingly low (Table 16.9).

16.8.3 Micronutrient content of the average urban diets across deciles

The micronutrient intake of the urban household in India at the average level has been computed to examine the adequacy of intake. The adequacy of intake has also been examined for the lowest two monthly per capita expenditure classes. It has been observed that as against the requirement of 600 mg of calcium, for all the income groups the consumption falls below and specifically, a deficiency of 419.95 mg for the lowest income group. The requirement of Iron is 29.538 mg per day and again the consumption of Iron by the urban population falls short for all the three income groups. In the case of Beta Carotene that contains vitamin A, the RDA is 2.340 mg per day. All the three groups again fall short of this requirement. The amount of Thiamine that needs to be consumed is around 1.30 mg per day. The lowest income group alone has fallen short of this requisite. The intake of Thiamine for the other two groups is well above the requirement level. In the case of Riboflavin, it is essential that 1.221 mg be consumed per day. However, all the three classes fail to meet this requirement.

The requirement of Niacin, also known as Nicotinic Acid, is 13.418 mg per day. The lowest-income group fails to meet this requirement, although by only a small amount. The other two classes again have been observed to consume more than the necessary level. The required consumption of Vitamin C or Ascorbic Acid is 42.778 mg per day. The lowest income group again fails to satisfy this condition. The intake is only around 31.76 mg per day for the lowest expenditure group, whereas the middle income group consumes about 43.570 mg per day on an average which is just about the RDA level. The highest income group consumes around 65.890 mg. that is much higher than the required levels.

The RDA for Folic acid is around 0.017 mg per day and all the three income groups have been observed to consume more than the required levels. The intake of Phosphorous must be around 1000 mg per day and it was observed that apart from the lowest income group, the other two income groups have been observed to consume the required amounts of Phosphorous. The RDA of Zinc is expected to be 15 milligram per day. However, it was observed that none of the classes satisfy the adequacy norms of Zinc. The high-income class consumed only around 8.38 milligram per day, the middle-income group consumes only 7.96 milligrams per day and the lowest income group consumes 6.83 milligrams per day (Table 16.10).

Zinc, Vitamin C, Vitamin A, Riboflavin, Calcium, Iron, thiamine and Niacin in that order are the most deficient micronutrients in the average urban diets of the country. More than iron, zinc and vitamin 'C' deficiency are seen in all the states.

References

CACP (2006) Government of India, Department of Agriculture and Cooperation, Ministry of Agriculture, *Reports of the Commission for Agricultural Costs and Prices, 2003–2004.*

FAO (Food and Agricultural Organization) (2001) *The state of Food security in the world-2001*, Rome.

Gopalan, C., Ramasastri, B.V., and Balasubramanian, B.V. (1989) *Nutritive Values of Indian Foods.* ICMR, New Delhi.

Kamala Krishnaswamy (1999) *Obesity in Urban middle classes in Delhi.* Nutrition Foundation of India, New Delhi.

MSSRF (M. S. Swaminathan Research Foundation) (2002) *Food Insecurity Atlas of Urban India*, Chennai.

NFI (Nutrition Foundation of India) (1999) *Investigation of current prevalence, nature and etiology of obesity in urban communities.*

NSS (National Sample Survey) (2001) *Nutritional Intake of India*, Report no. 471, 1999–2000.

Punjab Singh (2001) *Agriculture Policy: vision 2020*, IARI-FAO/RAP Planning Commission, 2001 (Mimeograph).

Rekha Sharma and Meenakshi, J.V. (2004) The Micronutrient Deficiencies in Rural Data, In "*Towards Hunger Free India, From Vision to Action*" (Ed. M.S Swaminathan and Pedro Medrano).

Toteja, G.S., & Padam Singh (2004) *Micronutrient Profile of Indian Population.* Indian Council of Medical Research, New Delhi.

Vepa, S.S. (2003) Impact of Globalization on food consumption of urban India, In "*Globalization and urban food consumption*, Seminar proceedings of Oct. 2003, published by FAO.

Vepa, S.S. (2005) *Micronutrient deficiencies in India – Need for Food Fortification.* Commissioned paper by Micronutrient Initiative (May 05 – September 05) (Mimeograph).

Chapter 17

Food chain dynamics and consumption trends: Implications for freshwater resources

Jan Lundqvist

Stockholm International Water Institute, Stockholm, Sweden & Department of Water and Environmental Studies, Linköping University, Linköping, Sweden

17.1 Introduction

At the dawn of the Green Revolution, at the beginning of the 1960s, the average crop yield in the world was about 1.4 tonnes/hectare. Thirty years later, in the mid-1990s, it had increased to about 2.8 tonnes/hectare, i.e. a doubling of yields. In the mid 1960s, total global cereal production was about 0.94 billion tonnes. In 1995, it was in the order of 1.7 billion tonnes. The 2 billion ton mark was passed with a margin in 2004, when total cereal production was estimated at 2,254.9 million tonnes (FAO; Food Outlook, June 2, 2005).

Thanks to a historically unprecedented increase in overall production, there has been considerable improvement in the average amount of food available on a *per capita* basis. Within one generation, i.e. during the last thirty years, the average *per capita* food supply in developing world has risen from 145 kg/year to about 175 kg/year (see, for instance, CA, 2007). With reference to the figures quoted above, the current level of annual, average global harvested grains *per capita* is about 375 kilos.

Indeed, recent developments seem to contradict Malthusian concerns.

There have been optimistic projections about continuous increases in the availability of food, e.g. in FAO (2003). Behind these projections lies the reality of an alarming depletion and degradation of water resources, which, to a large extent, are related to increased food production. The erratically available surface water in rivers was first harnessed and gradually the ground water has been lifted on a grand scale (CA, 2007; Briscoe, 2005; Shah *et al.*, 2000). Reduction in the volume in water bodies and changes in water flow have negatively affected aquatic ecosystems (CA, 2007; Smakhtin & Anputhas, 2006; Smakhtin *et al.*, 2004; Tilman *et al.*, 2001). As underscored by the headline for the World Water Day on March 22, 2007, "Coping with Water Scarcity", a serious water challenge is already a reality in parts of the world. Considering the likely demand for additional food but also demand for other water intensive products like bio-fuels, together with the likely impact from climate change, we are on the verge of a new water scarcity era (Lundqvist *et al.*, 2007; Falkenmark, *et al.*, 2007).

At this critical juncture, food security cannot only be interpreted in terms of conventional projections about production and supply of food. It is equally important to look into food security in terms of access to food and the actual consumption of food. A basic challenge is to promote an intake or consumption of food that enables people to lead "an active, healthy and productive live". In addition, food security must be seen in a context where demands for other goods and services will also increase. Strategies must be identified where various socioeconomic demands may be satisfied without jeopardizing aquatic and other ecosystems.

In this article, food and water security is discussed from a food chain perspective with a focus on consumption aspects. What would be the implications for water and society of an improved efficiency in the food chain, including a food intake at a level that would enable people to lead an "active, healthy and productive life"? In other words, what are the water implications if food intake is not too much and not too little? No doubt, the challenge to feed the world and to meet a range of human needs and wants is huge. It is therefore important that options to reach these challenges are identified and assessed.

17.2   Food security and water pressure: current situation and likely future

Food production is among the most water intensive activities in society. Roughly one litre of water is required to produce one kcal of vegetarian food items. With reference to a widely used norm for food security, food supply should be about 2,700 kcal per person per day. This implies a consumptive use of water of almost 3 m^3 per person per day (CA 2007; SEI, 2005). Based on information compiled by FAO in Food Balance Sheets, estimates of the consumptive water use vary between countries; from about 1 m^3/capita per day to more than 6.5 m^3/capita per day (Lundqvist et al., 2007).

Interpretations of these figures must be done with certain circumstances in mind. For instance, sophistication in water management and climatic zone are important. Much more water is required to produce animal calories as compared to vegetal calories. Depending upon these and other circumstances, scarce water resources need to be used more prudently. The fraction of animal calories is as high as 40–45% in some countries (Ibid.). Water productivity differs significantly between countries, varieties/breeds and agricultural practices (CA, 2007).

Producing food in countries where the majority of the world population reside, most of them poor, and where demographic change is rapid, i.e. mainly in tropical regions, is a major water challenge. Per capita availability of water is already comparatively low. At the same time, evaporative demand of atmosphere is high. According to de Fraiture et al. (2004), the amount of water depleted ranges from 315 to 750 mm per season mainly as a result of variation in evaporative demands from atmosphere in different parts of the world. Most likely, climate change will aggravate the water challenge.

Current annual water consumption in food production is in the order of 7.000 km^3 (CA, 2007; Lundqvist et al., 2007). Estimates of the water that will be required in the future do naturally vary. To bring down the number of people suffering from under nourishment by 50% by 2015 and to feed an expected increase in population, the additional consumptive water use could amount to 2,200 km^3, i.e. quite a substantial increase (Falkenmark & Rockström, 2004). This estimate is based on the assumption that 20% of the food comes from animal sources and 80% from vegetarian sources. Moreover, it is assumed that the water consumption to produce the animal calories is 8 times higher as compared to vegetarian calories. With reference to FAO (2003), average food supply is supposed to be 3,000 kcal/person, day. Postel (2005) estimates that 2,040 km^3 will be required to support the diets of the additional 1.7 billion people at today's average diet by 2030. Calculations made by Gleick (2000) suggest higher figures for water requirements, especially for production of beef, i.e. red meat. In the Comprehensive Assessment Scenario, where considerable investments are taken into account – and thus improvements in productivity - crop water consumption increases by 20% while the withdrawals increase by 13% till 2050 (CA, 2007). In another recent study, which considers likely increases in GDP and demographic trends, an additional consumptive use of water of about 4,000 km^3, from about 7,000 to about 11,000 km^3, till 2045 is estimated (Lundqvist *et al.*, 2007). Plausible improvements in water and land productivity will dampen the demand for additional water.

Variation in estimates underlines the difficulties involved in making these projections. Improvements in water productivity, agronomic practices and increases in "virtual water" flows i.e. trade in food items can reduce the huge gulf between projected water requirements for additional food and what volumes of water that can realistically be exploited with regard to social and environmental costs. Capturing a larger share of the precipitation and using the harvested rainwater for supplementary irrigation could also reduce the pressure on blue water resources (Falkenmark & Rockström, 2004; SIWI *et al.*, 2005; SEI, 2005).

The common denominator of these scenarios shows that future water requirements can be substantially reduced through various improvements in water management. But they also show that the pressure on water will increase, even with substantial investments and a better use of both blue and green water sources (CA, 2007).

17.3 Under-nourishment, obesity and food wastage

Paradoxically, imbalances in the food security complex refer to a shortage of food but also an abundance of food. Some 15% of the world's population, or about 850 million, are under-nourished, primarily because they are deprived of access to adequate quantities of food. The number of under nourished people decreased slowly but steadily over many decades. Recent figures (FAO, 2004a) indicate that the number might have increased during the first years of the new century (Figure 17.1).

Total level of under nourishment is most serious in sub-Saharan Africa where also poverty is rampant (Table 17.1) and water scarcity serious. It is noteworthy that the majority of the food insecure live in rural areas, about 70%, i.e. in areas where food is produced. Small holders and rural landless are particularly affected. The higher the incidence of subsistence farming, the higher the degree of malnutrition (von Brown *et al.*, 2004). The high number of under-nourished people is, in itself, a tragedy. It is even more disgraceful in a world where the number of overweight and obese is higher than the number of under nourished. According to the estimates by WHO, about 1,100 million people belong to this group.

Lack of access to food is obviously related to risk of being under-nourished. It is also relevant to mention that part of the food that is produced and could have been used for human consumption is lost through losses and wastage. In tropical, hot climates, post harvest technologies are often poorly developed. Perishable food items, i.e. fruits and vegetables are particularly affected (Swaminathan, 2006), but also animal products. Losses are also substantial in subsequent stages of the food chain. Some of the food is lost during storage and in the wholesale and retail segments, while some is lost or wasted in the households, restaurants and other units at the end of the food chain

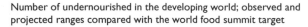

Number of undernourished in the developing world; observed and
projected ranges compared with the world food summit target

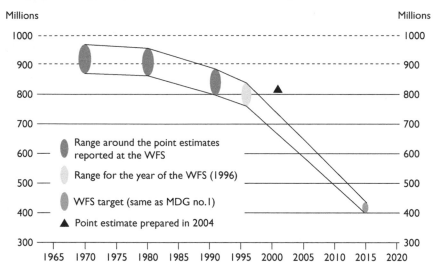

Figure 17.1 Observed and projected ranges of under-nourished people in developing world: The angle
of the dotted lines illustrates that reduction in number of undernourished must accelerate substantially
if the MDG targets for 2015 should be achieved. The point estimate for 2004 indicates that the number
of under-nourished may have rather increased during the first years of this century (Source: FAO,
2004a).

(Figure 17.2). It is likely that losses and wastage in the latter parts of the food chain
are comparatively small in poor societies where, at the same time, post harvest losses
are considerable. The opposite is presumably true in affluent societies, where wastage
among the consumers is significant (Milmo, 2005).

A certain amount of the food that is produced is converted, from vegetarian calories
to animal based food items (Figure 17.2). Animal based food items are much more
water intensive as compared to vegetarian food. From a resource perspective, however,
it is important to make a distinction between animal food items that come from animals
that get their feed from grazing and food that comes from animals that are fed on food
grains that could have been used for direct human consumption. Large tracts of land
in semi-arid regions are suitable for grazing but are not suitable for raising ordinary
crops, e.g. the case with parts of southern Africa and Mongolia (Lundqvist *et al.*,
2007).

About 40% of the beef is estimated to come from cattle raised on feedlots, whereas
the proportion of poultry and pork raised in the same manner may be higher
(Nierenberg, 2003). From a water perspective, it is also noteworthy that the trend
in developed countries seems to go from red meat to white meat. Renault & Wallender
(2000) estimate that the effect of this trend is a reduction in the water intensity in
producing the diet by ten percent, or 400 litres per capita per day.

Table 17.1 Figures illustrating the number of people living in extreme poverty, (income < 1 US$/day) and proportion of people who do not have access to the minimum dietary energy requirement.

	Percentage of population living below $1 per day		Population below minimum dietary energy requirements (%)	Total population, ×1,000
	1990	**2001**	**2000/02**	**2005**
Developing countries and transition economies	27.9	21.3	17	5,287,035'
Northern Africa and Western Asia	2.2	2.7	4	333,700'
Sub-Saharan Africa	44.6	46.4	33	713,843'
Latin America and the Caribbean	11.3	9.5	10	561,012'
Eastern Asia	33.0	16.6	11	1,396,296'
Southern Asia	39.4	29.9	22	1,451,004'
South-Eastern Asia and Oceania	19.6	10.2	13	564,480'
Commonwealth of Independent States	0.4	5.3	–	363,356'
Transition countries of south-eastern Europe	0.2	2.0	–	203,644'

[1] High-income economies, as defined by the World Bank, are excluded.
[2] Estimates updated by the World Bank in December 2004.

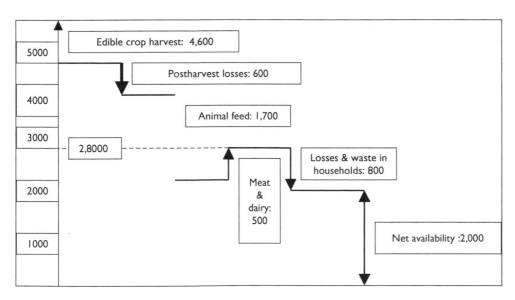

Figure 17.2 Estimated global per capita averages of food harvests, losses, waste and conversions, in kcal/day. (Based on information in: Smil, 2000).

17.4 What food for food security?

The foregoing discussion indicates that producing food is one thing, and accessing it is quite another. However, both are integral parts of food security. Due to socio-economic circumstances, access varies significantly. A complex and contentious issue refers to what and how much should be consumed in terms of diet composition and energy intake. This is particularly sensitive and problematic concerning food intake, which is part of our private sphere. At the same time, figures on food production are no longer considered adequate measures of food security. Today, the concern about chronic or transient food insecurity at household level seems to increase (e.g. Radhakrishna, 2003). Information about what is happening on the consumption side is, however, generally not related to production.

With a relative surplus of food being produced in combination with serious food security challenges, it is relevant to identify steps in the food chain and to analyse what can be done in order to improve food and nutrition security without a corresponding increase in the pressure on our land and water resources. In this process, it is relevant to make distinctions between:

(i)   the amount of food produced,
(ii)   the amount of food available on the market, i.e. the produce "at field level" minus losses before food items reach the food stand or supermarket, plus/minus changes in stocks, i.e. the food supply,
(iii) the amount demanded or bought by households, public institutions and other buyers,
(iv)   the actual intake of food, i.e. the amount of food eaten.

Strangely enough, most discussions about food security refer only to the second and third categories in the above typology. Information about the wide gap between gross food production and the actual intake of food does not figure in main stream analyses of food and water security.

17.5 Relation between food supply and food intake

As illustrated in Figure 17.2, there is a big difference between the amounts of food produced and the food intake. On a global level, the supply of food is currently about 2,800 kcal/capita, day (CA, 2007; cf. Figure 17.2). The actual intake of food is much lower than production and also lower than supply. Smil (2000) refers to figures from various societies, e.g. US, Canada, Kenya and Senegal and argues that the actual food intake is lower as compared to figures that are frequently quoted, "…no U.S. food consumption survey has returned an adult female mean higher than 1,700 kcal/day during the past generation" (*ibid.* p. 239). For urban India, Vepa (2002), in a comprehensive study on food insecurity in Urban India, argues that 1,890 kcal/capita per day is the acceptable level of calorie consumption for urban people. This figure represents 70% of the 2,700 kcal norm. It is, however, not clear if 1,890 refer to food intake or food accessed. From these and other studies (e.g. Schäfer-Elinder 2005), an average daily per capita food intake of about 1,800–2,200 kcal is considered adequate, depending on age, sex, activity, illness, etc. It should be added that energy intake is important but not the only factor determining under nourishment.

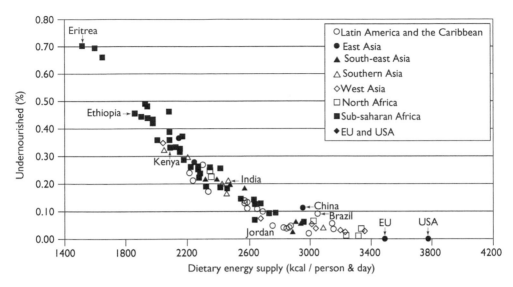

Figure 17.3 Percentage of undernourished people in society and average food supply. (Source: SEI, 2005 based on data from FAOstat and UNstat, 2005).

A distinction between the amount of food produced, food supply in society *versus* the net food intake at the household and individual level is important for many reasons. Since a certain fraction of the food produced is lost or converted in the food chain, there must be a certain buffer in society in order to reduce the risk of undernourishment. As indicated in Figure 17.3, there is clear empirical support for a hypothesis that the risk for undernourishment decreases with increasing supply. When supply is in the order of 2,700 to 3,000 kcal/capita, day, the fraction of people reported as under-nourished is quite low. Minimizing the risk for undernourishment is a very important objective, for instance, with regard to the Millennium Development Goals. On the other hand, a food supply at a high level seems to imply that also wastage is high and/or that the risk of an increasing number of people being overweight and obese is increasing.

While undernourishment is humiliating and often outside the control of those who are affected, it is, in some respects, resulting in similar social costs as overweight and obesity. Impairment of health, premature death, loss of ability to lead a productive life and pressure on the health care system, are all likely consequences of either too much or too little food. In one respect there is a big difference though; overweight and obesity implies a higher pressure on water and other natural resources as compared to under-nourishment.

Considering demographic trends and likely increases in purchasing power in the future, and thus an increasing demand for food, among other things, a high wastage means an extra pressure on water and other natural resources as compared to a situation where wastage is reduced. With a likely future of an increased need and demand for food, it is relevant to ask; which strategy is best to feed a growing world population properly while at the same time, not jeopardizing the functioning of natural ecosystems? Three types of approaches need further elaboration.

- One is to increase productivity, of water, land, labour, capital etc., i.e. produce more per drop.
- Quite another route is to reduce losses and wastage in the food chain.
- A third approach is to reduce over-consumption.

So far, only the first route is given attention.

17.6   Food security in an increasingly competitive water future

There is wide agreement that more food will be required in years to come. A doubling of annual food grain production, from about 210 Mill Tonnes to 420 Mill Tonnes in 10 years, has been mentioned as a target by the Prime Minister of India. Projections made in the Comprehensive Assessment of Water Management in Agriculture indicate that the demand for food and feed crops on a global basis will nearly double in the coming 50 years (CA, 2007). In addition, the agricultural sector will be stimulated and expected to produce other goods and services, notably biomass for bio-fuels and raw material for industry and commercial interests. Parallel with claims on water for production of more food and other goods, the concern about aquatic and terrestrial natural ecosystems and the conservation of environmental and esthetical values inherent in the landscape, mean that more food must be produced with less water. Constraints on an expansion of the area used for food are equally strong. Below, some principle components of a strategy for increased water productivity are noted.

17.6.1   Enhancing water and land productivity

Huge differences exist between different parts of the world in terms of yields, i.e. amount of produce per hectare. Whereas yields in the order of 8 tonnes/hectare, or more, are fairly common in the industrial countries, the yields are typically around 1–1,5 tonnes on sub-Saharan Africa. A low yield per hectare also means a low water productivity. Since the amount of water is erratically available, especially in the tropical region, the inter-annual variation in yield and variation from field to field is also considerable, especially in rain-fed systems but also in irrigated systems, which, of course, also depend on rainfall. Within irrigated agriculture, there is a noticeable variation in productivity between countries and regions.

A wide array of institutional and technological approaches is available for an enhancement of water and land productivity. There is no blue print for what specific approach and combination is "best". As so often stated, local circumstances will determine what approach is promising and which approach is not. A few principles may nevertheless be summarised.

Considerable gains in overall water productivity can be reaped if a larger share of the precipitation is harnessed and used to upgrade agricultural systems. Systems for rainwater harvesting are crucial (see, for instance, work done by Centre for Science and Development: http://www.rainwaterharvesting.org/index.htm) The harvested water will enhance soil moisture and can also be used for supplementary irrigation.

Such arrangements may increase yields substantially and, thus, also land and water productivity, particularly in areas where performance is poor today (CA, 2007; especially Part 4; pp 315–348).

Investments for building necessary arrangements will be required but of another sort as compared to the huge investments that were made in water storage arrangements during the Green revolution. A shift in thinking on water is crucial. Focus should not only be on the water that is available in rivers, lakes and aquifers, the so-called blue water. This part of the entire water resource is already heavily exploited and competition from the urban sector is intense.

Communities and individual farmers and other stakeholders in sub-catchments must be the drivers of this strategy. To enable and empower them to implement this strategy, support from outside is essential. Credit must be accessible and affordable also to those who venture to go ahead and build the required arrangements. Supporting institutions, e.g. the extension service, should be part of the package. In irrigation systems, simple measures like, for instance, better timing of water supply in relation to crop season, improved seeds and other agronomic inputs are important to enhance water productivity in more conventionally irrigated systems.

17.6.2 Improved efficiency to bring food to the market

Improvements in the chain from production to final demand require better post-harvest technologies and support to producers for investments and marketing, among other things. Current poor infrastructure in terms of roads and transport is hugely detrimental to bring the produce from field to places where demand is to be met. With urbanisation and a decline in the proportion of the farming population, the distance between areas where food is produced and where it is consumed increases. Poor linkages between the primary and secondary sectors, i.e. between producers and those who market and process food, are detrimental to small farmers. They face considerable difficulties to meet the demands imposed by supermarkets and other food processing industry (FAO, 2004a; Reardon *et al.*, 2003). To add value to the primary produce and thus enhance livelihood of farmers, it is essential that backward linkages are developed in terms of better access to credit and improved technologies for production, processing, storage and transport, but also improved forward linkages to markets (MSSRF, 2005).

In practice, some of the produce that could have come from the small holders is difficult or impossible to market. This is a waste of opportunity that should be added to the waste and loss that are implied in Figure 17.2. The implication is clear; a substantial fraction of the food that is produced – or could have been produced – never reaches the market or the consumer.

17.6.3 The critical link: Consumers

The most complicated part is probably related to the last part of the food chain, i.e. the consumers of food. With increasing purchasing power, consumer preferences and behaviour are vital bricks in the dynamic interplay between production, demand and consumption. Poverty alleviation as well as economic progress is high on the political as well as the social agenda and rightly so.

As consumers we are all intervening in the water cycle through our preferences for various goods and services. Being far away from where food is produced, few consumers are aware of this linkage.

The combination of demographic and economic trends implies increasing pressure on water and other natural resources. For this and future generations, it is of paramount interest that inefficiencies and wastage are reduced as much as possible, particularly in the last segment of the food chain, i.e. in the consumption of food. So far, the attention has been focused on production aspects in discussion about food security. With increasing consumer influence, it is logical and important to develop a perspective where this important driving force is taken into account. It is necessary to enhance consumer awareness about their footprint on water in order to stimulate sound preferences and smart consumption patterns. Without "water literacy" among those who are the main drivers behind water pressure, a vital component in water policy is being ignored.

17.7 Water implications of an improved efficiency in the food chain

Scenarios of consumptive water use in the future unanimously suggest that pressure on water will increase. Even with substantial investments in both irrigated and rain-fed systems and improvements in fisheries and livestock, it is likely that crop water requirements will increase. Estimates vary depending upon assumptions, but a plausible figure for additional withdrawals from blue water sources is between 10–15% till 2050 (see e.g. CA, 2007). The figure may be higher if the likely demand for other agricultural products, e.g. biomass for energy production is taken into account. On top of this, climate change will aggravate the water situation.

Faced with these challenges, it makes sense to consider what additional efforts need to be made in order to reduce the additional pressure on water and other resources. i.e. apart from improvements in productivity. A strategy to reduce wastage and losses in various segments of the food chain should be considered. If a larger fraction of the food produced would reach the consumer, i.e. if waste is reduced, the possibility will increase to feed the world on the same amount, or less, water. Reduction of losses and wastage, including household waste, by a proposed 50%, in combination with reasonable improvements in water productivity would result in a consumptive water use by 2045 that is at the same level as at the beginning of this century (Lundqvist et al., 2007). Improvements in efficiency in the various segments of the food chain will require a broad and long term strategy where representatives of the various segments of the food chain are mobilised. However, if we agree on the magnitude of the challenges, effective measures can be and must be gradually introduced. A strategy that considers what can be done through an improvement in both production and the food chain could result in a win-win-win outcome where water, environmental and public health concerns can all benefit.

Acknowledgement

Financial support from Sida and FORMAS for research on which this article is based is gratefully acknowledged.

References

Briscoe, J. (2005) *India's water Economy: Bracing for a turbulent water future*. Draft report, October 5. http://www.worldbank.org. in/WBSITE/EXTERNAL/COUNTRIES/SOUTH ASIAEXT/INDIAEXTN/0„contentMDK:20674796~pagePK:141137~piPK:141127~theSite PK:295584,00.html

CA (Comprehensive Assessment of Water Management in Agriculture) (2007). *Water for Food, Water for Life. A Comprehensive assessment of Water Management in Agriculture*. London: Earthscan, and Colombo: International Water Management Institute.

de Fraiture, C., Molden, D., Rosegrant, M., Amarasinghe, U. and X. Cai. (2004). *Does International Trade Save Water? The Impact of Virtual Water Trade in Global Water Use*. Comprehensive Assessment Report 4. Colombo. (http://www.iwmi.cgiar.org/assessment/)

FAO (Food and Agricultural Organization of the United Nations). (2003). *World Agriculture: Towards 2015/2030*. Bruisma, J., ed. London: Earthscan Publications.

FAO (Food and Agricultural Organization of the United Nations). (2004). *The State of Food Insecurity in the World, 2004. Monitoring Progress Towards the World Food Summit and Millennium Development Goals*. Rome.

FAO (Food and Agricultural Organization of the United Nations) (2005). World Cereal Production. *Food Outlook*, June 2. Statistical appendix. http://www.fao.org/documents/show_cdr. asp?url_file=/docrep/008/j5667e/j5667e00.htm

Falkenmark, M. and J. Rockström. (2004). *Balancing Water for Humans and Nature. The New Approach in Ecohydrology*. London: Earthscan Publications.

Falkenmark, M., *et al.* (2007) *On the Verge of a New Water Scarcity: A call for Good Governance and Human Ingenuity*. SIWI Policy Brief, SIWI, Stockholm.

Gleick, P. (2000). *The World's Water – The Biennial report on freshwater resources 2000–2001*. Washington D.C. Island Press.

Lundqvist, J., Barron, J., Berndes, G., Berntell, A., Falkenmark, M., Karlberg, L. and Rockström, J. (2007). "Water pressure and increases in food & bio-energy demand implications of economic growth and options for decoupling". Chapter 3 in: *Scenarios on economic growth and resource demand*. Background report to the Swedish Environmental Advisory Council, memorandum 2007:1. Swedish Environmental Advisory Council, Stockholm.

Milmo, C. (2005) "What a Waste". *The Independent*, April 15. http://www.findarticles.com/p/ articles/mi_qn4158/is_20050415/ain13608404

MSSRF (M.S. Swaminathan Research Foundation) (2005). "Fifteenth Annual Report: 2004–2005". Chennai.

Radhakrishna, R. (2003). "Food and Nutrition Security". *India Development Report* 2002. pp. 47–58. IGIDR & Oxford University Press.

Renault, D. and Wallender, W.W. (2000). Nutritional Water Productivity and Diets: From "crop per drop" towards "nutrition per drop". *Agricultural Water Management*, 45:275–296.

Reardon, T., Timmer, P., Barrett, C. and Berdegue, J. (2003). "The Rise of Supermarkets in Africa, Asia, and Latin America." *American Journal of Agricultural Economics*. 85 (5): 1140–1146.

Schäfer-Elinder, L. 2005. Obesity, hunger and agriculture: the damaging role of subsidies. *BMJ*. Vol 331, 3 December. Pp. 1333–1336.

SEI (Stockholm Environment Insitute) (2005). *Sustainable Pathways to Attain the Millennium Development Goals. Asessing the Key Role of Water, Energy and Sanitation*. Report prepared for the UN World Summit, 4th Sep. 2005, New York. Stockholm. (http://www.sei.org)

Shah, T., Molden, D., Sakhtivadivel, R. and Seckler, D. (2000). *The Global Groundwater Situation. Overview of Opportunities and Challenges*. International Water Management Institute, Colombo.

SIWI (Stockholm International Water Institute), IFPRI (International Food Policy Research Institute), IUCN (International Union for Conservation of Nature) and IWMI (International Water Management Institute) (2005). *Let it Reign. The New Water Paradigm for Global Food Security*. Stockholm. (http://www.siwi.org).

Smakthin, V., Revenga, C. and Döll, P. (2004). *Taking into account environmental water requirements in global scale water resources assessments*. Comprehensive Assessment Research Report No.2. Sri Lanka: Comprehensive Assessment Secretariat / IWMI.

Smakhtin, V. and Anputhas, M. (2006). An assessment of Environmental Flow Requirements of Indian River Basins. Research Report 107. International Water Management Institute, Colombo.

Smil, V. (2000). *Feeding the World. A Challenge for the Twenty-First Century*. MIT Press, Cambridge, Massachusets. London.

Swaminathan, M.S. (2006). *2006–07: Year of Agricultural Renewal*. 93rd Indian Science Congress, Hyderabad. Public Lecture, 4 January.

Vepa, S.S. (2002). *Food Insecurity Atlas of Urban India*. M.S. Swaminathan Research Foundation and World Food Program. MSSRF, Chennai, India.

WHO (World Health Organization). (2004). *Global Strategy on Diet, Physical Activity and Health*. Report by the Secretariat of the 57th World Health Assembly. Geneva.

von Braun, J., Swaminathan, M.S. and M.W. Rosegrant. (2004). *Agriculture, Food Security, Nutrition and the Millennium Development Goals*. IFPRI 2003–2004 Annual Report. International Food Policy Research Institute, Washington D.C. http://www.rainwaterharvesting.org/index.htm

Chapter 18

Micro-enterprises and food security

Paul Rigterink

Potomac Technical Advisors, Potomac, USA

18.1 Introduction

This paper describes how micro-enterprises can contribute to the food security and income of small-scale farmers by providing supplies and training in return for a share of the small-scale farmer's earnings. A table is given which provides the business planning required so that small-scale farmers and micro-enterprises jointly can become productive. With emphasis on arid and tropical regions, more information is given on how to develop the supply chains and training materials required.

The Rockefeller and Gates Foundations (2006) have started a pilot program in Africa using ideas similar to what are suggested in this paper. In particular, the Program for Africa Seed Systems (PASS) hopes to provide training, capital, and credit to establish a network of 10,000 small African agro-dealers. In turn, these agro-dealers can serve as conduits of seeds, fertilizers, chemicals, and knowledge to small-scale farmers to facilitate increasing productivity and income.

18.2 Food security for small-scale farmers

According to the UN Food and Agriculture Organization (FAO) (2006), food security exists when all people, at all times, have access to sufficient food to meet their dietary needs for an active and healthy life. The FAO Special Programme for Food Security (SPFS) estimates that there are currently 852 million people chronically hungry due to extreme poverty. Velarde (1992) has outlined what is necessary to increase food security for small-scale farmers. Her ideas apply equally well to home gardens in large populated areas and cultivated fields in rural areas. According to Velarde:

- "A primary goal, if not the primary goal of small scale-farmers is food production for household consumption. Yet both agricultural policies and research efforts have tended to focus on cash crops and ignore this goal or give it little more than token consideration.
- In rural areas with poorly developed marketing systems, household production is the main food procurement source for small-scale farmers. Households must produce both a wide variety of food items and sufficient quantities of these to ensure adequate supplies to cover needs.
- The transition from traditional subsistence farming to commercially oriented farming leads to permanent farming systems emphasizing mono-cropping and creates competition for farm resources often to the detriment of the household food security."

Many extension services have agricultural policies that focus on cash crops and ignore food and subsistence crops (Women in Development Service (2006)). Except for "black markets," the marketing systems are poorly developed and place emphasis on mono-cropping to the detriment of food security. However, many of these past policies are under review and may be changed. In particular, Velarde (1992) offers the following general objectives in developing a food security program:

- "Define the food base of the target area.
- Define the patterns of food availability and accessibility.
- Identify main sources of food procurement, with any seasonal variations.
- Determine if the food base is capable of covering general household energy and nutrient needs.
- Identify household types suffering from or at risk for food insecurity.
- Identify causal factors of any household food insecurity situations and their plausible solutions.
- Identify areas of priority for research by adaptive research planning teams which can lead to enhancing or improving the home food security situation."

Velarde's (1992) perspective on the importance of cash crops may differ to some extent with those of the former U.S. Secretary of Agriculture, Orville Freeman (1989). Freeman noted:

"The ability of small farmers to take the plunge into cash crops has an immediate effect on a village. Even if only marginally successful, the farmer needs help at harvest time so that casual labor is taken on. When farmers have a profitable year, their first expenditure is home improvement. Materials and labor come from the village. As farmers climb out of subsistence production, they take other members of their village with them. Perhaps one of the most important effects on a small farmer having just a little money to spare is the apparently universal instinct to spend money on educating his children.

The raising of cash crops introduces the farmer to modern methods of cultivation that constitute a triple-win proposition. It increases the income of the farmer and his family; it has an immediate effect on the local community, as well as the nation; and it contributes to the global food supply."

Freeman's perspective on agriculture's modernization was reviewed later by Rivera (2000). Rivera noted that Freeman considers agribusiness as orthodoxy with its tenets based on agribusiness economics. In addition, he noted that Freeman's view contrasts with those who view agriculture as best understood within the context of the environment (Edward et al., 1990). Rivera's opinion is that the modernization in agriculture resulted in large part from the technical development of agricultural inputs and processes. According to Rivera, the recent modernization of agriculture culminated in the contemporary dominance of what are known as Green Revolution technologies that include chemical fertilizers, pesticides, herbicides, chemically-treated seed varieties, and more recently, the development of bio-genetically treated seeds and plants and hormonally treated farm animals.

Green Revolution technologies have had only marginal benefit to people living in extreme poverty (struggling to survive on less that $1 a day) and to those who eke out an existence on $2 a day. The U.N. Development Program (2004) estimates that one third of the people in the world are in these conditions. The perspective of this paper is to show how we can:

- Improve agriculture business concepts for extremely poor subsistence farmers.
- Use micro-entrepreneurs to transfer the needed training and supplies to the field including "Green Revolution technologies" and other agricultural technologies that will help the extremely poor earn a better living.

18.3   Identifying what small-scale farmers need

In a recent conference (Rigterink, 2006), I outlined how major Non-Government Organizations (NGOs) and government relief organizations could help subsistence farmers. To aid in solving these extreme poverty problems, I suggested that NGOs and government relief organizations use processes similar to those used by large businesses to improve productivity and capital. In particular, I suggested that they:

1. Identify the business concept that small-scale farmers can use to solve standard business problems (see Table 18.1).
2. Identify the areas of change and discuss them with local personnel to see if these changes are possible.
3. Identify the requirements for changing the business.
4. Verify that local people can afford the costs and that the supplies and training needed can be obtained at the local level.
5. Document the business solution.
6. Review and upgrade the solution with a few local participants (such as local school teachers) to verify the solution.
7. Use local school teachers and other community leaders to promote the program at a broader level.

This technical approach requires that a thorough analysis be made of such business concept elements as those shown in Table 18. 1. For each business concept element, one needs to identify the steps and work required, the benefit mechanisms, performance metrics, and cost. In particular, the suggested steps and work required must be reviewed with the local stakeholders and local community leaders in detail so that fewer mistakes are made.

An examination of the column in Table 18.1 titled "Work Required" will show that investment, credit, and micro-entrepreneurship will be required. Improving the food security of a subsistence farmer will require investment and credit so that the farmer can improve pest and disease control practices, water use, and seed variety selection. In addition, it will require the identification of cash crops that will allow the subsistence farmer to repay the investments made. Investment and credit need to be supplied to micro-entrepreneurs for providing training, developing quality control procedures, maintaining standards, obtaining and using market data, negotiating the best deals both for themselves and for the subsistence farmer, providing transport, and improving business operations.

Even if investment and credit are available, creating a balance between cash crops and food production for household consumption will not be easy. The investment and credit required to make an arid or tropical region farm productive are generally much

Table 18.1 Definition of business concept elements.

Business concept element	Step	Work required
Requirements analysis	Preliminary analysis	• Assessment of product's capability • Assessment of market
	Strategy formulation	• Definition of objectives • Definition of priorities • Determination of alternative approaches • Selection of preferred approach
Design	Formulation of development plan	• Identification of customers • Identification of area to sell product • Identification of constraints • Definition of product • Analysis of costs • Definition of required actions • Sequence of required actions • Schedule
	Establishment of support required	• Definition of required resources • Acquisition of required resources • Organization of human and material resources • Acquisition of training
	Ensuring acceptable levels of quality and supply	• Selection of technology • Site selection • Variety selection • Feed selection • Pest and disease control practices • Water use • Determining when to sell
Operations	Maintaining quality	• Quality control procedures • Maintaining standards • Packaging procedures • Training workers on use of technology
	Capturing a market	• Obtaining and using market data (pricing analysis) • Making use of market intelligence • Selection of target markets • Identification of middleman • Negotiating the best deal
	Optimal transport	• Selection of means of transport • Packing for transport • Keeping quality control during transport
Maintenance	Maximization of income, profits, and production of product	• Analysis of costs • Analysis of returns • Analysis of business operations

higher than the investment and credit required in other regions of the world. It is not surprising that the most poor live in these regions as described by Sachs (2005). Gaining the confidence and trust of the population is key to the success of any program. The locals must be made to understand that cash crops are a way of not only attaining food security for themselves and their families, but also a way out of extreme poverty.

It is anticipated that individuals and small groups can help improve the following businesses by trying new ideas and using information that might not be available at the local level:

– Grain production
– Production of vegetables
– Fruit orchards and fruit nurseries
– Poultry production
– Small animal production such as goats and pigs
– General purpose micro-farms

While it may seem strange that the level of planning shown in Table 18.1 is required to improve a subsistence farm, engineers and successful businessmen consider detailed planning key to the success of a business. In contrast, many community development personnel do not develop their ideas to this level of detail. This results in the extremely poor taking huge "business risks" that would never be taken by successful businessmen. A complete set of supplies, training, and business processes, along with a stable supply chain for new supplies, must be available for a business is to succeed. The extremely poor cannot be expected to work their way out of poverty if only some of the supplies and training is available. In particular, one cannot expect a subsistence farmer to start a successful vegetable or fruit business with only seeds and tools (no stable water supply, fencing, pest control, or disease control). Similarly, the extremely poor cannot be expected to start a successful backyard poultry farm without such necessities as veterinary supplies and fencing.

The problems that must be solved at the micro-enterprise level are the same problems that must be solved at the macro level. The main difference is that personnel working from major organizations can bring in outside resources while individuals and small groups must use the resources that are available. Individuals or small groups, however, can better assess local conditions and take advantage of existing opportunities which would be difficult to accomplish at the macro level. In particular, individuals and small groups can better assess the problems of local people who live in poverty.

18.4 Providing the necessary supplies

One major difference in creating a food security program in a developed country as compared to a less developed country is the amount of supplies available. For example, better food storage equipment for grain and produce would help most subsistence farmers. Also, by providing improved sources of supplies for subsistence farmers in remote regions, micro-entrepreneurs can help subsistence farmers while earning income from the sale of the supplies. It is expected that most of the poor will not have access to a complete set of supplies necessary for conducting a business with a reasonable amount of risk. Lehman's "Non-Electric Catalog" (available at http://www.lehmans.com/) provides examples of supplies that may be needed and not currently available in many regions of the world. It is extremely frustrating to start a project and find that the supplies needed are not available. One must be careful of this problem when giving micro-loans to the world's poor in an effort to help them get out of the "poverty trap."

Micro-entrepreneurs may incur huge "business risks" if loans are given without ensuring first that necessary supplies for a project, such as veterinary supplies and fencing materials in the case of a backyard poultry farm, are available.

18.5 Providing the necessary training

There is an abundance of training material on farming available on the Internet and in books and pamphlets. However, this material must be tailored so that it fits the particular farming region. For example, extensive horticulture information is being supplied by Ecology Action (see http://www.growbiointensive.org/index.html). They have introduced the GROW BIOINTENSIVE method in 130 countries around the world including Kenya, Ecuador, Uzbekistan, Russia, Argentina, and the Philippines. The method (see http://www.growbiointensive.org/biointensive/GROW-BIOINTENSIVE.html):

- "Enriches the soil with compost produced from crops grown on the farm
- Uses organic fertilizers
- Prepares soil deeply to promote lush growth,
- Spaces plants within growing beds rather than in rows to maximize yields, and
- Grows diverse crops including vegetables, grains, herbs, and flowers for diet, income, compost, and a healthy mini-farm ecosystem.

The method is appropriate for the situation in many under developed regions because it:

- Brings food security to areas that are lacking in nearly all basic needs, including water and sewage systems, road, and electricity
- Generally uses only locally available resources
- Avoids use of chemical fertilizers and pesticides that can be difficult and expensive to obtain
- Makes optimal use of limited water and seed supplies
- Provides the farmer with a significant legal income while also producing nourishing food for families
- Relies on no machinery, thereby being accessible to anyone with basic hand tools
- Has proven capable of producing the highest yields per unit area compared with other systems, and
- Builds a healthy, fertile soil while achieving all this!"

It is important to remember that informational material on farming is only useful if it has been tailored to the region.

18.6 Working in arid and tropical regions

Arid regions and tropical regions can create special problems. Unfortunately, the extremely poor usually work on arid lands that could be characterized as too dry for conventional rain-fed agriculture. Yet, millions of people live in these regions, and if current trends continue, there will be millions more. These people must eat and earn a living; the wisest course would be for them to produce their own food. The local inhabitants usually have developed very suitable techniques for using the surface water that is available. However, in many regions, the ability to produce agriculture crops is restricted. In particular, there are areas in many countries where the amount of precipitation is too small to penetrate the soil sufficiently, or it may run through a porous soil too quickly, or it may run off too quickly. Furthermore, weedy species are so adept at utilizing scarce water that they rob water from crops. For these regions, special micro-irrigation equipment or runoff irrigation schemes will be needed. The use of runoff irrigation can provide substantial amounts of water in certain regions. Prinz and Singh (2003) have reviewed the potential of various runoff irrigation techniques. These techniques are generally not used in the United States for capturing water for high-value agriculture. In the United States, water is generally carried by wells, canals, or other means so that normal agriculture can exist in spite of the aridity of the climate. This allows for a much more consistent water supply if the water is used wisely. More use of the techniques suggested by Prinz and Singh (2003) is needed.

In some arid regions, water can be obtained from wells. The depth of a well necessary to obtain water may vary from a few feet to thousands of feet. In areas where the water table is within 50 meters, the use of inexpensive equipment by various means may be possible by including hand dug wells, percussion drilling, or auger drilling. These sources of water are limited and can be exhausted easily. Such water conservation techniques as reducing evaporation with windbreaks and light-shades, planting in furrows, pits, or swales, establishing plants in a nursery, using drought resistant crops, using mulch, irrigating efficiently, and keeping weeds down are needed to establish food security and provide the water needed for a cash crop. This requires money, equipment, and training that the extremely poor do not have. Community development workers will need to identify the micro- irrigation equipment, training, and funding that are needed and help local personnel develop cash crops to cover the cost of ensuring a steady water supply.

International Development Enterprises (2006) has developed inexpensive micro-drip irrigation systems for many water scarce areas. These systems include drum kits, bucket kits, kits made with double layered plastic bags (called Family Nutrition Kits), as well as customized micro irrigation systems. In particular, International Development Enterprises India (IDEI) (2006a) has developed customized systems, costing from $5 for 20 sq.m. to $80 for 1000 sq.m. that will enable a family to grow produce for their own use or for a cash crop. In addition, IDEI sells a variety of treadle pumps for use by small farmers in areas where water is available at or near the surface.

Equally important, IDEI (2006b) has developed a program called "Integrating Poor into Market Systems (IPMAS)" that provides help in overcoming market constraints as well as water constraints. IDEI seeks to achieve this through capacity building and training of the farmers in successful crop management and use of high quality seed, water technology, fertilizers, pest control, and training. In particular, IDEI:

- Identifies relevant technologies
- Creates the necessary supply chains
- Develops diversified crop portfolios
- Provides training for small holders, suggests crop options, and provides best practices required to grow various crops

- Provides linkages to public crop experts and private sector service providers of seed, fertilizer, technology, micro credit, and on-farm agronomic training.
- Introduces and promotes Vermin wash technology as a high quality inputs for micronutrients.
- Links farmers with nursery growers that provide saplings for high value crops.

Programs of this nature need to be developed for major refugee centers so that the most poor can get back on their feet after an economic or political disaster. It is suggested that major NGOs should buy large quantities of IDE equipment as well as other needed supplies and use it to develop nurseries and small gardens at refugee centers to demonstrate to the poor that the technology can significantly improve their income. In addition, major NGOs should supply some of this equipment to villages used for resettlement, schools, and community centers so that the refugee and other poor people can begin to work their way out of poverty. These ideas will only work if the proper supply chains and training also are provided.

Tropical regions also present problems that may require additional funding, training, and supplies to aid the subsistence farmer in achieving food security and potential cash crop income. Meitzner and Price (1996) have provided extensive ideas for growing foods under difficult tropical conditions. It is suggested that community development workers use this book and the planning material provided in Table 18.1 to establish improved food security programs both in tropical and arid regions.

18.7 Advantages and disadvantages that subsistence farmers possess

The extremely poor do possess potential business advantages that should allow them to make a better living. In particular, the advantages that the extremely poor possess in running a subsistence farm or starting a micro business are:

- Low labor costs
- Low fixed and variable costs (not capital or labor intensive)
- Few resources subject to theft
- Lack of attention by the local governance
- Ability to provide food, water, and energy while earning profit for work on a subsistence farm
- Ability to control costs and make appropriate decisions regarding their businesses
- Few barriers to entry into a new micro-enterprise

The disadvantages that the extremely poor face are:

- Needed supplies and training not readily available
- Undereducated and often unwell
- Little or no investment funds available
- Successful business plans and business processes not provided at the local level
- High crime-rate affecting business profits (theft of equipment/livestock)

It is clear that establishing processes for the extremely poor to develop food security and income is no easy task. However, financial and engineering analyses indicate that given the low cost of labor in underdeveloped countries, a profit margin greater than fifty percent for any investment can be expected. If this truly is the case, then these profits will allow program expansion with the result that micro-entrepreneurs (and NGOs) will earn profits while aiding in the reduction of poverty.

References

Edwards, C.A., R. Lal, P. Madden, R.H. Miller, and G. House (eds.) (1990) *Sustainable Agricultural Systems*, Ankeny, IA, Soil and Conservation Society.

Freeman, Orville (1989) Reaping the Benefits: Cash Crops in the Development Process, *International Health and Development*, March/April 1989, pp. 21–23.

International Development Enterprises (IDE) (2006) *IDE Technical Library*, available at http://www.ideorg.org/page.asp?navid=224.

International Development Enterprises India (2006a) *IDEI Products and Technologies*, available at http://www.ide-india.org/ide/pt/photo_gallery/aditi/familynkits/fnk.shtml.

International Development Enterprises India (2006b), *IDEI Integrating Poor into Market Systems*, available at http://www.ide-india.org/ide/pt/photo_gallery/aditi/familynkits/fnk.shtml.

Meitzner, Laura S. and Price, Martin, L. (1996) *Amaranth to Zai Holes*, published by ECHO, 17430 Durrance Road, North Fort Myers, FL, 33917, 404 pp.

Prinz, Dieter and Singh, Anupam K. (2003) Technological Potential for Improvements of Water Harvesting, *Technical Report on the World Commission on Dams*, Cape Town, South Africa, available at http://www.wcainfonet.org/servlet/BinaryDownloaderServlet?filename=1066992525558_harvesting.pdf

Rigterink, Paul V. (2006) A Revolutionary Approach to Reducing Poverty and Increasing the Agriculture Productivity of Subsistence Farmers, *Product and Market Development for Subsistence Marketplaces: Consumption and Entrepreneurship Beyond Literacy and Resource Barriers Conference*, August 2006.

Rivera, William, M. (2000) The Changing Nature of Agricultural Information and the Conflictive Global Development Shaping Extension, *Journal of Agricultural Education and Extension*, 7(1), 31–41.

Rockefeller Foundation and Gates Foundation (2006) *Major Effort to Move Millions out of Poverty and Hunger Begins with a $150 Million Investment to Improve Africa's Seed Systems*, available at http://www.rockfound.org/Library/agra1.pdf

Sachs, J.D. (2005) *The End of Poverty: Economic Possibilities for our Time*, Penguin, New York, 396 pp.

U.N. Development Program (2004) *Human Development Report*, available at http://hdr.undp.org/2004/.

UN Food and Agriculture Organization (2006), *Special Program for Food Security*, available at http://www/fao.org/spfs.

Velarde, Nancy (1992) Putting Household Food Security Aspects into Farming System Research. *Forest, Tree and People Newsletter*, 11, 12–16.

Women in Development Service (SDWW) FAO Women and Population Division, (2006), "*Towards Sustainable Food Security, Women and Sustainable Food Security*", available at http://www.fao.org/sd/fsdirect/fbdirect/FSP001.htm

Chapter 19

Rainwater harvesting: Resources development and management

Sekhar Raghavan
Rain Centre, Chennai, Madras, India

N. Parasuraman
M.S. Swaminathan Research Foundation
Taramani Institutional Area, Chennai, Madras, India

19.1 Introduction

In many regions of the world, particularly those with monsoon climate, rainfall is confined to three or four months in a year. Consequently, from time immemorial, many cultures practiced the collection, storage, treatment and distribution and use of rainwater and other forms of precipitation (e.g. fog drip and dew are collected in Namibia). During 3000 – 1500 B.C, Dholavira, a major site of the Indus valley civilization, had several reservoirs to collect monsoon runoff. Wells were probably a Harappan invention. A recent survey revealed that every third house had a well. Sophisticated examples of rainwater harvesting have been found in the ruins of the Palace of Knossos (1700 B.C.) in the island of Crete, Greece, the Alhambra Palace in Granada, Spain, and Chihuahua area of Mexico, etc. (Vide historical accounts described in UNEP, 1983; Agarwal and Narain, 1997; Aswathanarayana, 2001; Mukundan, 2005; Raghavan, 2004, 2005, 2006).

Rain Water Harvesting (RWH) systems existed in India for the collection of water for immediate use as well as recharge. At the micro level, rainwater was collected directly from rooftops and stored it in below ground level (bgl) tanks called *tankas*. Both at the micro and macro levels, the runoff rainwater was collected from courtyards and open community lands respectively and stored in another kind of bgl tanks called *kundis*. Huge tanks had also been built within forts situated on hilltops for storing rainwater for use during wars. These are found in the states of Rajasthan and Gujarat in Northwest India.

- The monsoon runoff was also harvested by capturing water from swollen rivers and stored in earthen tanks as surface storage called by different names in different states – *zings* in Ladakh state, *ahars* in Bihar state and *johads* in Rajasthan and *eris* in Tamil Nadu to name a few.
- Water from flooded rivers was also harvested in places like North Bihar and West Bengal state.

There existed a tank system of harvesting rainwater in South India since 4th Century A.D. These tanks are earthen bunded reservoirs constructed across slopes by taking advantage of local depressions and mounds. The three states of South India – Andhra Pradesh, Karnataka and Tamil Nadu have 150,000 such tanks among them and most of them are still functional. Though called by different names – *cheruvu*, *kere* and *ery* respectively in these three states, they are identical in structure and performance. These are primarily meant for irrigation and most of them are inter-connected meaning that the overflow of one will go to fill up the neighbouring one at the lower level and so on. These tanks are of two types called system and non-system tanks. The former also known as riverfed tanks gets water diverted from rivers by means of embankments called *anicuts* and the non-system tanks have their own catchment and are known as rainfed tanks.

During the period of British rule, the traditional management and cultural practices of the communities in respect of water resources were disrupted, with the government taking the responsibility. Unfortunately, the same beauracratic system of management of tanks was continued in India after India's Independence in 1947, with disastrous results.

19.2 Rainfall pattern in India

The rainfall in India varies from 100 mm in the deserts to 14000 mm in northeast India. Nearly 12% of the country receives an average rainfall of less than 610 mm per annum, while 8% receives more than 2500 mm. But more than 50% of this rain falls in about 15 days and less than 100 hours out of a total of 8760 hours in a year. The total number of rainy days can range from a low of 5 days to a high of 150 days in the northeast. As India has a land area of 329 million hectares (Mha), and has an average rainfall of 1050 mm, the total annual precipitation in the country is about 345 million hectare-metres (M ha m), with evaporation of 59% (203.55 Mha m) and percolation of 12% (41.40 M ha m). The average runoff is has been estimated to be 180 M ha m, out of which 69 M ha m is considered utilizable. Presently, there is storage capacity of 14.3 M ha m in the form of large reservoirs, and 3 M ha m in the form of minor irrigation tanks. Plans are underway to develop 8 M ha m of additional storage. The annual per capita of use of water in India is 612 m^3 (594 for agricultural and industrial purposes and 18 for domestic use) (Aswathanarayana, 1995, p. 82).

19.3 Strategy for rainwater harvesting in rural areas

19.3.1 *At the macrolevel*

Collection of rainwater into tanks can be done by the state, village councils, or farmer cooperatives. Where a tank already exists, the community bodies concerned will have to maintain it. The tank could serve the dual purpose of improving the infiltration into local wells, and also provide water for irrigation. Irrespective of the extent of spread of water, the top 1–1.2 m layer of water is invariably lost due to evaporation. Therefore it is necessary that the tank is as kept as deep as possible. This will also mean that for a given volume of water stored, the surface area of the spread of water will be less pro rata. It is evidently necessary to desilt the tank regularly to ensure that sedimentation does not reduce the volume of water that the tank could hold.

The catchment area should have as much grass cover as possible. The rain drops in the monsoon climate are large and hit the ground with great force. The impact of rain drops detaches particles from the soil clods and moves them by splashing, followed by runoff when the material thus loosened is transported by turbulent water. The grass cover will cushion the rain drop impact, promote percolation, and reduce the amount of runoff and its turbidity. This will reduce sedimentation in the tanks.

In order to harvest flood waters, the customary practice is to construct a dam at the mouth of a gully to divert the muddy water to the flats on either banks of a stream. Alternately, intercepting ditches and canals can be constructed along contour lines to intercept the overland flow on slopes and the water thus intercepted can be utilized to irrigate the terraced fields (UNEP, 1983).

19.3.2 *At the micro-level*

Farmers are encouraged to dig farm ponds in the downslope side of the farm. It is possible to collect about 40 ha.cm. ($4000 m^3$) of water from a farm of 1 ha, with rainfall of 800 mm, and collection efficiency of 50%. This water can be stored in one or more farm ponds of 5 m depth, with total surface area of $800 m^2$. To reduce evaporation, the pond may be covered with bamboo or casuarinas poles, and viny vegetables (such as, gourd) with large leaves can be trained on them. If the soil porosity is high, the bottom and sides of the pond should be covered with a suitable sealant, such as an embedded membrane of plastic sheeting and bentonite clay. Animal-powered sprinkler or drip irrigation, could be made use of to use the water efficiently for irrigation. The stored water (40 ha.cm) will be more than enough to irrigate low water-need crops, such as maize, groundnut, jowar, gram and bajra.

19.4 Strategy for rainwater harvesting in the urban areas

19.4.1 *At the macro level*

The floods in Chennai in 2005 have drawn attention to the disastrous consequences of neglect of maintenance of water bodies in the urban areas. River Cooum which flows through the city is now a foul-smelling sewer. It was not always like this. It is said that in the eighteenth century, the brahmins of the Chintadripet area of Chennai used to bathe in the clear waters of the river Cooum, and offer "arghyam" (obeisance to Sun god). Collection of storm water and storing it in the form of surface water bodies, or for recharging the groundwater through recharge pits, lead to a number of benefits:

(i) Improvement of groundwater quality and rise of water-table. In Chennai city for example, the residents had implemented GWR about two years back, in order to comply with the law enacted by the government. This in addition to the record rainfall that the city received during October to December 2005 resulted in the groundwater level rising in the entire city by almost 6 to 8 metres. This was revealed in a survey conducted by the Rain Centre.

(ii) Renovated water bodies, such as temple tanks, could serve as decentralized fresh water sources for the immediate neighbourhood and also facilitate GWR. For example, Chennai still meets 40% of its fresh water needs from four such decentralized water bodies located in its suburbs.

(iii) Flooding will be mitigated.

(iv) The excess storm water could be used to flush water ways such as Cooum river and Buckingham Canal, and prevent saline incursion. If saline incursion has already taken place, the storm water could be made use of to push back the saline water (this has been achieved in Long Island, New York).

In the case of inland cities, even though the rain water is not harvested efficiently *in situ*, it is bound to be around in some other way. In the case of coastal cities, uncollected rain water ends up in the sea, besides creating floods on land (Raghavan, 2006).

19.4.2 At the micro level

A family of five can be assured of clean water for domestic use at the rate of ~60 L/d/capita, assuming a rooftop with an area of 200 m², rainfall of 1000 mm, and collection efficiency of 50%. The island of Bermuda in the Caribbean constitutes an excellent example of efficient application of rainwater harvesting technology. Bermuda depends wholly on the harvesting of rainwater (1430 mm) for providing 80 L/d/per capita of water for its citizens. Rainwater is collected and stored in underground cisterns. Necessity, tradition, financial incentives and administrative procedures have led to this highly desirable situation (Aswathanarayana, 2001, p. 138).

19.5 Database required for rainwater harvesting

1. Pedo-topo-geological profiles of the area,
2. Geoelectrical and hydrological study of the city and its environs, to understand the disposition and dynamics of movement surface water and groundwater,
3. Porosity, permeability, infiltration capacity and water retention capacity, etc. of different layers,
4. Construction of a three-dimensional map of aquifers,
5. Modeling the dynamics of movement of surface water and groundwater,
6. Techno-socio-economic analysis of various options of harvesting, storage and distribution of water at macro and micro levels.

19.6 A case history of rainwater harvesting in the MSSRF Estate, Chennai

19.6.1 Biophysical setting of the Estate

Chennai (previously known as Madras) is located at 13.04°N; 80.17°E. The annual high average temperature is 32.9°C, and a low average temperature is 24°C. The mean annual precipitation is ~1290 mm.

The estate of the M.S. Swaminathan Research Foundation (MSSRF) is spread out in an area of 3.54 acres (~1.4 ha). MSSRF is located in the institutional area of about 20 sq.km., where the groundwater is brackish. Recourse has been taken to rainwater harvesting to provide fresh water to the estate, as rain water is naturally soft (unlike well water), contains almost no dissolved salts, does not need any chemical treatment and is a relatively reliable source of water for households.

The Estate is located about 5 km. inland from the sea. The soil profile of the Estate consists (from the top) of the following units:

Desiccated crust of 2–6 m. depth of medium to high plastic clay
Alluvial soil
Kankar
Basement (charnockite – hypersthene granulites).

The thick clay layer at the top impedes the infiltration of rain water in the shallow aquifer.

There are about 60 rainy days in a year, but 70% of the precipitation occurs in 15 rainy days. These factors have to be taken into account in designing the rainwater harvesting system. If this potential can be tapped even to the extent of 30%, it can reduce the demand on groundwater and prevent seawater ingress.

A geoelectrical and hydrogeological investigation of the Estate was made to start with, in order to understand the disposition of the groundwater, and to help in deciding upon the location of the recharge pits. Three recharge pits were constructed with a dimension of 3 m × 2 m × 2 m and with a hand bore to a depth of ∼2 m from the bottom of the recharge pit. The recharge pit was filled with gravel and sand in alternating layers and topped with charcoal at depth of about 0.6 m. below the ground level. It is covered with sand and gravel leaving a freeboard of 500 mm below the ground level. The cost per recharge pit worked out to Rs.10,000/–(∼USD 200) per pit. These recharge pits are meant to increase the infiltration rate of the rainwater through the soil strata to recharge the potential under ground aquifer. The recharge pits not only increase the sub-surface storage of fresh water but also reduce the brackishness of the groundwater. After leveling the collection area with an inward slope from North to South and from East to West, channels be dug along the bunds to make the final collection point of surface storage of water at southwest corner of the collecting zone.

19.6.2 Economic viability

There are about 60 rainy days in a year, but 70% of the precipitation occurs in 15 rainy days. These factors have to be taken into account in designing the rainwater harvesting system. If this potential can be tapped even to the extent of 30% alone, it can reduce the demand on groundwater and prevent seawater ingress. About 3000 m^3 of runoff was harvested during the rainy days (Sept '95–Nov '96) by constructing harvesting structures viz. roof water collecting system, recharge pits and surface water collecting channels. At the then market price of Rs. 33 per kL of water, the market value of the harvested water per annum works out to Rs.112,500/–(∼USD 2250). During the last two rainy days (June 13th–14th 1996) roughly around 924 m^3 has been collected. Since the construction cost of the harvesting structures is about Rs.200,000 (∼USD 4000), the costs of construction can be recovered in about two years. An additional benefit is the improvement in the quality of groundwater.

19.6.3 Water Quality improvement

The water harvesting work commenced during the month of April '95. That there was discernible improvement in the quality of groundwater after the construction of the recharge pits is evident from a comparison of the physico-chemical characteristics of the water in the borewell in the Estate, before and after the construction of recharge pits.

It may be noted that after the construction of recharge pits, the TDS (mg/l) of groundwater decreased markedly from 2490 from 398.

Observation from the tube well shows an increase in the height of water table from 9 feet (∼3 m) below G. L during summer season to about 4 feet (∼1.3 m) below G.L.

Table 19.1 Comparative study of the physico-chemical parameters of bore well water before and after the construction of recharge pits.

S. No.	Parameters	Before construction June'93	After construction of recharge pits		
			Sep '95	Jan '96	Nov '96
1.	E.C.	4180.0	1820.0	630.0	600.0
2.	pH	7.8	8.2	7.9	7.6
3.	Calcium (mg/l)	12.0	48.0	24.0	24
4.	Magnesium (mg/l)	24.0	41.0	20.0	18
5.	Sodium (mg/l)	902.0	274.0	81.0	79
6.	Potassium (mg/l)	1.0	19.0	2.0	2.0
7.	Bicarbonate (mg/l)	647.0	249.0	244.0	198
8.	Carbonate (mg/l)	0.0	0.0	0.0	0.0
9.	Sulphate (mg/l)	470.0	257.0	66.0	60.0
10.	Chloride (mg/l)	738.0	231.0	50.0	48
11.	Nitrates (mg/l)	0.0	2.5	1.0	1.0
12.	Total dissolved salts(mg/l)	2490.0	1063.0	403.0	398
13.	Total Hardness (mg/l)	130.0	290.0	140.0	123

during monsoon season. This may be due to increase in sub-surface storage level which after a period few years can be utilized for tapping good quality water from the ground.

19.6.4 Other benefits of rainwater harvesting – Fisheries

About 2,000 fingerlings of Catla, Rohu, Common carp and Grass carp was stocked in the water collected in the surface storage system on 20/11/96. The water was turbid at the level of 1500 (1.5 cm Secchi disc transparency Method). The stocked fishes were monitored at intervals for testing their growth. Rice bran, groundnut cake, tapioca flour, *Cajanus cajan* husk, and black gram husk were used as feed for the fishes. The survival rate of Catla and Common carp was higher than of Rohu and Grass carp. On an average the depth of the water in the system during the culture period was about 30–60 cm. The total harvest (16th May '97) was 56.3 kgs at the end of sixth month. The weight of individual fishes harvested ranged from 200 grams to 725 grams.

The case study demonstrates that the harvesting of the rainwater improves the water quality of the groundwater, raises the water-table and is economically viable.

References

Agarwal, A., and Narain, S. (1997) (Ed.) *Dying Wisdom – Rise, fall and potential of India's traditional water harvesting systems*, Centre for Science and Environment, New Delhi.

Aswathanarayana, U. (1995) *Geoenvironment : An Introduction*. Rotterdam: A.A. Balkema.

Aswathanarayana, U. (2001) *Water Resources Management and the Environment*. Lisse (The Netherlands): A.A. Balkema.

Mukundan, T.M. (2005) *The Ery Systems of South India – Traditional Water Harvesting*, Akash Ganga Trust, Chennai, India.

Raghavan, S. (2004) *Rainwater Harvesting in Urban Areas – The Chennai Experience*, Arid Lands Newsletter 56, University of Arizona, Tucson, USA.

Raghavan, S. (2005) *Rainwater Harvesting in India with Special Reference to Urban Areas and the Chennai Experience*, Proceedings of the Tokyo-Asia Pacific Skywater Forum, People for Rainwater, Tokyo, Japan.

Raghavan, S. (2006) *Rainwater Harvesting – Need, Relevance and Importance of Groundwater Recharge in Urban Areas with Particular Reference to Coastal Cities*, Proceedings of the International Workshop on Rainwater Harvesting, Kandy, Ministry of Urban Development and Water Supply, Government of Srilanka.

UNEP (1983) *Rain and Stormwater Harvesting in Rural areas*. Dublin: Tycooly.

Chapter 20

Techno-socio-economic dimensions of food policy in India

M.C. Varshneya & R.H. Patel

Anand Agricultural University, Anand, Gujarat, India

20.1 Current status of food security in India

The food policy and agricultural development strategy adopted by India to improve food security situation paid rich dividends, and the ensuing improvements in food security can be assessed from several angles. The most significant change was the increase in the domestic output of food grains, particularly cereals (Table 20.1). The production of cereals increased from 72.1 million tonnes during the triennium ending (TE) 1964/65 to 130.2 million tonnes during TE 1984/85 and further to 186.4 million tonnes during TE 2003/4. Increase in the production of staple food (cereals) has kept pace with the population growth. Per capita net output of cereals, which had increased from 110.4 kg in 1951 to 130.9 kg in 1964, went up further to 166.1 kg in 1984 and has hovered around that level for the last 20 years.

The long-term growth rate of all cereals, which was 2.61 per cent per annum over the period 1967/68 to 1980/1, and 2.77 per cent per annum over 1967/68 to 2001/02, has exceeded the Indian rate of population growth (Table 20.2). Owing to the increase achieved in the production of cereals, the dependence on imports for meeting the staple food needs of the population dropped considerably. Net imports as a percentage of net domestic output had increased to unprecedented levels during the mid-1960s. For example, in 1966 the net import of cereals at 10.3 million tonnes represented 19 per cent of net production. Reviewed on quinquennial basis, cereal imports totaled 8.2 per cent of net output during 1961–65 and 9.6 per cent during 1966–70, declining to 4.4 per cent of net production during 1971–75, 1.5 per cent during 1981–85 and only 0.4 per cent during 1986–90. Since then, India has become a net exporter, accounting for

Table 20.1 Production of food grains in India (in million tonnes).

Period	Cereals				Pulses	Total food grains
	Rice	Wheat	Coarse	Total		
TE 1951/2	21.8	6.3	16.1	44.2	8.3	52.5
TE 1964/5	36.5	11.0	24.6	72.1	11.3	83.4
TE 1974/5	41.0	23.5	26.0	90.5	10.0	100.5
TE 1984/5	55.2	44.1	30.9	130.2	12.2	142.4
TE 1994/5	78.1	60.8	32.6	171.5	13.4	184.9
TE 2003/4	84.3	70.0	32.1	186.1	13.2	199.3

Note: TE = Triennium ending.
Source: Acharya (2002a); Govt. of India (GoI) (2004; 2004/5).

Table 20.2 Total growth rates of production of cereals in India.

Period	Rice	Wheat	Coarse cereals	All cereals
1949/50 to 1964/50	3.50	3.98	2.25	3.21
1967/68 to 1980/81	2.22	5.65	0.67	2.61
1980/81 to 1989/90	3.62	3.57	0.40	3.03
1990/91 to 1999/2000	1.90	3.81	1.48	2.10
1957/58 to 2001/02	2.78	4.34	0.54	2.77

Source: Gol (1999,2000,2000a, 2003).

0.1 per cent during 1991–95, 1.3 per cent during 1996–2000 and 4.0 per cent during 2001–03 of net cereal production.

In addition to the increase in domestic cereal production, the inter-year instability in production was reduced considerably. This happened for two reasons. First, the irrigated area under cereals expanded considerably, reducing the dependency on uncertain rainfalls. Out of total cereal area, irrigated areas increased from 23.1 per cent in 1964/65 to 50.1 percent by 2000/1. And second, the share of more stable grains (wheat) increased while unstable grains (coarse cereals) decreased. Wheat had accounted for 15.2 per cent of total cereals in TE 1964/65, increasing to 37.6 per cent in TE 2003/04. On the other hand, the share of coarse cereals declined from 34.1 per cent to 17.2 per cent during this period.

Another noteworthy feature of India's advancements in macro food security is that 97.4 per cent of the incremental output of cereals between TE 1964/5 and TE 2003/04 were due to improvements in the per hectare productivity (yield); area expansion accounted for only 2.6 per cent. For example, during this period, the area under cereals increased from 93.7 million hectare to 97.3 million hectare and the average yield per hectare went up from 770 kg during TE 1964/65 to 1,946 kg during TE 2003/04. The improvement in yield resulted from advancements in technology, irrigation, and the diversion of low-yielding crops to high value produce. There has been considerable improvement in the physical access to food in different parts of the country, helped by several initiatives and measures. First, the share of rice, the production of which is more geographically dispersed, has continued to be quite considerable. Rice contributed 42 per cent of the increase of 114.3 million ton in cereal production between TE 1964/65 and TE 2003/04. Moreover, rice itself became geographically more dispersed. Second, the network of public distribution system was expanded, enabling food grains to reach the deficit, geographically difficult regions (hilly or desert) and tribal dominated areas. Finally, systematic measures to expand food-marketing infrastructure increased physical access to food. These included the creation of market yards in rural areas, storage and warehousing facilities, expansion of the road network, transport and communication facilities, and incentives to promote food processing and packaging industries.

Yet another important development has been the continuous improvement in the economic access of consumers to food. The increase in retail prices of the two staple food items (rice and wheat) has been lower than the increase in per capita income, and thus the proportion of consumer income required to buy a unit quantity of rice or wheat has continued to decline. For example, the proportion of annual per capita

Table 20.3 Production of other food products in India.

Items	Total production (million tonnes)			Per capita production (kg p.a)	
	1980/81	2003/04	CGR % p.a	1981(688.5)[f]	2002 (1050.6)[f]
Edible oilseeds/oil[a]	9.4	25.1	4.36	6.4	8.6
Fruits	23.8[b]	47.7[c]	3.07	34.5	45.4
Vegetables	45.4[b]	97.5[c]	4.33	65.9	92.8
Spices	1.4	2.9[d]	3.91	2.0	2.8
Sugar	5.1	19.9	6.10	8.2	16.2
Milk	31.6	88.1	4.56	45.9	82.2
Eggs[e]	10.1	40.4	6.21	14.7	39.7
Meat	0.8	4.9[c]	8.59	1.2	4.7
Fish	2.4	6.4	4.36	3.5	5.9

Notes: (a) Production of oilseeds and per capita production of edible oils; (b) Pertains to 1984/5; (c) Pertains to 2002/3; (d) Pertains to 1999/2000; (e) Production in billions and per capita production in number; (f) Population in millions.
Source: GoI (2003/04,2004,2004/05); Singhal (2003).

income needed in the rural areas to purchase a quintal of wheat has declined from 15.4 per cent in 1973/74 to 8.7 per cent in 1983/84, 5.9 per cent in 1990/91, 5.0 per cent in 1994/95 and finally to 4.4 per cent in 1999/2000. A similar declining trend is noticed for urban communities, as well as in the case of rice for both rural and urban areas (Acharya, 1997, 2002a, 2004a).

A related development needs to be mentioned: in addition to the expansion in the availability of cereals and the decline in their relative prices vis-à-vis incomes, the per capita consumption of cereals has also tended to drop in recent years (Dev, 2003), going down from 173.6 kg per year in 1987/8 to 160.8 kg in 1993/4 and further to 152.6 kg in 1999/2000. The decline in consumption has been sharper in coarse cereals, and has occurred even among the lowest 30 per cent of consumers, reflecting a shift towards more nutritive foods like fruits, vegetables and livestock products. Long-term data from National Sample Survey Organization also indicate a declining trend in the per capita consumption of cereals in both rural and urban areas from the early 1970s to 1999/2000, accompanied by a decrease in the proportion of expenditures on cereals and an increase of that on milk, meat, eggs, fruits and vegetables (Selvarajan and Ravishankar, 1996; Dev, 2003).

Improved availability of staple food at declining real prices has contributed to improved nutritional security. Farmers have shifted from the low-yielding coarse cereals to non-cereal food products since the middle of the 1980s (Acharya, 2003a), a fact which has inter alia helped to increase production and availability of edible oils, sugar, fruits, vegetables, spices, milk, eggs, meat and fish/fish products. During the last two decades, the output of fruits and spices increased at a total rate of 3.07 to 3.91 per cent per annum, while the production of vegetables, edible oilseeds, milk and fish recorded increase of 4.33 to 4.56 per cent per annum during this period. The annual production rates of sugar, eggs and meat were even higher: sugar increased at the rate of 6.10 per cent, eggs 6.21 per cent and meat 8.59 per cent during this period (Table 20.3). As the

Table 20.4 Incidence of poverty in India.

Year	Poverty ratio (%)			Number of poor (millions)		
	Rural	Urban	Total	Rural	Urban	Total
1977/78	53.1	45.2	51.3	264.3	64.6	328.9
1983/84	45.7	40.8	44.5	252.0	70.9	322.9
1987/88	39.1	38.2	38.9	231.9	75.2	307.1
1993/94	37.3	32.4	36.0	244.0	76.3	320.3
1999/2000	27.1	23.6	26.1	193.2	67.1	260.3
2007*	21.1	15.1	19.3	170.5	49.6	220.1

Note: *Projected.
Source: Gol (2003/04:2004).

production growth of all these food items was considerably higher than the population growth, per capita production of nutritive foods went up substantially in India.

In addition to the advancements made in macro food security, there has been considerable improvement in food availability, and a reduction of hunger at the household level. In the rural areas, which account for nearly three-fourths of the poor in India, the percentage of households reporting sufficient food availability every day throughout the year for all family members increased from 81.1 per cent in 1983 to 96.2 per cent in 2000. The percentage of households with at least one member not getting enough food daily during some months declined from 16.2 per cent in 1983 to 2.6 per cent in 2000, and the percentage of households with at least one member without sufficient daily food throughout the year came down from 2.4 per cent in 1983 to 0.7 per cent in 2000 (NSSO, 2001).

Economic poverty is an important factor affecting food security at the household level. Over the years, the incidence of both rural and urban poverty has declined considerably (Table 20.4). The percentage of population below the poverty line declined from 51.3 per cent in 1977/78 to 38.9 per cent in 1987/88 and finally to 26.1 per cent in 1999/2000 which, according to some scholars, may have been even lower (Bhalla, 2003). However, the absolute number of poor or food-insecure people continues to be large. Another disquieting aspect of food security is nutritional status, particularly with regard to children and women. Based on the reports of National Nutritional Monitoring Bureau (NNMB), Radhakrishna and Ravi (2004) have observed that 47.7 per cent of the children (under 3 years) still suffer from malnutrition and the incidence of child malnutrition is higher in the rural areas. Even among adults, the incidence of chronic energy deficiency (CED) is quite high, with 37.4 per cent of males and 39.4 per cent of females suffering from CED in 2000/01.

20.2 Agricultural price support policy

20.2.1 *Policy objectives and framework*

Price support for foodgrain producers has been an important instrument of the agricultural and food policy pursued by India since the mid-1960s. These instruments included minimum support prices, subsidized farm inputs, food marketing system improvements, and direct food assistance and employment generation programmes. The broad policy framework was initially outlined in the terms of reference of the Agricultural Price Commission (APC), set up in 1965 to advise the government on a regular basis for developing a balanced and integrated price structure. In formulating price policy, the APC was to recognize, on the one hand, the need to provide incentives to farmers for adopting new technology and maximizing production, and on the other hand, to the likely effect of price policy on the cost of living, wage levels and industrial cost structure. In 1980, when the demand and supply of food grains appeared to be in balance, the framework of the policy was modified. The emphasis of the APC policy (later renamed Commission for Agricultural Costs and Prices, CACP), shifted from maximizing the production of food grains to developing a production pattern consistent with the country's overall economic needs. The issue among farmers and consumers of a fair split of the gains accruing from technology and public investment was also explicitly recognized, and CACP was to monitor the terms of trade for the agricultural sector. Policy was reviewed again in 1986 when its long-term perspective was emphasized. This implied that in order to make the farm sector more vibrant, productive and cost effective (Acharya, 2004b), policy should be extended to major factors, which would influence agricultural prices in the long term.

20.2.2 Salient features of support prices

For almost 20 years until 1991, the distinction between the support price and procurement price of wheat and rice (paddy) was blurred. Each year, the procurement prices were announced and procurement targets fixed. To meet the procurement targets, government imposed movement restrictions in the surplus-producing areas, which was a disincentive for the farmers. Procurement targets also affected farmers' price support; once the target quantity had been procured, public agencies exited the market and farmers were left without support price for their produce. Thus, target-based procurement was becoming an obstacle to the support prices of rice and wheat growers in major producing areas. Therefore, in 1991, the government decided to eliminate the system in favor of fixed minimum support prices only, also for rice and wheat (as was being done for other farm products).

Currently, minimum support prices in the nature of a price guarantee to farmers, are applied by the Indian government to 25 farm products, which include paddy, wheat, maize, pearl millet, sorghum, ragi, barley, chickpea, pigeon pea, black gram, green gram, lentils, groundnut, mustard, sesame, soybean, sunflower, safflower, Niger and dried coconut. If market prices fall below the support level, government agencies buy the quantities offered at support prices, but the support is linked to specified quality standards. Farmers also have the option to sell in the open market. On the other hand, while there is no obligation on the part of farmers to sell to government agencies, these are bound to buy all quantity offered by the farmers at guaranteed prices.

Determination of the support levels is governed by the cost of production; changes in input prices; input-output price parity; trends in market prices; emerging demand and supply situation; inter-crop price parity; effect on the cost of living; effect on general price level; effect on industrial cost structure; international price situation; and parity between prices paid and prices received by the farmers (terms of trade). Support prices, usually announced at the time of sowing, are fixed for the year and applied

Table 20.5 Share of government purchases in total output of rice (in million tonnes).

Particulars/ period	Total production	Marketed surplus	Purchased by		% Share in total marketed surplus	
			Public agencies	Private trade	Public agencies	Private trade
Wheat						
TE 1992/93	53.6	28.1	8.4	19.7	29.9	70.1
TE 2002/03	72.9	53.4	18.7	34.7	35.0	65.0
% Increase	36.0	90.0	122.6	76.1	–	–
Rice						
TE 1992/93	73.9	31.6	11.7	19.9	37.0	63.0
TE 2002/03	83.7	61.6	19.9	41.7	32.3	67.7
% Increase	13.3	94.9	70.1	109.5	–	–
Total						
TE 1992/93	127.5	59.7	20.1	39.6	33.7	66.3
TE 2002/03	156.6	115.0	38.6	76.4	33.6	66.4
% Increase	22.8	92.6	92.0	92.9	–	–

Notes: for the TE 1992/93, MS to out put ratio was taken as 42.7 per cent for rice and 52.4 per cent for wheat (Acharya, 2004b). For the TE 2002/03, the ratio was taken as 73.6 per cent for rice and 73.3 per cent for wheat (GoI, 2004).

across all areas of the country. Inter-year changes in support prices are essentially non-negative (medium-term guarantee to farmers). A central nodal agency is designated for each commodity or group of commodities to undertake support operations, and for cereals this agency is the Food Corporation of India (FCI) and for pulses and oilseeds, the National Agricultural Cooperative Marketing Federation (NAFED). The resulting stockpiles are used in various ways. For example, on the part of rice and wheat, these are used for (i) meeting the requirements of the public distribution system and food assistance programmes; (ii) creating buffer stocks to even out inter-year fluctuations in supplies and prices; and (iii) open market operations, including supplies to flourmills and exporters.

20.2.3 Public-private share in grain trade

Farmers have the option of selling to private traders or to public agencies, but generally deal with the public agencies only when the support price is more favorable than that offered by private traders. In the last 10 years, there has been positive development in the level of price support operations of two major cereals, viz., rice and wheat, with price support purchases increasing from 20.1 million tonnes during TE 1992/93 to 38.6 million tonnes during TE 2002/03. The increase was more pronounced for wheat (Table 20.5). Public agencies purchased greater quantities, but so did private trade and these handled 76.4 million ton of rice and wheat during TE 2002/03 compared to 39.6 million ton a decade earlier. If viewed in terms of the total market surplus of these staple food grains, there was no change in percentage shares of purchases made by the public and private sectors.

20.2.4 *Impact and issues*

An assessment of the impact of price support policies should be based on the achievement of specified objectives, incentives or disincentives to farmers as well as the distortions, if any, created in the marketing system. The policy has been instrumental in reducing price uncertainty for farmers, thus inducing them to adopt new technology and thereby increase the output of food grains and attain macro food security. The price support programme, in conjunction with other policy instruments, has helped to improve physical and economic access to food. Despite these positive impacts of the price support policy, certain other issues have been raised and debated.

The first issue is the level of support prices: farmers consider the level inadequate, but consumer groups feel differently. The conflicting interests were reconciled with complementary instruments of input subsidies on the one hand and distribution of subsidized food on the other. Further, the level of support prices is a political-economic decision, and the government has relied mostly on recommendations made by CACP, the autonomous expert body. Whenever the government has deviated substantially from their recommendations, distortions emerge. For example, during 1999–2002, the government fixed minimum support prices (MSPs) of rice and wheat at levels much higher than recommended by CACP (Acharya and Jogi, 2003). This lack of prudence led not only to excessive stocks, but also increased public cost for the food grain policy. Apart from undue hikes in the levels of MSPs for rice and wheat, relaxation of the fair-average quality norms, inappropriate timing of price rises for grains for the public distribution system (PDS), and improper meshing of export-import policy contributed to the accumulation of government stocks in 2002. To overcome similar difficulties, there have been suggestions to declare CACP a statutory body.

The second issue relates to operational incentives for the private sector's grain trade. The often-cited example is the fact that the intra-year price rise for rice and wheat has been considerably lower than storage costs, discouraging private sector investment in storage and trading activities. Private sector involvement in the food grain trade continues to be large, and a curb on intra-year price rise has benefited both food grain producers and consumers. Petty traders, who generally operate in short-term markets, have not been adversely affected, and it is most likely the large food grain traders or trading giants, who cannot operate profitably. The question that arises is whether a country that is facing serious food shortages should prioritize its concern on farmers, consumers and petty traders, or on the large-scale trading companies.

The third issue relates to the efficiency of the Food Corporation of India vis-à-vis private trade in price support operations and subsequent distribution of food grains. The efficiency of FCI has been questioned on the ground of its economic cost and subsequent outlay on food subsidies. Several aspects of FCI's operations need to be noted. First, both the purchase price (support price) and the issue price are determined by the government. Second, around 70 per cent of FCI's total expenditures for procurement and distribution are spent on items over which it has no control (Acharya, 1997; GoI, 2002b). The same costs would also have to be incurred by private trade unless it is able to evade some of the statutory taxes/charges (Acharya, 1997). Third, losses occurring during storage and transit are estimated to be around one per cent, which, in comparison to private channels, is not unduly high. Fourth, FCI's establishment charges and administrative overhead are estimated to be 2.8 per cent of its economic costs, and thus are no higher than private-trade net margins. Fifth, a recent study commissioned by the Union Ministry of Consumer Affairs, Food and Public Distribution (Chand, 2003) has pointed out that in order to encourage the private sector to purchase wheat and paddy from the markets of surplus-producing states, retail prices in deficit states during lean months should be approximately at the same level as FCI's economic costs on wheat/rice.

Another important issue relates to the problem of ineffective implementation of price support operations for rice and wheat in certain states where, despite surplus yields of the last decade, farmers cannot get the minimum support prices. This situation has evolved mainly because the nodal agency (FCI) and state agencies in these new surplus states are not fully geared to undertake price support operations, as FCI continues to focus on large volume purchases from the traditional surplus-producing states. Decentralization of procurement and a refocusing of FCI operations towards the non-traditional cereal states, measures that are being currently pursued, may help in this regard.

20.3 Farm input subsidies

Farm input subsidies were used in conjunction with support prices to reconcile the conflicting interests of food grain producers and consumers. Input subsidies in India's agricultural sector constitute (i) direct or explicit subsidies, and (ii) indirect or implicit subsidies. Direct subsidies are payments to farmers intended to cover a part of the cost of inputs or equipment (improved seeds, plant protection equipment and improved farm implements). Direct subsidies are provided to well-defined target groups such as small or marginal farmers or those belonging to scheduled castes or tribes. The implicit or indirect subsidies arise as a result of the pricing policy for certain inputs such as fertilizers, electricity and canal water. There is no direct payment to producers, but as the inputs are supplied at less than the cost of production or supply, this amounts to an implicit subsidization of the input for the farmers. Implicit or indirect subsidies on fertilizers, electricity for irrigation and canal irrigation water account for more than 99 per cent of total subsidies in Indian agriculture.

Major input subsidies to the country's agriculture sector at current prices were estimated to total Rs 14.1 billion in 1980/81, Rs 114.6 billion in 1990/91 and Rs 504.4 billion in 2000/01, or in dollar terms for 2000/01, US$11.2 billion (Table 20.6). At constant (1993/94) prices, the input subsidies increased at the rate of 13.5 per cent per annum during the 1980s and 7.0 per cent per annum during the 1990s. As a proportion of agricultural GDP, farm input subsidies accounted for 2.6 per cent in 1980/81, 6.4 per cent in 1990/91 and 10 per cent in 2000/01.

Agriculture in India is basically smallholder farming. According to the 1995/6 agricultural censuses, there are 115.6 million farm holdings with an average operational area of 1.41 hectares. Nearly 62 per cent (71.2 million) of these holdings operate on less than one hectare (average 0.4 hectares) and 19 per cent (21.6 million) on 1–2 hectares of land (average 1.42 hectares) (Table 20.7). Marginal and small farmers account for 36.4 per cent of total subsidies, and this is slightly higher than their share (36 per cent) of total cultivated area. In contrast, the so-called large farms (average cultivated area 17.2 hectares) accounted for 11.8 per cent of total subsidies, which is lower than their share (14.8 per cent) of cultivated area. The average subsidy of

Table 20.6 Major input subsidies in Indian agriculture.

Particulars	1980/81		1990/91		2000/01	
	Current prices	Constant prices (1993/94)	Current prices	Constant prices (1993/94)	Current prices	Constant prices (1993/94)
Subsidy (Rs billion)						
Fertilizer	5.0	15.4	43.9	59.6	138.0	83.9
Electricity	3.3	10.2	46.0	62.5	269.5	163.9
Canal water	5.8	17.9	24.7	33.5	96.9	58.9
Total	14.1	43.5	114.6	155.6	504.4	306.7
Subsidy (billion US$)*	0.3	0.9	2.5	3.4	11.2	6.8
Subsidy as % of GNP	1.1	–	2.3	–	2.7	–
Subsidy as % of GDP ag.	2.6	–	6.4	–	10.0	–

Note: * Rs 45 per US$.
Source: Acharya and Jogi (2004a; 2004b).

Table 20.7 Farm input subsidies according to farm size, 2000–01.

Size	Total subsidy (US$ billion, Current price)	No. of farms (in millions)	Subsidy per farm (US$)	Subsidy per ha. of operated area (US$)	% of total operational area	% of total subsidy
Marginal (less than 1 ha.)	1.9	71.2	26.7	68.9	17.2	17.3
Small (1–2 ha.)	2.1	21.6	97.2	69.7	18.8	19.1
Semi-medium (2–4 ha.)	2.8	14.3	195.8	70.8	23.8	24.6
Medium (4–10 ha.)	3.1	7.1	436.6	73.8	25.3	27.2
Large (above 10 ha.)	1.3	1.4	928.6	54.6	14.8	11.8
Total	11.2	115.6	96.9	68.6	100.0	100.0

Source: Computed from Acharya and Jogi (2004b).

US$68.60 per hectare of cultivated area applied across the different farm-size groups, except for large size farms where it was considerably lower. In access to fertilizers, electricity or canal water, there is no preferential treatment with regard to size of the farm.

Input subsidies are mainly focused on food crops (Table 20.8), constituting as much as 95.6 per cent of the total. Breakdown by different crops was rice 32.1 per cent,

Table 20.8 Crop-wise subsidy in Indian agriculture, 2000–01.

Crops	% of total subsidy	Subsidy per ha. (Current price)		Subsidy per ton of output	
		Rupees	US$	Rupees	US$
Rice	32.1	3587	79.7	189	4.2
Wheat	27.5	5039	112.0	186	4.1
Gram	1.8	1495	33.2	201	4.5
Groundnut	2.5	1827	40.6	187	4.2
Mustard	4.0	3306	73.5	354	7.9
Sugarcane	5.1	6099	135.5	90	2.0
All food crops	95.6	2661	59.1	NE	NE
Cotton	4.4	2573	57.2	451	10.0
Total	100.0	2658	59.1	NE	NE

Note: NE = Not estimated.
Source: Acharya and Jogi (2004b).

wheat 27.5, sugarcane 5.1, mustard 4.0, groundnut 2.5 and gram (chickpea) 1.8 per cent. The subsidy was, on average, US$59 per hectare of cropped area, but varied from around US$33 for gram to US$136 for sugarcane. If subsidy is compared to output, it was close to US$4 per tonne for wheat and rice, around US$8 for mustard, and only US$2 per tonne for sugar.

Farm input subsidies, particularly their rising levels, have remained one of the most debated aspects of the agricultural policy since the launch of economic reforms in 1991. However, the withdrawal of subsidies has been cautious and gradual for several factors. Input subsidies have been considered not only from the fiscal perspective of the electricity or irrigation department, but more importantly, their overall role in food security and agricultural development of the country has been recognized. Input subsidies have enabled the country to improve its food security and to keep food prices low, improving access to food for the population, while providing reasonable returns to farmers. Furthermore, the subsidies are not net transfers to the farmers. For example, of the total fertilizer subsidies, farmers receive an estimated 62 per cent and 38 per cent goes to the industry (GoI, 2004/05). A considerable portion of power and canal water subsidies is wasted because of inefficient production and distribution systems. Thus, the burden of subsidies can be reduced with more efficient production and distribution systems of key farm inputs. With better supply systems of electricity and canal water, farmers would be willing to pay higher user charges. As already mentioned, the share of marginal and small farmers in total input subsidies is quite significant. Their food security depends on self-production, and thus the option of compensating this group for increased user charges with higher support prices is not feasible, as the marketed surplus of the small and marginal farmers is either negligible or very low. Furthermore, many crops are not covered by the support policy. It is also being argued that subsidies on farm inputs cannot be seen in isolation of the multiple subsidies in other sectors of the economy, and consequently their withdrawal is less painful. Total subsidies in the union budget alone (un-recovered cost of non-public goods) were estimated at Rs 1158 billion in 2003/4 (GoI, 2004/05), including subsidies on LPG and kerosene.

20.4   Direct food and other assistance programmes

Apart from food production and agricultural development programmes, the problem of food insecurity and malnutrition at the household level was tackled through measures such as direct food assistance, wage employment, food-for-work, and some other welfare schemes. Historically, poor households in India have relied on traditional family and community-based mechanisms of social protection to cope with deprivation. However, the process of change has eroded many of these traditional systems. Therefore, the country's post-independence history of social development has highlighted food as a cornerstone of the national strategy to accord some measure of social protection to vulnerable citizens. India's development policy after independence has always had a niche for food-based anti-poverty and social protection programmes (Medrano, 2004). Once a fairly satisfactory situation with respect to macro food security had been achieved, even the Apex Court (Supreme Court) intervened with a series of directives to central and state governments for implementing programmes to eliminate hunger and malnutrition within a stipulated timeframe. The joint government-judiciary endeavor received an additional boost with the 'right to food' campaign launched by civil society and non-governmental organizations (NGOs). Government's food assistance and related intervention have sought to address food insecurity on three fronts. Chronic food insecurity is alleviated through subsidized food distribution, food-for-work, and employment generation programmes. Nutritional insecurity, primarily of pregnant and nursing women and children, is addressed through supplementary nutrition and school feeding programmes. Transitory food insecurity is covered with food assistance as part of disaster relief and long-term disaster preparedness and prevention programmes. Over the years, the nature of food assistance has changed considerably. In the 1960s, national assistance policies focused on food production, and food assistance was mainly intended to augment food availability. The goal was to target the entire population through undifferentiated generalized food distribution as well as to build food reserves. Over the years, food assistance has been motivated more by the need to alleviate poverty and hunger, and currently by the prevention of malnutrition. According to UNWFP (2002), food assistance strategy in India has moved from 'food for the nation' to 'food for the people' and most recently to 'food security for the vulnerable'.

Current direct food and other assistance programmes in India fall broadly into five groups, viz., (i) distribution of subsidized food grains; (ii) supplementary nutrition programmes for children and women; (iii) food-for work and wage employment programmes; (iv) self-employment augmentation programmes; and (v) welfare or social assistance programmes for specific vulnerable groups.

20.5   Strategies

To achieve food security, emphasis is to be laid on improved agricultural production but it is a multifaceted problem, therefore multi-pronged strategies will have to be developed to address the problem. Outlines of the aspects, which lead to be attached for overall improvement in food security, are listed bellow.

- Fast growth in GDP.
- High growth in employment.

- Priority to effective population and migration policies.
- Development and proper maintenance of rural infrastructures by unified investment plan by all the development departments.
- Accelerated investment in generation and transfer for agricultural technology and increased participation of private sector.
- Promoting cutting edge science like biotechnology, molecular biology, system analysis, information etc.
- Environmental protection and sustainable use of natural resources.
- Human resource development including farm-women and youth.
- Market development and appropriate trade liberalization.
- Adequate and timely inputs (*e.g.* credits, technical assistance, etc.) to farmers.
- Effective implementation of poverty alleviation programmes and targeted distribution of food grains.
- Making available rural housing, safe drinking water, primary health care and primary education.

References

Acharya, S. S. (1997) Agricultural Price Policy and Development: Some Facts and Emerging Issues. *Indian Journal of Agricultural Economics*, 52 (1): 1–47.

Acharya, S. S. (2002a). Food Security and New International Trade Agreement: Perspectives from India, In G. S. Bhalla, J. L. Racine and F. Landy (eds), *Agriculture and the World Trade Organization: Indian and French Perspectives*. Paris: Editions de la Maison des sciences de l'home, 151–78.

Acharya, S. S. (2004b) Agricultural Marketing and Rural Credit , In *Millennium Study of Indian Farmers*, v. 17. New Delhi: Academic Publishers for the Indian Ministry of Agriculture.

Acharya, S. S., and R. L. Jogi (2004a) Farm Input Subsidies in Indian Agriculture. *Agricultural Economics Research Review*, 17 (January-June): 11–41.

Acharya, S. S., and R. L. Jogi (2004b) Input Subsidies and Agriculture: Future Perspectives. Paper presented at the IRMA Silver Jubilee Workshop on Governance in Agriculture, 18–19 December.

Bhalla, S. S. (2003) Recounting the Poor: Poverty in India, 1983–99. *Economic and Political Weekly*, 38 (4): 338–49.

Chand, R. (2003) *Government Intervention in Foodgrain Markets in the New Context*. Policy Paper 19. Study report prepared for Ministry of Consumer Affairs, Food and Public Distribution. New Delhi: National Centre for Agricultural Economics and Policy Research.

Dev, S. M. (2003) *Right to Food in India*. Working Paper No. 50. Hyderabad: Centre for Economic and Social Studies.

Government of India (GoI) (1999) Agricultural Statistics at a Glance. New Delhi: Ministry of Agriculture.

GoI (2000) *Agricultural Statistics at a Glance*. New Delhi: Ministry of Agriculture.

GoI (2002a) *Agricultural Statistics at a Glance*. New Delhi: Ministry of Agriculture.

GoI (2002b) *Report of the High Level Committee on Long-term Grain Policy*. New Delhi: Department of Food and Public Distribution, Ministry of Consumer Affairs, Food and Public Distribution.

GoI (2002c) *Report of Inter-Ministerial Task Force on Agricultural Marketing Reforms*. New Delhi: Ministry of Agriculture.

GoI (2003) *Agricultural Statistics at a Glance*. New Delhi: Ministry of Agriculture.

GoI (2003/4) *The Economic Survey*. New Delhi: Ministry of Finance.

GoI (2004) *Agricultural Statistics at a Glance*. New Delhi: Ministry of Agriculture.

GoI (2004/05) *Agricultural Statistics at a Glance*. New Delhi: Ministry of Agriculture.

Medrano, P. (2004) *The Experience with Food Safety Nets in India*. Paper presented at National Food Security Summit, 4–5 February. New Delhi.

National Sample Survey Organization (NSSO) (2001) *Reported Adequacy of Food Intake in India 1999–2000*. Report No. 474 (55/1.0/7). New Delhi: Ministry of Statistics and Programme Implementation.

Radhakrishna, R., and C. Ravi (2004). Malnutrition in India: Trends and Determinants. *Economic and Political Weekly*, 39 (7): 671–6.

Selvarajan, S., and A. Ravishankar (1996) Foodgrain Production and Consumption Trends in India. *Agricultural Economics Research Review*, 9 (2): 142–55.

Singhal, V. (2003) *Indian Agriculture*. New Delhi: Indian Economic Data Research Centre.

United Nations World Food Programme (UNWFP) (2002). *Tackling Hunger: UNWFP's Efforts to Help Eliminate Food Insecurity in India: A Review of Strategic Actions*, New Delhi: UNWFP, 1–32.

Chapter 21

Overview and integration

U. Aswathanarayana (Editor)

Section 2: Socio-economic dimensions of food security

In Chap. 15, Prajapati dealt with the role that fermented foods could play in combating malnutrition. Fermentation is the oldest form of food preservation. In the ultimate analysis, fermentation is a process of bioconversion of organic substances by microorganisms and /or enzymes of microbial, plant or animal origin. According to FAO, fermented foods contribute to about one-third of the diet worldwide Cereals, pulses, root crops, vegetables, fruits, milk, meat and fish can be preserved by one or other method of fermentation. Fermentation improves the nutritional and therapeutic value and digestibility of food items. The following R. &.D areas need to be pursued in order to promote greater utilisation of fermented foods: (1) Identification of new raw materials that can be fermented (e.g. ways of making bread from coarse grains, like sorghum and pearl millet), (2) Understanding and optimizing the biochemical and physicochemical mechanisms of fermentation, (3) Preparation of library of known and biogenetically alterable microorganisms and enzymes to customize groups of them for fermentation of specific substances for specific fermentation protocols, (5) Assessment of nutritional improvements, (6) Clinical trials to evaluate therapeutic properties, and (6) Technology for preservation, packaging and distribution of fermented foods.

In Chap. 16, Vepa traces the changing food basket in India and its dietary implications. Food consumption patterns depend upon a variety of factors, such as changes in tastes and preferences, affordability and availability. Food basket is definitely changing for the country as a whole. Cereal consumption declined sharply with affluence. This is a clear indication of increasing affordability of other foods by the urban middle class and rich in both rural and urban areas. The consumption of cereals is high in poorer, less urbanised States, such as Bihar, Orissa, Madhya Pradesh and Uttar Pradesh, where local availability of cereals is good. There are reasons to believe that consumption out of homegrown produce, contributes to poverty reduction and higher cereal intake. Higher output subsidy going to wheat than rice in production has made wheat more profitable to grow than rice. Rice and pulses combination in the rice eating states leads to a better protein amino acid combination and easier digestibility of superior rice protein. Processed foods have become more important in the urban diets than rural diets. The calorie intake of almost all the states was above 2200 kilo calories per consumer unit per day. The consumption of micronutrients in the rural areas was found to be inadequate for all the classes, especially in the case of Calcium, Iron, Beta-carotene, Riboflavin and Zinc. In the average urban diets of the country Zinc, Vitamin 'C' Vitamin 'A', Riboflavin, Calcium, Iron, thiamine and Niacin in that order are the most deficient micronutrients. More than iron, zinc and vitamin 'C' deficiency are seen in all the states.

In Chap. 17, Lundquist traces the implications for freshwater resources arising from the changes in food consumption trends. For several decades, the availability of food in the world has improved, both in terms of total production and on a per capita basis. The number of undernourished has steadily, albeit slowly, been reduced. Recently, some disturbing shifts in these trends have occurred. One is that the gradual decrease in the number of undernourished has been reversed. Another problem is a rapid increase in the number of overweight and people suffering from obesity. In addition, a comparatively large fraction of the food produced is lost or wasted in the food chain, from the field to the end user. As a result of demographic trends and an increase in purchasing power, the demand for food, among other things, will increase and the composition of the diet will change. Water intensive food items are likely to be increasingly demanded with a commensurate increase in the pressure and inter-sector competition on land and water resources. In this new context, consumption patterns and resource and environmental implications are major issues. Efforts to achieve food and nutrition security need to address a triple challenge: (i) improvements in productivity in the use of water and land resources, (ii) reduction of losses and wastage in the food chain, i.e. from field to consumption, and (iii) promotion of a food intake that enables people to lead 'an active, healthy and productive life'.

Rigterink (Chap. 18) explains the methodologies of using micro-enterprises to address food security. Micro-enterprises can contribute to the food security and income of small-scale farmers. One view is that the primary goal of small-scale farming should be growing of enough food for the family of the farmer. Another view is that growing of cash crops is to be promoted, as this will not only enhance the income of the farmer, but would also introduce the farmer to modern methods of agriculture. Poor farmers in developing countries did not get the benefits of the Green Revolution as they did not have access to the needed inputs (high-quality seeds, irrigation water, fertilizers, pesticides, etc.). A micro-enterprise needs to be operated on the same business principles as is the case with large enterprises. The chapter provides the business planning required so that small-scale farmers and micro-enterprises jointly can become productive. Special attention is devoted to ways of developing the supply chains and training materials in arid and tropical regions, where most of the subsistence farmers live.

In Chap. 19, Raghavan and Parasuraman provide a case history of water harvesting in Chennai (Madras), India. India has a long tradition of rainwater harvesting - for instance, reservoirs to collect monsoon runoff have been found in the Indus Valley civilization sites. Riverfed and rainfed tanks for harvesting rainwater have been known in South India since 4th Century A.D. These were primarily meant for irrigation and most of them were inter-connected such that the overflow of one will go to fill up the neighbouring one at the lower level and so on. Several of them continue to be in use, but most of them got silted up due to lack of maintenance. The chapter describes the strategies for rainwater harvesting at macro and micro levels in rural and urban areas. The heavy rains during October to December 2005 in Chennai (Madras) flooded the city, but resulted in the groundwater level rising markedly in the entire city. Flooding and salt water incursion can be avoided, and the groundwater level improved by a proper system of rainwater harvesting in coastal cities. The Chapter provides a case history of rainwater harvesting in MSSRF Estate, Chennai. Construction of recharge pits in the Estate raised the water-table and improved the quality of groundwater. Also, fish were grown in the surface water ponds. Based on the market value of the harvested water, the construction costs of recharge pits were recovered in a matter of two years.

In Chap. 20, Varshneya and Patel delineate the techno-socio-economic dimensions of food policy in India. During the last forty years, the food security situation in India improved markedly. Since 1990, India became self-sufficient in food grain production, and became a net exporter. The general improvement in incomes led to reduced consumption of cereals and increased consumption of eggs, meat and fruit. Improved availability of staple food at declining real prices has contributed to improved nutritional security. Price support for food grain producers has been an important instrument of the agricultural and food policy pursued by India since the mid-1960s. These instruments included minimum support prices, subsidized farm inputs, food marketing system improvements, and direct food assistance and employment generation programmes. The policy has been instrumental in reducing price uncertainty for farmers, thus inducing them to adopt new technology and thereby increase the output of food grains and attain macro food security. Farm input subsidies were used in conjunction with support prices to reconcile the conflicting interests of food grain producers and consumers. Food security at the aggregate level may not translate into food security at family level – for instance, one in five Indians is food-insecure. Current direct food and other assistance programmes in India fall broadly into five groups, viz., (i) distribution of subsidized food grains; (ii) supplementary nutrition programmes for children and women; (iii) food-for work and wage employment programmes; (iv) self-employment augmentation programmes; and (v) welfare or social assistance programmes for specific vulnerable groups.

Governance of Food Security in Different Agroclimatic and Socioeconomic Settings

Chapter 22

Role of knowledge in achieving food security in India

P.M. Bhargava & Chandana Chakrabarti*

ANVESHNA, "Furqan Cottage", Tarnaka, Hyderabad, India*

22.1 Introduction

A country can consider itself having achieved food security when it has adequate and sustainable agricultural production, its agricultural production keeps pace with its increase in population, when food grains are available at affordable prices, and food is accessible to its poor.

Statistics of 2005 show that, in India,

- 200 million are underfed (roughly one-fifth of its population)
- 50 million are on the brink of starvation
- more than 38% children under three years in the cities are underweight
- more than 35% children in the cities are stunted
- 21% live in slums
- 8% urban dwellers have no safe drinking water
- <30% BPL families have access to the public distribution system
- 80–90% farmers are in debt.

The above facts are a clear pointer to the fact that India is today far from being food-secure.

Food security remains a far-fetched dream in many parts of the world, for instance, in several African countries. We also know that, the world over, famines have <u>not</u> become events of the past.

Farmers are the primary food producers in our country. Unless and until the needs of the farmers are adequately met, there can be no sustainable food security in India. Therefore, food security can be achieved only through farmers' security. And farmers' security would mean that farmers' welfare be given the top-most priority and all their needs met, such as,

- Seeds, Agrochemicals, Pest management, Water, Power, Connectivity, Credit
- Storage, Marketing, Fair Price
- Rights
- Value addition through post-harvest technologies
- Knowledge (General, Specific, New).

In fact the last item could cover all the preceding ones. We will, therefore, lay emphasis on knowledge.

22.2 Knowledge-based society

A knowledge-based *society* is not the same as a knowledge-based *economy* and the former is not necessarily implied by the latter. While a knowledge-based economy can be exploitative, a knowledge-based society is not easily exploited. A knowledge-based society inevitably leads to a knowledge-based economy but not vice-versa. So far around the world, emphasis has largely been on knowledge-based economies and rarely on knowledge-based societies.

Today, knowledge dissemination, not only in the developing countries but also in the developed countries is extremely uneven. This unevenness provides, perhaps, the largest base for exploitation. Indeed, many parts of the world are already knowledge-based economies, but no part of the world can claim to be a knowledge-based society. India, a developing country with its much strength in various areas of human endeavor, is a fine example of an emerging knowledge-based economy, yet it is far from being a knowledge-based society, given the rampant exploitation of the ignorant by those who are armed by the power of knowledge. On the other end of the spectrum is the United States, a highly developed nation with a sound knowledge-based economy but a far-from-sound knowledge-based society.

There is a hierarchy between data, information, knowledge and wisdom. Data is a prerequisite for information, information for knowledge, and knowledge for wisdom. Knowledge requires connecting and collating information of multiple kinds from multiple sources. Conversion of knowledge into wisdom requires weaving of experience (own and collective) into the fabric of knowledge. Our rural folks such as farmers are adept at converting information into knowledge and knowledge into wisdom–but they are generally extremely short of information.

Having a certain minimum amount of knowledge, becomes pertinent for any citizen of a country to protect his rights and discharge his responsibilities. Besides, surviving with dignity and without being exploited in an increasingly competitive world demands that one be both a specialist and a generalist.

Therefore, for farmers' security, it becomes essential that our farmers are made knowledgeable and have access to knowledge on a continuing basis.

22.3 How do we make our farmers knowledgeable?

The rural sector needs knowledge of the following three kinds:

(a) General knowledge that would make them reasonably informed citizens of the country and the world.
(b) Specialised knowledge, e.g. on seeds; optimal use of fertilizers, water and power; management of live-stock; output channels; market information; water management such as rain water harvesting; technologies for value addition; time and labour saving devices; and energy plantations. One would also need to establish a system by which one could update this knowledge in real time.
(c) Knowledge for immediate use – that is, updated information, for example, on weather and the movement of fish, which one can obtain using satellite imagery.

We will deal with each one of the three categories above, one by one, keeping in mind that what we suggest is practical.

Sheer numbers would make it difficult to reach every farmer directly. However, it should be possible for farmers to have access to knowledge if we empower the *Panchayats* that are directly elected by and are accountable to their respective *Gram Sabhas* of which all farmers are automatically members.

A few words about *Panchayats*. *Panchayats* are comprised of elected representatives of our rural communities that account for some seventy percent of Indian population; it is in these communities that we have the largest amount of deprivation – be it educational, social or financial. They are the lowest tier of local self government in India. [India has 601 districts in 35 States (big and small) that together constitute the Republic of India.] The functions, powers and duties of the *Panchayats* includes every possible aspect of living be it construction, repair, maintenance of community assets, agriculture, animal husbandry, fisheries, housing, education, development programmes, judicial powers and what have you. *Panchayats*, therefore, need to be empowered with information, both general and specialised, that would equip them to handle their above responsibilities through *Gram Sabhas*.

Local governance and rural development would remain an unfulfilled dream without empowering *Panchayats* with knowledge, even if the necessary devolution of powers to them becomes a reality, and the required resources are placed at their command.

We, therefore, need to develop and disseminate, using an appropriate medium, information packages to *Panchayats*. The basic steps involved could be the following:

(a) Identification of the areas/topics in which knowledge, both general and specialized, would need to be communicated to *Panchayats* over 40 hours (one working week of getting together) at the venue of one of the *Panchayats*.

(b) Identification of those who would prepare the detailed modules (approximately 40). One would need to make a provision of adequate payment to them.

(c) Identification of people (including, where possible, those who have prepared the module) (150 for Punjab, as an example) who would convey this information to the required number of individuals employed for this purpose (the "trainers") on a reasonable wage by the State Government.

(d) Working out the mechanism by which the trainers would then interact with the various *Panchayats*. It would be a three to five year programme for a State.

Financial calculations have been made and show that the entire project is financially viable.

The detailed steps would be the following:

i. Decision on the messages and the information – both general and specialized – to be communicated to the *Panches* and *Sarpanches* (the members and head of the *Panchayats*) in 40 hours (five days at a stretch).

ii. Decision on who would make these information packages. One would need to involve high-level experts who would also be good communicators, for this purpose.

iii. Recruitment of, say, for the State of Punjab, 150 young people of a high level of intelligence who are good communicators, at an attractive salary.

iv. Training (over a period of two weeks) of these carefully selected individuals by high-level experts who prepare the information packages.

v.   The trainees would then communicate with *Panches* and *Sarpanches* through 40-hour sessions as above, over a period of three years during which they can cover, for example, all the 12,000 and odd *Panchayats* in Punjab. For Punjab the expenditure over a period of three years would not exceed Rupees 50 crores. (INR one crore = ≈USD 220,000).

vi. The *Panches* and *Sarpanches* will be given material which would help them communicate this knowledge, through discussion, in course of time, to the *Gram Sabhas*. This will complement and supplement what the children in the rural sector will learn in schools.

The above persons (trainers) can then continue to be involved with the villages and would be responsible for setting up programmes for keeping the information updated and communicating it to the various *Panchayats*. If need be, a separate all India or State service could be created for these persons. Continued employment of such persons will not cost the Government more than Rs. 10 crores a year for Punjab as an example.

Another suggestion that has received much support is the setting up of knowledge clubs – 50,000 such clubs for the country, each club taking care of approximately 15,000 people including men, women and children in our villages – with adequate finances to subscribe to newspapers/magazines and to invite people from outside to talk to them. The school premises could be the venue of such clubs. An annual allocation of Rupees 500 crores to be given to the clubs plus Rupees 100 crore as other expenses, should be made for such clubs for the whole country. They should be managed by the Local Self Government (perhaps, a consortium of *Panchayats*).

All rural development programmes must be integrated/networked. This will reduce cost of administration and make more money available for the programme. Notice boards should be put up in every village that would indicate the money received for various programmes and the use to which it is being put to.

The Prime Minister has announced the setting up of Krishi Vigyan Kendras in every district as well as a National Horticulture Mission. One of the responsibilities of the trainers mentioned above could be to integrate the following: the National Horticultural Mission; the Krishi Vigyan Kendras of the Indian Council of Agricultural Research (ICAR); the various schemes of the Government; the above scheme of empowerment of *Panches* and *Sarpanches* and, through them, in the long run, the *Gram Sabhas*, through information packages delivered over 40 hours; the Knowledge Clubs mentioned above; the computerised kiosks set up by MSSRF (M S Swaminathan Research Foundation); and the school libraries. Such integration would reduce the cost of administration and make each mission/activity more productive and useful. The school premises could be used for many of these activities.

22.4 Specialised Information

We believe that specialised information in the following areas must be a part of the information package. (1) *Panchayat* Raj Act, (2) Pesticides and alternatives, (3) Fertilizers and alternatives, (4) Seeds, (5) Understanding business needs (e.g., marketing), (6) Farmers rights (WTO, TRIPS, Convention on Biodiversity, and Indian Acts that are relevant, etc), (7) Water harvesting and water conservation, (8) Soil, (9) Food storage, preservation and processing, (10) Disaster management, (11) New materials,

(12) Importance of statistics, (13) Energy Conservation, (14) Wasteland and grassland development (15) Techniques to improve breed of cattle, poultry and other livestock, (16) Fisheries development, (17) Fuel plantations, (18) Forestry, (19) Indigenous and low-cost construction (20) Non-conventional energy sources, (21) Dairy/ animal product development, (22) Post-harvest technologies for value addition (food processing), (22) Organic agriculture and vermiculture, (23) Integrated pest management and biopesticides, (25) Better weather prediction models (DST; MST Radar Facility, Tirupati, DOS), (26) Prevention and management of epidemics, (27) Marketing of our creative, cultural and heritage industries, (28) Additional employment.

22.5 Conclusion

We believe that if the above knowledge is conveyed appropriately, it will, taken together, also convey the following important messages, which will empower our rural sector of which farmers are the most important constituent, with important and lasting values without which knowledge ceases to be knowledge.

Empowerment with knowledge can yield optimal results only when means are provided to apply and use the knowledge. In our context, the following steps must be taken concurrently with knowledge empowerment, to ensure sustainable farmers' (therefore food) security.

1. All seed business must be Indian business. If we wish to protect the sovereignty of our country, besides our food security, we must prevent neocolonialism through control of our agriculture by multinationals who are in the business of seed production.
2. Encourage organic farming and integrated pest management, and discourage excessive use of fertilizers and pesticides which have played havoc with our environment and our export potential. Therefore, learn to deal with the powerful fertilizer and pesticide lobby.
3. Safeguard ourselves from the potential hazards of genetically modified seeds/crops by setting up adequate and appropriate biosafety measures in the country to thoroughly assess such seeds and crops before introducing them in the environment.
4. Find local solutions for water management such as rain-water harvesting and water-shed programmes, instead of launching mega projects such as interlinking of rivers. Prevent exploitation of underground water by soft-drink industries, such as Coca Cola.
5. Make assured and quality power available to farmers, instead of free power.
6. Make credit available to farmers with easy terms of repayment.
7. Provide adequate storage of food grains.
8. Ensure easy access of farmers to markets.
9. Ensure fair price to farmers for their produce.
10. Set up a mechanism by which farmers have an attractive stake in the quality of their produce.
11. Ensure that the provisions of the Plant Variety Protection and Farmers Rights Act are implemented.

Chapter 23

Governance of food and water security in China, with reference to farming in Northwest areas

Zhang Zhengbin & Xu Ping

Center for Agricultural Resources Research, Institute of Genetics and Developmental Biology, Chinese Academy of
Sciences, Shijiazhuang, China

23.1 Introduction

China has the largest population (1.3 billion) in the world. It produces 500 million
tonnes of food, which works to 400 kg per capita. It has to feed 22% of the population
of the world by using 7% of the world's land and 8% of the world's freshwater. China's
total population is expected to reach 1.37 billion in 2010, 1.46 billion in 2020 and 1.5
billion around 2033, and to stabilize at 1.6 billion later (Hu, A.G., 2006). In 2020, the
projected population of 1.5 billion would need 600 million tonnes of food grains, i.e.
100 million tonnes more than the present food production. When the Chinese population will reach 1.6 billion, 640 million tonnes of food will be required. It is estimated
that there would be deficit of 1300–2600 billion m^3 water by 2030. Therefore, the
only option for sustainable development of agriculture and economy is to undertake
water-saving agriculture, to develop the Blue Revolution and comprehensive drainage
systems. As the total extent of land and water endowment of the country remains
the same, the per capita endowment of land and water would decrease pro rata with
time as the population increases. Increasing population would need more food. Also
more land would be needed for the construction of habitations and establishment of
industries and social facilities. The challenge facing China is how to grow more food
with lesser amount of land and water.

About 70% of the poor in China live in arid and semi-arid areas of Northwest China
(Fig. 23.1). In such areas, problems of slow economic growth and food security are
likely to persist for some time to come. In order to address the problem, China has
implemented the strategy of developing the western region, through measures such as,
reducing and exempting agricultural taxes, increasing financial input, building water
conservancy and other construction projects. All these measures will play key roles in
solving the poverty problems of northwest region.

23.2 The achievements and challenges of poverty alleviation in China

China has the largest number of poor people in the world. Reducing and eradicating
poverty is not only the foremost but also the most difficult task in China's development.
The total number of people in the world living below the Absolute Poverty Line (one

Figure 23.1 Incidence of rural poverty in China – 1996 data from "China:Overcoming rural poverty", World Bank Poverty Assessment – 2000.

USD per capita per day) got reduced due to marked reduction in the number of the poor in China and India. Chinese absolute poverty population living in rural areas has decreased from 80 million in 1993 to 29 million in 2003. The ratio of the poor to the total rural population fell from 8.7 % to 3.1%, thereby improving upon the goal of halving the number of extremely poor population in "Millennium Development Goals" (Millennium Development Goals, MDGs) suggested by the UN (Wang, Q.J., 2004).

At present, China has the ability to produce 450 million to 500 million tonnes of food grains. Grain consumption, energy and protein intake per capita per day in China is catching up with the world average level. Historically, the long-term shortage of grain and other agricultural products gave place to sufficiency and later abundance. During the last 20 years, over 200 million poor Chinese in the rural areas have overcome the shortage of food and clothing, and water conservation made this possible (Sun, J., 2002; Wang, S.C., 2005; Zhang, Z.B., 2006a,c). After being in existence for 25 years, the World Food Aid to China was discontinued in 2005 (Xu, J.Y., 2005).

By the end of the year 2005, China has 23.65 million BPL (Below the Poverty Line (BPL of one USD per capita per day) population, which is next to India (Xiao, X., 2006). The "Key Indicators 2005" of ADB (Asian Development Bank) pointed out that in 2003 there were more than 600 million undernourished people living in Asia, 93% of them being distributed in India, China (173 million) and South Asia.. According to the survey of the Chinese Poverty Alleviation Office, 76% of poor farmers who suffer from inadequacy of food and clothing live in mountainous areas (Fig. 23.1–23.2), and 46% of poor families possess arable land of less than 1/15 ha per capita. At present there are about seven million poor people living in region with harsh conditions (Deng, Y.W., 2005). The "Tianjin Binhai Declaration" of May, 2006, has elucidated the sustainable development strategy in China's poverty areas.

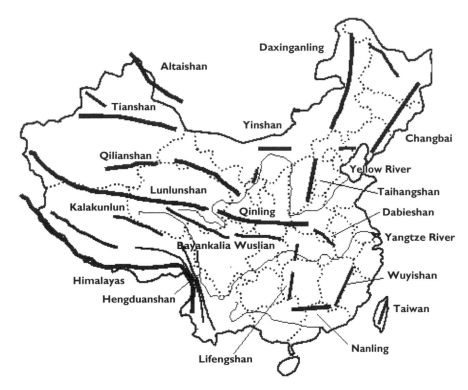

Figure 23.2 China's mountains and rivers.

23.3 The poverty problem in Western China

The majority of the Chinese poor is concentrated in rural areas, primarily in the six areas of the West China as follows: 1) high and cold mountain areas of western desert, 2) sandy region eroded by wind in the southeastern fringe of Mongolia plateau, 3) ravine areas of loess plateau which have suffered serious water scarcity and soil loss, 4) Qinba mountain, 5) southwest hilly area of Karst plateau, and 6) alpine gorge area of Hengduan mountains. It is not an accident that nearly 80% of state-level poverty counties in China and the majority of the poor population are concentrated in these areas of infertile soil and water scarcity (Fig. 23.1).

There is a strong contrast in regard to the availability of arable land and water per capita between the southwest and northeast regions. In the southwest, the terrain is rugged, and rich in water resources, i.e. with more water (5,500 m³ per capita) and less arable land (410 m² per capita). In contrast, in the northwest, the terrain is relatively flat and subject to severe drought – it has less water (2,500 m³ per capita) and more land (3200 m² per capita) (Figs. 23.2, 23.3 & 23.4). These are the key constraining factors in the agricultural, economic, cultural and ecological development of the regions (Wang, J., 2001).

Therefore, poverty arising from the local biophysical conditions is sought to be eliminated through new industrial and agricultural technologies or through some other

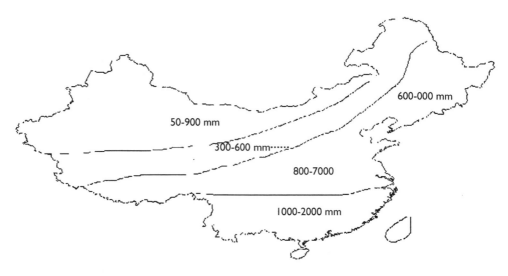

Figure 23.3 Precipitation in China.

Figure 23.4 Map of China.

means such as emigration. For example, poor farmers are encouraged to migrate from the mountainous areas to plains, from the water-deficient areas to the irrigated areas. In recent years, Gansu Province and other regions tried to solve poverty by such solutions and have achieved very good results. The project of "south-north water transfer", involving the transfer of the water of the Yangtze River (Chang Jiang) and Yellow

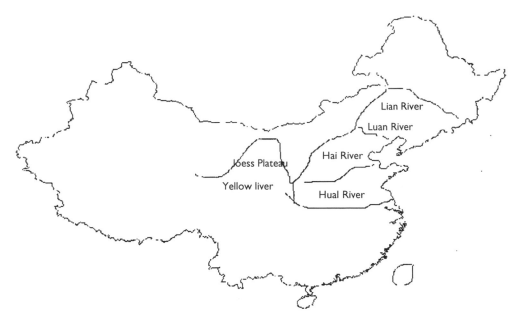

Figure 23.5 Rivers of North China.

River (Huang He)(Fig. 23.5), and extending irrigation area, has been effective. There is still a big potential to get more food production in western arid and semi-arid areas. Currently, China carries out the strategy of exploiting western region, which has an important and realistic significance in solving poverty problem.

23.4   Outlook of water resource security in China

China is a big country with severe water deficit. The precipitation in 50% of the land area is less than 400 mm. The total water resource in China is only 2.8 trillion m³, ranking the sixth in the world. With per capita water resources of less than one-fourth of the world average, China is one of the 13 water-poor countries in the world. The apportioning of water resources to farmland is only three-quarters of the world average (Wang, S.C., 2005; Zhang, Z.B. *et al.*, 2001; State Environmental Protection Administration, 2006). However, the total water consumption in China is the largest in the world. In 2002, the freshwater consumption reached 549.7 billion m³ in China, which accounts for about 13% of the world water consumption.

Parts of China have been facing water scarcity since 1970s. It has gradually spread to the whole country after the 1980s, and the situation is becoming increasingly serious, with significant impact on agriculture and economy. For example, in the past 20 years, there have been several water transfer projects to alleviate water shortages in the cities in Xian of Shaanxi province in loess plateau, Tianjin and Beijing and other cities in north of China. The water shortages are caused by paucity of water resources in northern china, poor water quality in southern China, as well inadequacy of water conservation projects in western China (Fig. 23.5).

At present, the water resource shortage problem is serious in China. The total amount of water shortage amounts to about 30 billion m³ to 40 billion m³. During 1991 to 2005, the average annual land area suffering from drought has been 26.02 million ha. From 2003, the extents of drought-affected areas in northern region have become relatively less, while the frequency of drought occurrence in the southern region has been increasing. In the past five years, drought reduced the annual grain output by 35 million tonnes, and denied safe drinking water to 320 million rural populations. 400 out of 669 cities in the whole country have inadequate water supply, and 110 cities of them have serious water shortages. Drought reduced the industrial output by more than 2,000 billion Yuan. About 70 million people do not have adequate drinking water in China. Water scarcity has thus a serious impact on environment and human health (Wang, S.C., 2006).

There has been excessive exploitation of water resources in China. At present, China's water resources exploitation and utilization rate has reached 19% in most areas, and even higher in some areas. This is nearly three times that of the world average level. Most rivers in the north China have been over exploited. The utilization rate of water resources in the Huai River, Liao River and Yellow River basins is about 60%. It is more than 100% in the Hai River basin (Fig. 23.2; Fig. 23.5). These utilization levels are far in excess of scientifically accepted rate of exploitation of water resources recommended by international water organizations. The groundwater exploitation also has been excessive. By the year 2050, the average utilization rate of groundwater will reach 64%, and will be lower in the case of inland rivers (27%). The utilization of groundwater in other basins (excluding southwest rivers) will be greater than 56%. In the case of Hai River, Luan River, Huai River and the Yellow River (Huang He) it will be as high as 100%, 74% and 93% respectively (Fig. 23.5). Excessive withdrawal of groundwater causes ground subsidence and salinization of coastal aquifers due to intrusion of seawater (Wang, S.C., 2005).

North China is not only an important political and economic center but it is also the principal grain production base. More than half of wheat and one-third of corn are produced in North China Plain. However, the per capita water resource endowment of this region is only 15% of that in China. This region has long relied on groundwater for irrigation. Some surveys show that North China uses more than 100 billion m³ of groundwater, which is twice the quantity of water resources of the Yellow River. The area of composite cone of depression reaches 7.28 million km² in the North China Plain (Hubei Pingyuan) around the Bohai (Yellow Sea) gulf, covering Hebei and Shandong province, Shijiazhuang, Tianjin and other cities, which occupy 52% of total area of North China Plain (Fig. 23.5) (Zhang, 2006b). Many urban areas have developed cones of depression. In the past 50 years, in north China's major cities such as Beijing, the underground water level has dropped by 50 m, the annual fall rate of underground water being about 1 meter. Consequently, the groundwater is no longer renewable. Meanwhile, in order to get good quality drinking water under conditions of serious pollution, two ultra-deep wells of 1,800 m and 1,543 m depth were drilled in Beijing. In the past 40 years, there were more than 200 earth ruptures/ surface collapses in Hebei Province, with consequent damage to farming. Most of them were caused by excessive extraction of groundwater. The excessive utilization of water resources in the Yellow River (Huang He) and Huai River and Hai River in North China Plain, caused them to dry up in the dry season. The water shortage seriously impacts industry and agriculture in North China (Fig. 23.5).

The poor quality of water in many parts of China is caused by pollutants diffusion, poor management, widespread drought and the excessive economic expansion. Water shortage caused by pollution is more severe than water shortage caused by scarce resource, especially in economically developed coastal regions in southeast China. Seventy percent of rivers and lakes have been polluted to different degrees in China, while the Yangtze River (Chang Jiang) and the Yellow River (Huang He) are facing survival crisis. The three major rivers, namely, Yellow River, Huai River and Hai River in north China, face serious problems of water shortage and water pollution (Fig. 23.2; Fig. 23.5). It has been reported that China is spending USD 1.39–1.89 billion every year to control pollution in the Yellow River (Huang He). This does not include the public health spending for cleaning up rivers or annual medical treatment of people affected by drinking polluted water. As much farmland is irrigated with polluted water in the Yellow River basin, the health costs are more than 330 million U.S. dollars each year in this area. 360 million rural residents lack clean drinking water in China (Bai, J., 2006). Water crisis is becoming a hazard for the health of millions people and economic development.

Natural disasters, such as floods, typhoons, landslides, and mud-rock flows, have had a significant adverse impact on food security of China, particularly south China The World Bank estimates that China's losses averaged 10 billion U.S dollars due to annual flood, of which only 30%–40% due to the floods and 60%–70% due to the secondary effects of floods (Li, L.F. et al., 2006).

23.5 The impact of water security on food security in China

Water security for supporting China's food security is facing four sets of challenges: shortage of water resources, increasing need of industrial and urban water, extensive water management and serious water and soil loss.

There is a mismatch between soil and water resources in China. 81% of the water is concentrated in the Yangtze River and southern China, while northern China accounts for 64.1% of the arable land in the country (Fig. 23.6).

Irrigated agriculture plays an important role in Chinese food security. China has the largest irrigated area (53.33 million ha) in the world. In 1999, irrigated land area occupied 40% of the total farmland area but produced 75% of the Chinese total grain, 80% cotton output and 90% vegetables production. In the past 20 years, the proportion of agricultural irrigation water as a percentage of total water use amount dropped from 85% in 1980 to 65% in 2006 (Zhang, Z.B. et al., 2001; Zhang, Z.B., 2006a,c).

Meanwhile agriculture water resource is increasingly being diverted to urban and industrial development, leaving lesser quantity of water for food production. From now upto 2010, the population of China is likely to increase by 126 million people. The World Bank estimates that the urban water demand will increase by 60%, from the current 50 billion m^3 to 80 billion m^3. Meanwhile industrial water consumption is expected to increase by 62%, from 127 billion m^3 to 206 billion m^3. Consequent upon excessive withdrawal, the underground water level continues to drop. So that

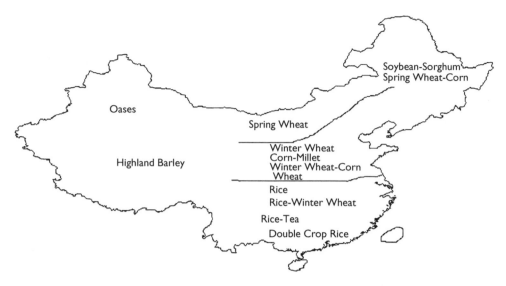

Figure 23.6 China's agriculture region.

increasing pumping costs will lead to many farmers not being able to afford the cost of underground water for irrigation. With the drying up of rivers and depletion of underground aquifers, the increasingly serious water shortage is likely to lead to lower grain production capacity, requiring import of food from other countries (Wang, S.C., 2005; Li, L.F *et al.*, 2006).

Wastewater problem in urban and industrial areas is also very serious. In the case of Beijing, Tianjin, Dalian and Qingdao cities, the wastewater recycle rate reaches 70%, but in a large number of urban areas, the water recycle rate is only 30%–50%, and in some cities even lower. In contrast, wastewater treatment in the industrialized countries is about 75%, several times higher than that in developing country, sometimes even more than 10 times. The water loss due to faulty water supply involving theft, dripping and leakage in most cities is estimated to be 15%–20%. The water saving potential should be exploited in cities (Wang, S.C., 2005; Li, L.F *et al.*, 2006).

Water use efficiency in China is 30% to 40%, compared to 70% to 90% in the Developed countries. Crop water use efficiency averages 0.87 kg/m^3 in China, while it is 2.32 kg/m^3 in Israel (Wang, S.C., 2005; Zhang Z.B., 2006a). So the potential of agricultural water saving is considerable.

China has 53.3 million ha of irrigated farmland. By planting water-saving, high-yielding, and drought-resistant varieties, and by reducing the frequency and scale of irrigation, it is possible to save 1500 m^3/ha of water annually. Therefore annual total water saving could reach the 80–100 billion m^3 entirely. For example, rice is the major food crop in South China, and its cultivation accounts for 70% of agricultural water. If the water-saving irrigation techniques (named as "thin-water-layer irrigation often disclosed fields") are used in 30.7 million ha of rice land in China, there will be annual saving of 1500–4500 m^3 per ha rice land. Thus more than 46–100 billion m^3 water could be saved in the whole Chinese rice land each year, which is more than the amount of the

Yellow River (Huang He) water resource. In wheat-corn-growing areas in the North China Plain (Fig. 23.2; Fig. 23.6), land is irrigated 8–10 times, the irrigation amount being about 900 m³/ha each time. If the frequency of irrigation is reduced to 3–6 times (1–3 times for wheat and maize respectively), the total water saving amount would be more than 1500–3000 m³/ha. Therefore, there is great potential for water saving in irrigation (Zhang, Z.B., 2006c). China must launch a revolution for enhancing water use efficiency. It is estimated that if we adopt the scientific water-saving agriculture technology, the utilization rate of irrigation water in 2030 will reach 60%–70%; water productivity will reach 1.5 kg/m³. In sum, if during the next 30 years, the irrigation water utilization rate is enhanced by 30%, there will be a saving of 120 billion m³ which could be used to produce 120 million tonnes of grain, assuming water productivity is 1.5 kg/m³. This strategy is very important to ensure food security in future.

For a long time, China's grain growth, agriculture and rural economy development was based on predatorily using natural resources. For instance, the thickness of the black soil layer which produces crops, decreased quickly from 1–2 m to 3–5 cm. in Jilin province in Northeast China (Fig. 23.6). In some places the loess is exposed, indicating the seriousness of soil erosion. At present, soil erosion has degraded 3.56 million km² of area, which constitutes 37% of the total Chinese land area. Total annual soil loss amount reaches 5 billion tonnes. Soil and water loss in the Loess Plateau region is particularly serious, as it accounts for half of the total area of soil and water loss in China. Serious soil and water loss leads to land degradation, grassland desertification and ecological deterioration, enhanced river and lake sedimentation, and exacerbation of flooding in rivers downstream (Wang, S.C., 2005; Li, L.F. et al., 2006).

23.6 China's food security and its solutions

Despite dire prognoses, the annual food output in China during the last 25 years has increased from 300 million tonnes to 500 million tonnes, with annual per capita of food availability exceeding 400 kg. According to a report of WFP (World Food Programme), in 2005, China became the third largest food-aid contributor, after US and EU.

In some regions, the food crop area has decreased markedly due to the cultivation of economic crops and acceleration of urban and industrial development. According to a latest investigation made by Ministry of Land and Resources of China, the area of arable land in China has decreased from 130 million ha in 1996 to 126 million ha in 2002. At present, the availability of farmland area per capita is only 0.094 ha, which is 40% of the world average. In China, there are currently 666 counties (or regions) where per capita farmland availability is less than 0.053 ha which is the minimum area stipulated by FAO (World Food Organization). It is predicted that by 2020, China will have about 1.5 billion population, which would need about 600 million tonnes of food i.e. about 100 million tonnes more than at present (Wang, S.C., 2005; State Environmental Protection Administration, 2006; Zhang, Z.B. et al., 2004). The challenge thus involves higher food production on limited arable land.

Recently, the Ministry of Agriculture enacted the 11th Five-year Plan (2006–2010) for the development of agriculture and rural economies in China. This document classified China's agricultural regions into three types: advantageous development areas, pivot development areas and moderate development areas, in terms of natural resources, industrial bases and development potentials. The pivot development areas mainly include north Xinjiang, where focus will be on the settlement of field irrigation, and the construction of China's strategic alternative food zone; the low-yield and middle-yield fields in northeast region where we will focus on improving arable land productivity and quality, enhancing modern agricultural material equipments, and constructing modernized bases for food production with a high jumping-off point; dryland farming areas in the Yellow River and Huai river as well Hai river basin regions where we will focus on popularizing dryland farming and water saving technologies in a large scale, improving water use efficiency, and adopting the policies of resources saving for agricultural development; tropical agriculture areas in South China where we will focus on strengthening the construction of national bases for natural rubber, and actively establishing complex rubber forest ecological systems; the west coast of strait where we will focus on the acceleration of modern agricultural construction; grass hills and grass slopes featured areas in South China where we will focus on accelerating grassland improvement, and developing grassland-based stockbreeding. The moderate development areas mainly include agriculture and stockbreeding alternative areas, Qinhai-Tibet Plateau areas, Loess Plateau areas, Karst areas in Southwest China, desertification areas in Northwest China and wet land areas in Northeast China (Ministry of Agriculture, 2006) (Fig. 23.6).

In China, 30.8% of whole country land area is arid, and it is mainly distributed in Northwest China, especially in Xinjiang (Fig. 23.4), which has the largest arid area in China. The applicable areas for agricultural, forestry and stockbreeding development in Xinjiang amounted to 68.53 million ha, and the per capita arable land area in Xinjiang is about 4 ha; both rank the top in China. In Xinjiang prefecture, uncultivated lands account for 59.2% of the area. Although located in extremely arid belt, Xinjiang is endowed with extensive glacial water resources, which have great potential for exploitation. In the mountain areas of Xinjiang, there are approximately 200 billion tonnes of rain and snow water resources, capable of generating annually about 90 billion m^3 of river water from the mountains, which is 1.8 times of the annual runoff amount of the Yellow River (50 billion m^3). The per capita water resource in Xinjiang is more than 6,000 $m^3/$ year, ranking the fourth largest in China (Office of West regions exploiting of Xinjiang Uygur Autonomous Region, 2005). With an annual water resource amount of 58 billion m^3, the Yellow River basin can support more than 100 million population; while despite having an annual water resource of more than 40 billion m^3, the Tarim River basin only feeds about 10 million population (Figs. 23.2–23.4). Therefore, compared with the Yellow River (Huang He), the water shortage in Xinjiang is caused by the lack of more water conservancy projects, rather than the problems of less of water. So we should farther increase water conservancy projects construction and strengthen water resources management in the Tarim river basin. As belonging to typical non-irrigation areas, namely non-agriculture inland provinces, total amount of water used in agriculture use is 77.8 billion m^3 in Xinjiang, covering 90% of its total amount water used; and its irrigation level is about 12000 $m^3/$ha, which is twice that of North China. Currently, Xinjiang adopts such new method as under-mulch drip-irrigation in planting 0.2 million ha of cotton and other crops, thus recording extremely prominent achievements in both water-saving and improvements in economic benefits, such as making the irrigating quota in Xinjiang only more than 3000 $m^3/$ha, which represents a 1/3 to 1/2 water-saving compared with conventional irrigation methods. Therefore, the agricultural water-saving in irrigated areas of

Northwest China represents one of the important ways in improving the productive capacity of water resource.

In addition, the poverty areas of southwest China, such as Guizhou and Yunnan Provinces, where annual precipitation is about 1000–1300 mm, has numerous rivers and are rich in water resources. Because of the ruggedness of the topography, water resources utilization remains undeveloped (Figs. 23.2–23.4). Therefore, the investment on the construction of water conservancy projects in this region constitutes the best way to achieve poverty alleviation through the development of agriculture and industry in the poverty areas of southwest China.

Of all the arable lands in China, the extent of fields with low-yields, middle yields and high yields amount to 41.6%, 29.6% and 28.8 % respectively. Through increasing inputs such as water conservancy, fertilizer use and other factors, we will reclaim 2 million ha of low-middle-yield fields every year. Altogether about 20 million ha of such fields will be improved during the next 10 years, whose output will be equal to 40 million tonnes (or the like increase output from 60 million ha of fields). The cultivation of super rice can increase the yield by 30 million tonnes; and that of the super maize can also increase the output by 40 million tonnes. In addition, China's 2.67 million ha of water areas will yield 30 million tonnes of food, and more forest areas will yield 20 million tonnes of food. According to the statistics given by Chinese Ministry of Agriculture, in 2005, China's agricultural science and technologies contributed 48% of agricultural output growth, i.e. nearly half of the increase of more than 10 million tonnes of food were attributable to science and technologies. If China wants to improve food output to more than 600 million tonnes by 2020, the contributions of science and technology to food production should be at least 50%. Therefore, it is essential to actively pursue the technologies for decreasing ecological cost while safeguarding food security, and to effectively strengthen the investment on food science and technologies (Zhong, Y., 2006).

Our local governments at all levels should give high priority to food production in order to achieve food self-sufficiency. The imbalances in regional development and regional food shortages may be addressed through domestic interregional and trans-regional regulative measures such as "North-South food transfer" and "South-North water transfer", which have played and will continue to play important roles in solving the problems of China's food and water resource security.

Acknowledgement

This work was supported by the China National 863 Water-saving of Important Item (2006AA100201). We are grateful to Prof. U. Aswathanarayana for improving our manuscript.

References

Bai, J. (2006) China's water crisis is a fan strobe that causes other problems. June 9, 2006. http://finance.sina.com.cn/g/20060609/15522639344.shtml\

Deng YW. (2005) Why Chinese poverty population is more with shrinking? *Chinese Youth Daily*, Sept. 13, 2005.

Hu, A.G. (2006) How to treat China's population conditions. *People Forum*, 4, 19.

Li, L.F. and Lan, Z. (2006) China's water crisis. Oct.23, 2006 http://www.huanjing65.com/viewthread.php?tid=1281.

Ministry of Agriculture. (2006) Agricultural and rural economic development in the 10th Five-Year Plan of China 2006–2010 , Aug. 3, 2006 http://www.xyjj.org.cn/html/xwgg/2006/0807/479.html.

Office of West regions exploiting of Xinjiang Uygur Autonomous Region. (2005) General Situation of the Xinjiang Uygur Autonomous Region. Apr.19, 2005 http://www.xjwd.gov.cn/xbhj.php.

State Environmental Protection Administration. (2006) Chinese environment situation in 2005. *Economic Environment*, 7:15–31.

Sun, J. (2002) Reading speedily national figures: 500 billion kilograms grain, the world acclaiming for China. *Beijing Evening News*, Oct. 14, 2002, 2(Page).

Wang, J. (2001) Poverty problem in China's western region in 2020. July 31, 2001 http://www1.cei.gov.cn/forum50/doc/50hgjj/200107311541.htm

Wang, Q.J. (2004) China advanced to achieve the goals of halving the proportion of people in extreme poverty. *People Daily*, Sept.8, 2004 :1(Page).

Wang, S.C. (2005) Solving the problem of water to safeguard China's grain security. *Study Times*, 6:12–18.

Wang, S.C. (2006) Follow the path of a water-saving society. June 23, 2006 http://www.habc.cn/houqin/html/jl0623.htm

Xiao, X. (2006) China's poverty population lies second in the world. *Financial Times*, 6: 50–52.

Xu, J.Y. (2005) China will end up the history of accepting foreign food from next year. Sept. 28, 2005 http://politics.people.com.cn/GB/1026/3734346.html.

Zhang, X.F. (2006) Water shortage is serious in north China, which that is equivalent to two times of the Yellow River water resources. *Science and Technology Daily*, Sept. 11, 2006 1(Page).

Zhang, ZB, Xu P. (2004) Prospects and strategy of food production and agricultural development in China. *Science News*, 18: 11–13.

Zhang, Z.B, Zhang, J.H, Liu, M.Y, Zhong, G.C and Yuan P. (2001) Growing up of blue revolution in the 21st century. *Chinese Journal of Eco-Agriculture*, 6:18–23.

Zhang, Z.B. (2006 a) Research and development of dryland and high water use efficiency agriculture. *Science Press*, **May, 2006**, 65–80(Pages).

Zhang, Z.B. (2006 b) Speaking with facts, biological water saving will become bright prospects. *Science News*, **20**, 11–13.

Zhang, Z.B. (2006 c) The successful experience of Chinese agricultural development. *Science News*, **15**, 30–31.

Zhong,Y. (2006) Food security remains austerity, China must rely on science and technology. May 18, 2006.

http://finance.sina.com.cn/g/20060518/16132578828.shtml.

Chapter 24

Adverse impact of green revolution on groundwater, land and soil in Haryana, India

S.K. Lunkad & Anita Sharma

Department of Geology, Kurukshetra University, Kurukshetra, Haryana, India

> *Bubhuksitah kim na karoti pāpam*
> *(There is no crime that a hungry man would not commit)*
>
> *–A Sanskrit saying*

24.1 Agroclimatic setting

The Green Revolution enabled the small state of Haryana to become the 'wheat granary' of India – the state, though occupying 1.3% geographical area and 2% of the population of India, produces 13% wheat and about 3% quality rice of India besides other cereals, oil seeds, sugarcane and cotton. Haryana paid a heavy price for this impressive agricultural development in the form of serious degradation of basic geo-resources, i.e., water-soil-land. The purpose of the chapter is to bring out the linkages between the geomorphological and hydrogeological setting, agroecological zones and cropping patterns, irrigation from canal water and from groundwater and use of fertilizer, and their impact on the degradation of soil and groundwater.

The Yamuna which flows along the eastern border of Haryana is the only perennial river in the state. Ghagghar is the west flowing ephemeral stream forming the northern and western boundary of the state with Punjab and Rajasthan. The 'upland plain' (Fig. 24.1) between these two streams is the most fertile alluvial terrain of Haryana, termed 'Ghagghar Yamuna Do Aab' on which intensive cultivation of wheat, rice and sugarcane is done (Fig. 24.2 A,B&C). The western sandy terrain (millet-belt) known for cultivation of millet, cotton and other coarse crops such as barley, maize, pulses and oilseeds (Fig. 24.3 A & B) (Sharma,2007).

The average precipitation varies across districts from 350 mm in Hisar, Sirsa, Fatehabad and Bhiwani to as high as 1560 mm in the Himalayan foothills of Ambala and Panchkula. However, leaving these two extremes, the average rainfall in the State varies between 500–600 mm. while the normal rainfall is less than the average (400–450 mm). The rainfall in western arid region is 350–400 mm (Fig. 24. 4). Thus, Haryana has a subtropical monsoon climate with scanty and aberrant rains, hot summers with bright sunshine and dry-cold winters. The average number of rainy days in the eastern semi-arid district of Kurukshetra is 35 only . It is for this reason that the irrigated land in Haryana remains desiccated for 10–11 months in a year, leading to the deposition of salts in the top soil. The summer temperatures rise upto 48°C and the same drop to 1°C in January.

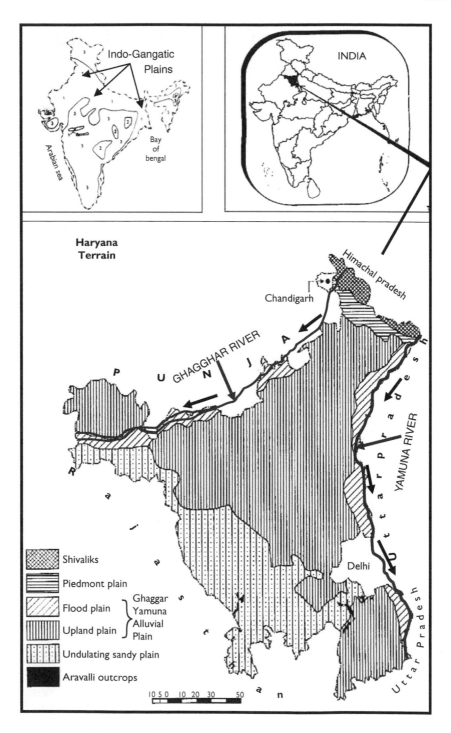

Figure 24.1 Geomorphic Terrains of Haryana (The location of Indo-Gangatic Plains and Haryana shown in the insets).
Source: Modified after Sharma, 2007.

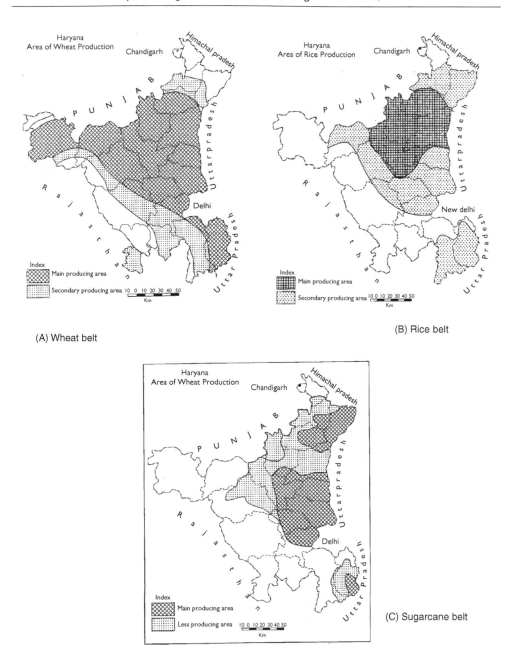

(A) Wheat belt

(B) Rice belt

(C) Sugarcane belt

Figure 24.2 Principal crop belts of Semiarid Haryana (A)Wheat, (B) Rice & (C) Sugarcane.
Source: Sharma (2007).

On the basis of climate, rainfall, groundwater quality and cropping pattern Haryana can be divided into three agro-ecological regions, namely, western arid region, eastern semi-arid region and a transitional central region in between the two (Table 24.1; Fig. 24.4). With decrease in the average rainfall from semiarid eastern region to arid

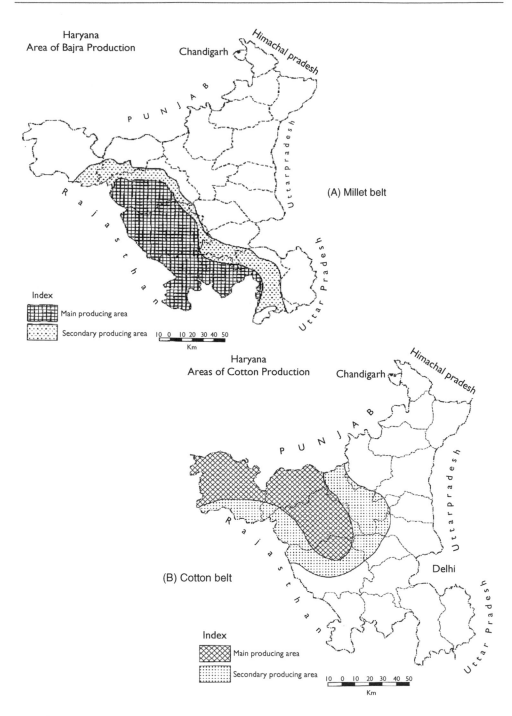

Figure 24.3 Principal crop belts of Arid Haryana (A) Millet (B) Cotton in Arid Region.
Source: Sharma (2007).

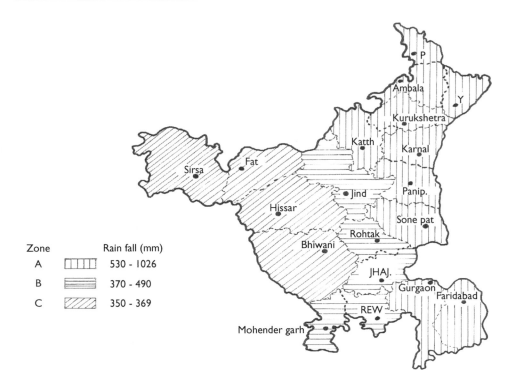

Figure 24.4 Average rainfall and climate zones of Haryana.
Source: Lunkad (2006).

Table 24.1 Agro-ecological regions of Haryana.

Region	Climate & soils	Average rainfall (mm)	Average groundwater quality TDS (mg/l)	Districts	Crops (major in bold letters)
Arid (Western)	Hot and Arid with desert and saline soils	~350	>4000	Sirsa, Fatehabad, Hisar, Bhiwani, Mahendragarh, Rewari	**Millet, Barley, Pulses, Cotton***, Maize, Sorghum, Oilseeds, Wheat, Rice, Sugarcane
Semiarid (Central)	Hot, semiarid with alluvial derived soils	350–500	>1500–4000	Kaithal, Jind, Rohtak, Jhajjar, Gurgaon	**Wheat & Rice** with maize Sorghum Cotton**, Oilseeds, Pulses, Sugarcane
Semiarid (Eastern)	Hot, semiarid with alluvial derived soils	500–600	<1500	Panchkula, Ambala, Yamunanagar, Kurukshetra, Karnal, Panipat, Sonepat, Faridabad	**Wheat, Rice, Sugarcane**, Maize, Sorghum, Oilseeds

*Not cultivated in Mahenderagarh and Rewari districts.
**Not cultivated in Gurgaon and Jhajjar districts.

Figure 24.5 Canal network in eastern Haryana.

western region the groundwater also becomes progressively more saline from east to west. The deep, loamy alluvium of the semi-arid belt of the state supports a variety of crops such as wheat, rice, maize, pulses, millet, sugarcane and cotton under irrigated conditions. *Per contra*, the sandy and sandy loam seric soils of the western arid belt are largely sown in millets and pulses under rainfed conditions.

24.2 Irrigation

Most of the eastern and central semiarid region of Haryana is irrigated by western Jamuna Canal (WJC) and its distributaries (Fig. 24.5) and some area in the north and northwest by Habra canal system bringing Satluj river water through Punjab. The arid western and southern region has no canal network reflecting lopsided development of irrigation in the state. This rainfed region has remained largely dependent on supply of scarce and low quality groundwater for extensification and intensification of agriculture. However, groundwater exploitation has been more in the semiarid fresh water region where surface water from canals did not reach. There is stagnation in the area of irrigated area, as there has been no expansion of irrigated land in the State since 2000–2001. During the first decade of Green Revolution (1970–71 to 1980–81) the groundwater use increased tremendously from 1.0 BCM to 7.0 BCM but remained fluctuating between 5.5 to 7.0 BCM during next decade (upto 1990–91). In 2000–01 the conjunctive use of groundwater and surface water for irrigation reached 50–50%. However, the groundwater use crossed over in 2001 and increased to 52.6% in 2003–04, as there

Table 24.2 Fertilizer use and number of tube wells in Haryana (1966–67 to 2004–05).

S.No.	Year	Total chemical fertilizer (NPK)(kg/ha)	Nitrogenous fertilizer (N)(kg/ha)	Number of tubewells & pumpsets ($\times 10^3$)
1	1966–67	4	3.5	25
2	1970–71	17	20	104
3	1980–81	64	52	332
4	1990–91	124	164	498
5	1995–96	164	202	549
6	2000–01	260	200	589
7	2002–03	278	204	602
8	2003–04	285	215	607
9	2004–05	318	239	612

Note: *Calculated by authors over net sown area.
**Data source Statistical Abstract of Haryana 2004–05, Govt. of Haryana, Chandigarh.

has been no enhancement in the canal irrigated land (Anonymous, 2006). The number of tube wells increased from 25×10^3 in 1966 to 612×10^3 in 2005 (Table 24.2).

24.3 Hydrogeology

The subsurface geological information on the topmost alluvial sediment cover available through tubewell logs upto a depth of about 100–300 m suggests that the sediment comprises alternating layers of coarse and gravelly sand, medium sand, fine sand and silty-clay. Sometimes the silty–clay layers contain lime concretions. The sediments depict a "fining upward sequence" deposited by braided as well as meandering rivers (Lunkad, 1988). The nature of sedimentation and the aquifer geometry around Kurukshetra have been described by Lunkad and Gupta (2003). The groundwater occurs in abundance in the gravelly coarse and medium sand between 50 to 110 m depth (Fig. 24.6). The fine sand layers occurring between 0–45 m depth from the surface constitute relatively poor aquifers yielding water to shallow wells and hand pumps. The first two good aquifer zones occur at 45–60 m and 90 to 110 m depth. The sediment in the central basin probably continues upto a depth of 1000 to 2000 m. No bedrock is encountered upto a depth of 600 m in Kurukshetra and Karnal area where deep drilling has been carried out by Central Groundwater Board (CGWB) and Haryana State Minor Irrigation and Tubewell Corporation (HSMITC) (Tanwar, 1997, 1998).

24.4 Groundwater quality

The salinity of groundwater in Haryana is due to multiple causes – mode of deposition of alluvial sediments, semi-desertic climate since middle-Holocene times, surface evaporation of precipitation water when shallow aquifers are being recharged and long residence time of water in aquifers containing large amount of fine sediment, i.e. clays and silty-clays, (Lunkad, 2006).

The groundwater quality aspects of Haryana have been critically reviewed by Thussu (1999) and Lunkad (2006). Almost 43 to 47% of groundwater of the state (both shallow as well as deep) is either marginally or highly saline (Tanwar 1997, 1998).

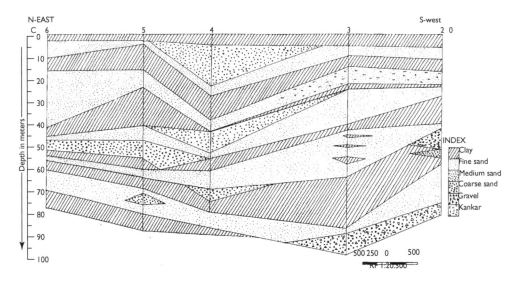

Figure 24.6 Groundwater aquifers around Kurukshetra, Haryana (Lunkad & Gupta, 2003).

The quality of groundwater fluctuates with depth depending on lithology of the aquifer and the climate. In general water associated with fine, clayey and silty sediments is comparatively more saline than in sandy aquifer. The water is generally fresh in semiarid eastern Haryana (TDS < 1500 mg/l), marginally saline (>1500–4000 mg/l) in central districts and highly saline (>4000 mg/l) in arid western Haryana (Table 24.1). The water quality also fluctuates with climate, becoming more saline in arid region relative to that in eastern Haryana (Fig. 24.7). From the agricultural point of view, the groundwaters in western Haryana in the district of Sirsa, Fatehabad, Hisar, Bhiwani and Mahendragarh are more charged with sodium, potassium, chloride and sulphate ions along with carbonate and bicarbonates of calcium and magnesium. Due to evaporation, these cyclic salts precipitate and deposit in the topsoil causing soil salinity and thereby reducing soil porosity and crop yield. However, soil salinization in Haryana is not restricted to western arid region alone. **In fact one third of the irrigated land in entire Haryana is salinity affected.**

24.5 The green revolution

At present (2001–04), 85% cultivated land in Haryana is irrigated against overall 40% in India. Of the total irrigated land 53% is irrigated by groundwater and 47% by surface water (Anonymous,2006). The current production of wheat and rice is ~9.5 and 2.72 million tonnes, respectively. The total production of cereals exceeds 13.0 million tonnes which is more than 6% of India's total food grain production.

It is interesting to note that there has been no extension of cultivated land in Haryana since its creation (Table 24.3). The net sown area has fluctuated between 80–81% upto the year 2000. In recent years (2001–04), the percentage of cultivated land has decreased to 79–80% as a result of diversion of arable land towards fast sprawling urbanization and industrialization particularly in Gurgaon, Faridabad, Sonipat and

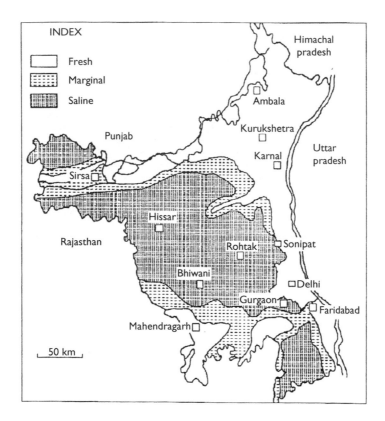

Figure 24.7 Generalized deep groundwater quality map of Haryana (Tanwar, 1997).

Table 24.3 Agricultural land use (Net area sown) in Haryana, 1970–71 to 2003–04 (in hectares $\times 10^3$).

1970–71	1980–81	1990–91	1995–96	2000–01	2002–03	2003–04
3565.4	3602.0	3575.0	3586.0	3526.0	3458.0	3534.0
(80.64)	(81.47)	(80.86)	(81.10)	(79.73)	(78.21)	(79.90)

Note: The numbers in parentheses indicate percentages of the area sown to the total area of Haryana 44212 $\times 10^3$ hectares.

Panipat districts. Hence the agriculture sector for its growth remained dependent on intensification of inputs in terms of irrigation, chemical fertilizers, pesticides and high yielding varieties of seeds.

The production of wheat and rice increased in both arid and semiarid regions during two decades (1971–91). The cropped area of rice doubled in semiarid region (9.0 to 18%) and tripled in arid region (0.9 to 2.6%). The area of wheat doubled in arid zone (11.7 to 22.9%) and increased in semiarid zone from 31 to 40%. However the area of millet, the crop environmentally suitable and nutritive, showed a considerable decline (12 to 5%) (Table 24.4). This is a reflection of market forces at work, and changes in the food habits of the people as their incomes increase. The production of wheat and

Table 24.4 Area under cereal crops (Rice, Wheat & Millet) in Haryana, 1971–1991 ($\times 10^3$ hectares).

Region	Rice			Wheat			Millet		
	1971	1981	1991	1971	1981	1991	1971	1981	1991
Arid	19.2	41.2	63.6	237.8	396.9	560.7	521.1	535.4	399.5
	(0.9)	(1.6)	(2.6)	(11.7)	(14.9)	(22.9)	(25.6)	(20.1)	(16.3)
Semiarid	271.8	463.4	573.7	939.2	1,165	1245.1	361	316.5	157.1
	(9.0)	(14.6)	(18.4)	(31.2)	(36.8)	(39.9)	(12)	(10)	(5)

Note: Figures in parentheses are the percentage each crop occupies of the gross cropped area.
Source: Statistical Abstract of Haryana (various issues).

Table 24.5 Cereal production in Haryana during 1971 and 1991 (Tonnes $\times 10^3$).

Crop	Year		Percent change
	1971	1991	
Wheat	2,402	6,496	170 (+)
Rice	536	1,803	236 (+)
Coarse grains	923	506	45 (−)
Total cereals	3,861	8,805	128 (+)
Pulses	682	273	60 (−)
Foodgrains	4,545	9,078	100 (+)

rice increased by 170 to 236%, respectively, but the production of pulses and coarse food grains declined by 60% and 45% (Table 24.5). Pulses being the main source of proteins for the largely vegetarian population of Haryana, this decline should be considered the cause of worry.. The area of oil seeds increased substantially in arid and semi arid region while those of cotton and sugarcane crops fluctuated.

24.6   Impact of green revolution on water, land and soil

During Green Revolution main phase (1971–91) the rice-wheat rotation became predominant in Haryana, as well as in rest of the Indo-Gangetic plains. Rice is sown during warm season (*Kharif* crop) and wheat during winter months (*Rabi* crop). The last three decades in Haryana have seen intensification and diversification in the cropping pattern for cereals, oil seeds and cotton, largely as a result of government policies, but the markets and population dynamics have also played an important role in the change. The adverse environmental effects of these pressures on the land have been felt mostly in the arid zones of the state. The satellite image of Haryana clearly shows the degraded arid land of sandy and saline soils as white and light-toned areas, concentrated mostly in the western, north-western and southern part of the state (Plate 24.1; HARSAC, 2006).

Intensification of agriculture has led to a water table decline (Table 24.6) and soil salinization in both the semiarid and arid regions, but more so in the latter (Table 24.7). Water logging has also degraded vast areas of land in west-central and

Plate 24.1 A satellite view of Haryana showing degraded land on the west.
Source: HARSAC (2006). (See *Colour plate 24.1*)

western Haryana. Equally worrisome is the excessive use of fertilizers and pesticides, which has led to a deterioration of land, soil and water quality.

The trend of increasing N-fertilizer consumption and groundwater use for irrigation in Haryana vis-a-vis rising nitrate levels (114–1800 mg/l, as against the permissible

Table 24.6 Water table – recession in 24 and rise in 19 districts of Haryana during 1974–1998.

Water table recession (in meters)	No. of districts	Remarks
1.0 to 2.0	4	Mostly in semiarid
3.0 to 4.0	4	(Eastern and
6.0	1	Southeastern)
10 to 15 m	3	Haryana
Total No. of districts	12	
Water table rise (in meters)		
3.0	2	Mostly in arid
4.0 to 5.0	2	(Western and
6.0 to 7.0	2	Southwestern)
9.0	1	Haryana
Total No. of districts	7	

Data source: Condensed from Tanwar (1999).

Table 24.7 Soil degradation in arid and semiarid Haryana (Modified after Vashistha *et al.* 2007).

Region	Area		Degraded area		
	(Sq. Km.)	(as % of all Haryana)	(Sq. Km.)	(as % of all Haryana)	(% of region)
Arid region	18,937	42.83	6,410	14.50	33.85
Semiarid region	25,275	57.17	8,267	18.70	32.71
All Haryana	44,212	100.00	14,678	33.20	33.20

level of 50 mg/l) in shallow groundwaters are indicative of high nitrate pollution risk in these two green revolution states (Lunkad & Bhatt, 1990; Lunkad, 1994 and Agrawal *et al.* 1999).

Policy failure (e.g., incentives, subsidies, imports etc.) and socio-economic factors (e.g. partitioning of land among farmers' families, enhanced input cost and poor government support pricing) have rendered agriculture in Haryana an unprofitable venture for 80% of its farmers having small (<2 ha) and medium (<10 ha) holdings particularly during past one decade or so. Hence, the state is presently facing a crucial and challenging task of reversing and ameliorating these adverse impacts.

24.7 Unsustainable groundwater development

The number of hand pumps and tube-wells in Haryana has increased from 25,000 in 1966 to 414,000 in 1984–85 (Lunkad, 1988) and then to 564,000 in 1998, i.e. a rise of more than 22 times in 32 years (Tanwar, 1999). From the data given by Limaye (2003), Lunkad has calculated that in Haryana the ratio of deep (>50 m depth) to shallow tube wells (0–50 m depth) has increased from 1:100 to 1:178 in 2003. Thus the utilization of shallow groundwater is 100–200 times more than that of deep groundwater. Similarly, the number of deep tube wells upto 300 m has also increased from 1000 to more than 5000 during three decades.

During the first decade of Green Revolution (1970–71 to 1980–81) the groundwater use increased tremendously from 1.0 BCM to 7.0 BCM but remained fluctuating between 5.5 to 7.0 BCM during next decade (upto 1990–91). In 2000–01 the conjunctive use of groundwater and surface water for irrigation reached 50–50%. However, the groundwater use crossed over in 2001 and increased to 52.6% in 2003–04, as there has been no enhancement in the canal irrigated land (Anonymous, 2006). The number of tube wells increased from 25,000 in 1966 to 612,000 in 2005 (Table 24.2). The present draft of 8 BCM has been projected to reach 9 BCM (85%) by the year 2010 if the development growth rate is kept slow (0.2 BCM/year).

Thus Haryana is likely to face an acute shortage of groundwater by 2020 when the State would be drawing more than cent per cent of its annually rechargeable groundwater (11 BCM maximum) and the population would have increased to 30 million from 21.1 million in 2001.

24.8 Declining water table

According to NABARD and CGWB a maximum development of 85% of replenishable water is permissible leaving 15% for other uses, viz., domestic, potable industrial and ecological. The CGWB has divided Haryana (19 districts) into 108 blocks. In 32 blocks (6 districts) the groundwater development exceeds 85% by far. These areas (dark zone) recorded a decline of 3 to 12 m. in water level during the period from 1974 to 1998. The maximum decline was recorded in Mahendergarh (12 m) and Kurukshetra (10 m). The maximum rise of water level (9 m) was recorded in Hisar district (Table 24.6)

In Haryana, more groundwater could be harnessed because only 68 per cent of the annually replenishable resources (~70 BCM) was being utilized. However, the distribution of remaining resource in terms of quantity and quality is not uniform. Farmers in arid western zone would not be interested to invest in deep tube wells in saline water zone. On the other hand, in semiarid region the average fall in the water table was 0.3–0.6 m annually indicating that groundwater utilization exceeded replenishment.

24.9 Waterlogging and soil salinization

Seepage through unlined canals and excessive irrigation in areas of shallow groundwater table are the main causes of waterlogging. An area where water table comes up within 0 to 1.5 m depth range from the surface, is said to be fully waterlogged, between 1.5 to 3.0 m is waterlogged and 3 to 10 m is potentially waterlogged. Safe areas are those where water table is at >10 m depth from the surface.

Unscientific canal irrigation in the western Haryana, where groundwater was at a shallow depth, has resulted in waterlogging over vast areas. In 1974 about 1059 sq. km. land (2.5% of the state's total area) was lying in the danger zone (0–3 m. depth from the surface) but the same increased to 4418 sq. km. (10% of the state total area) in 1983. Thus, in less than a decade, the area of land prone to waterlogging quadrupled, which is alarming.

Table 24.8 Extent of waterlogged areas in Haryana (in June 2005) (Sq.km.).

Geographical area of Haryana (sq. km.)	Hilly area (sq. km.)	Area in sq. km. under various depth range in meters			
		Fully water logged 0–1.5 (m)	Water logged 1.5–3 (m)	Potential water logged 3–10 (m)	Safe area >10 (m)
44212	890	251	2346	18513	22212
% 100	2.01	0.56	5.30	41.87	50.23

Source: CSSRI, Karnal, Haryana, India (in abstracted form).
Note: Of the total Geographic area of Haryana about 6% is waterlogged, 42% is potentially waterlogged and 50% is safe.

A recent and more precise study carried out by CSSRI, Karnal in June 2005 has brought to light state's 251 sq. km. area (0.56% of the total) was fully water logged, 2346 sq. km. (5.30% of the total) was waterlogged and 18513 sq. km. (41.87% of the total) was potentially waterlogged (Table 24.8). In Hisar district alone 105 sq. km. area was fully waterlogged, 392 sq. km. was waterlogged and 2300 sq. km. was potentially waterlogged.

Similarly, the absence of proper drainage has increased soil salinity affected land in Haryana (climate being dry and desiccating over 10 to 11 months in a year). According to a World Bank study carried out in 1985, about 0.5 M. ha land was affected by salinization in the State and the same has since been increasing. However, a recent wastelands atlas of Haryana (HARSAC, 2006) estimates that Haryana has 7.38% (3258 sq km or 0.3258 M. ha) of total wastelands under different categories.

According to the National Bureau of Soil Survey and Land Use Planning, one-third of the land area of Haryana has degraded soil of which 23% suffers from medium level degradation and 5% from high-level degradation. Chemical degradation, probably from the excessive use of chemical fertilizers and pesticides, accounts for 6 per cent of soil degradation. About one third of both regions are categorized as degraded area. In the 'Rice Belt' (Fig. 24.2B), 39% land is degraded.

24.10   Urbanization and depletion of cultivated land

The net sown area has remained static at 81% of the total area in Haryana since 1971 indicating that the potential for expansion of agriculture was already exhausted more than three and a half decade ago. Another striking feature of the land use pattern in Haryana is very small proportion of forest cover, 2.3% in 1970 and 2.5% in 1995 (much below 33% national norm), which also indicates extensive use of land for cultivation. In land use pattern, an increase in the area devoted to non-agricultural uses is the direct consequence of urbanization, industrialization and the state Governments policy of creating Special Economic Zones (SEZs) (Vashishtha et al., 2007).Between 1970 and 1990, 29,000 ha land has been diverted for housing, industrial estate and infrastructure (roads). Similarly, the land classified as permanent pastures and grazing land declined by 30,000 ha between 1970 and 1990 (from 1.2 to 0.5% of the total land).

Fallow land increased by 20,000 ha from 1970 to 1990. Net sown area declined by 25,000 ha in 1980s. More than 17,300 ha (0.47% of net sown area) land has been defaced by soil pits for brick kiln industries in Haryana (Lal, 1998). In May 2006, more than 10,000 ha land in Jhajjar and Gurgaon districts has been leased by the Government to the Reliance Company for the creation of SEZ (Special Economic Zone under WTO). *In this way food would no longer be grown in about 1,31,300 ha land, which constitutes about 3% of the total land.*

24.11   Artificial recharge possibilities

Artificial recharge of aquifers in the dark-zone of eastern Haryana such as that around Kurukshetra, depends upon the availability of surplus (silt–free) surface water (Lunkad and Gupta, 2003). There is a scarcity of surplus surface water in Haryana due to (i) lack of any perennial stream in its hinter lands (except Yamuna on its eastern border which has no surplus water), (ii) semi-desertic climate & general rainfall deficiency – on a decadal scale and (iii) long persisting inter-State dispute over Satluj-Yamuna Link Canal (SYL) between Punjab and Haryana. The shallow aquifer zone (0–50 m) holds poor prospects for artificial recharge around Kurukshetra because it is prone to pollution from surface sources and has mostly fine sand mixed with silt and clay which may not have a good water holding and yielding capacity. The depth zone 50–110 m with medium and coarse sand mixed with gravel and pebbles could be a better horizon for recharge. However, any effective and large scale artificial recharge at this depth will be costly compared to community based surface rainwater harvesting methods such as those suggested by Tanwar (1999).

Some experiments of recharge in Markanda and Saraswati (ephemeral streams) beds and Khanpur drain done by HIRMI (Haryana Irrigation Research and Management Institute) have failed. However, experiments by CGWB of injecting silt-free canal water above atmospheric pressure into under ground aquifers have shown initial success. At present there is no noteworthy and functional artificial recharge system in Haryana capable of raising the water level on a regional or even on a local scale over a time-span of ten years.

24.12   The challenges ahead

From the preceding it emerges that making optimum use of saline and marginally saline groundwater in the western zone, conservation and recharge of groundwater in the depleted eastern zone (dark zone), reducing waterlogging and salinization of soils in the affected areas are some of the challenges facing the State of Haryana in the management of its precious groundwater resource.

The irrigation water requirement varies from crop to crop. One hectare of sugarcane needs 300 ha.cm (3 ha. m or 30,000 m^3) of water. The same amount of water would irrigate about 2.5 ha of vegetables, 5 ha each of cotton, onion and paddy, 8 ha of wheat and maize, 12 ha of groundnut, 15 ha of jowar (Sorghum), 21 ha of gram and 30 ha of bajra (pearl millet) (Falkenmark *et al.*, 1990). Thus the use efficiency of water depends on crop selection and the method of irrigation. There is a need for a paradigm shift from the conventional 'flood irrigation' and 'gravity irrigation' methods to sprinkler, micro-spraying and drip irrigation practices to ensure more efficient water use (Hillel, 1987). Drip irrigation, although capital intensive has been found to cut water use by 30–70% relative to standard irrigation procedure and efficiencies as high as 95% have been achieved in some cases (Aswathanarayana, 2005).In saline water zone, salt-tolerant strains of crops like millet, barley, sorghum, gram, and maize, and vegetables like spinach and carrots and fruit trees such as dates, could be grown. Thus the problem of salinity of irrigation water can be managed by proper crop selection and crop alternation.

24.13 Suggested corrective measures

Groundwater is a critically important natural resource of Haryana, and its use needs to be regulated by the adoption of the following integrated soil-water-land and crop management strategies:

1. Recourse to sprinkler, micro-spraying and drip irrigation in place of current practice of gravity irrigation.
2. Encouraging practice of 'Irrigation with drainage' in place of 'Flooding the fields' method. This can be done in large sized farm holdings (10 ha and above).
3. Encouraging rainwater harvesting through 'johads' (small earthen dams) where topography allows.
4. Obtaining surplus surface water of Satluj river by early completion of Satluj–Yamuna Link Canal (SYL) which has long been in dispute with the adjoining Punjab State. This would enhance the possibility of artificial recharge of groundwater in the dark zone of Eastern Haryana.
5. The use of biofertilizers and green manures should be encouraged and only optimum nitrate fertilizer doze (120 Kg/ha maximum) should be applied. In the rural hinterlands where nitrate in groundwater is 50 mg/l (maximum permissible safe limit) and above, nitrate free or low nitrate water from alternative source should be made available in polythene containers for bottle-fed infants to avoid infant mortality due to Blue Baby Syndrome (death risk above 90 mg/l).
6. Taking up integrated soil & groundwater development projects; Gypsum treatment and flushing out mono-valent sodium salts in areas of saline and alkaline soils, mostly in South-Western Haryana.
7. Changing the presently practised wheat-rice crop alternation cycle to wheat-pulses/beans or rice-pulses/beans and reducing area of sugarcane, a water intensive crop (Aswathanarayana, 2005). In eastern Haryana (groundwater 'dark zone') declining water levels can be checked considerably by reducing cultivation of sugarcane and paddy in favour of wheat, maize, gram and jowar (Sorghum). Also, cultivation of salinity-tolerant crops, vegetables and fruit trees.
8. Water is a mobile resource which can be used and reused many times (e.g., municipal water supply → waste water → bio-ponds → irrigation water → groundwater recharge → groundwater withdrawal etc.) Therefore, recycling of water must be made mandatory by installation of sewage treatment plants in all municipal towns and district headquarters generating waste water in substantial quantities. Reuse of treated water for irrigation, washing, and cleaning depending on quality.

References

Agrawal, G.D., Lunkad, S.K. & Malkhed, T. (1999). Diffuse agricultural nitrate pollution of groundwaters in India. *Wat. Sci. Tech.* **39**(3), 63–65, Elsevier Science Ltd., Great Britain.

Anonymous. (2006). 'Statistical Abstract Haryana. (2004-05)', Economic and Statistical Adviser, Planning Department, Govt. of Haryana, Chandigarh (India).

Aswathanarayana, U. (2005). 'Water Resources Management: How to use science to make informed decision' *J. Appl. Geochem.*, **7**(2), 129–133.

Falkenmark, M., Lundquist, J. & Widstrand, C. (1990). *Water scarcity – an ultimate constraint in the Third World Development.* Tema V., Rept. 14, Univ. of Linkoping, Sweden.

HARSAC, (2006). 'Wastelands Atlas of Haryana, 2006', Haryana State Remote Sensing Application Centre (HARSAC), CCS HAU Campus, Hisar-125004 (Haryana), India.

Hillel, D. (1987). *The efficient use of water in irrigation.* World Bank Tech. Paper No. 64, Washington, D.C.

Lal, R. (1998) *Soil Quality and Agricultural Sustainability*, Ohio State University, Columbia, USA.

Limaye, S.D. (2003). *Some aspects of sustainable Development of Groundwater in India..* Eighth IGC's Professor Jhingran Memorial Lecture delivered at 90th Indian Science Congress , I.G.C. Roorkee, India.

Lunkad, S.K. (1988). Water resources and ecosystem in Haryana: An overview. Bhu-Jal News, Quar. J. Central Groundwater Board (C.G.W.B.), **3**(3), 18–21.

Lunkad, S.K. & Bhatt, S.K. (1990). Water resources and Ecosystem Management in Haryana. In *Growth Development and Natural Resources Conservation*; NATCON Publication-3, 221–232, Nature Conservators, Muzaffarnagar (India).

Lunkad, S.K. (1994). Rising nitrate levels in groundwater and increasing N-Fertilizer Consumption. BHU-JAL News, Quarterly Journal of Central Groundwater Board, Vol. **9**(1), pp. 4–10, New Delhi.

Lunkad, S.K. & Gupta, Y. (2003). Aquifer geometry around Kurukshetra: A synthesis from bore-hole data, In Khurshid, S. and Umar, R. (eds.) *Proceedings of National Conference on Groundwater Management-Future Challenges and its impact on Environment,* March 28—29, Department of Geology, A.M.U., Aligarh, India, 54–66.

Lunkad, S.K. (2006). The Present and the Future of Groundwater Resources of Haryana, India, In Venkateswararao, B. *et al.* (eds.) *Proceedings of International Conference on Hydrology and Watershed Management* (ICHWAM-2006), Dec. 5–8, Jawahar Lal Nehru Technical University, Hyderabad, India, **1**,165–181.

Sharma, B.L. (2007) *Haryana at a Glance*, Sahitya Bhawan Publications, Agra, India.

Tanwar, B.S. (1997). Water related problems of Haryana: Need for Science Technology inputs for water management and geohydrology. In pre-proceedings volume on *National meet on Science and Technology inputs for Water Resources Management,* I.I.T., New Delhi, 60–68.

Tanwar, B.S. (1998). Groundwater Management in Haryana. BHU-JAL News, Quar. J. Central Groundwater Board, **13**(3&4), 1–4, New Delhi.

Tanwar, B.S. (1999). Haryana me BHU-JAL ka upyog evam recharge ki aavashyakta: Haryana Sinchayee Patrika (in Hindi), **1**(7), 5–11, HIRMI Publication, Kurukshetra (India).

Thussu, J.L. (1999). Geo-environmental Appraisal of Haryana, *J. Geol. Soc. India,* **54**, 621–640.

Vashistha, P.S., Sharma,R.K., Malik, R.P.S. & Bathla, S., (2007). Population and Land Use in Haryana, In INSA-CAS (Eds), *Growing Populations, Changing Landscapes; Studies from India, China and the United States*, National Academies Press, Washington DC, 77–144.

Chapter 25

Governance of food security, with reference to farming in the tropical rainforest areas of Amazonia, Brazil

Alfredo Kingo Oyama Homma[1], *Ana Rita Alves*[2], *Sérgio de Mello Alves*[3] *& Avílio Antônio Franco*[4]

[1] *Embrapa Amazônia Oriental, Belém – Pará, Brazil;* [2] *Instituto de Desenvolvimento Sustentável Mamirauá, IDSM, Tefé-Amazonas, Brazil;* [3] *Embrapa Amazônia Oriental, Belém- Pará, Brazil;* [4] *Directorate of Research Institutes Coordination, Ministry of Science and Technology, Govt. of Brazil – Basília-Distrito Federal-Brasil*

25.1 Introduction

Food availability in Brazil is more than sufficient to feed the country's entire population. Excluding exports and adding domestic food production to imports, availability of grains is over 340 kg/per capita/year, which represents almost one third more than the minimum nutritional needs. If one considers that 200 kg/per capita/year of grains is sufficient to meet energy needs of 2,000 kcal/day for an adult with 70 kg and, considering that 21 million inhabitants live in Amazonia, it would take an estimated 4,200,000 tonnes of grains to ensure self-sufficiency. Considering that there are some 600,000 smallholders that adopt slash-and-burn based migratory agriculture, who manage to produce a maximum of 1,500 kg of hulled rice per hectare, the maximum area needed to sustain the population of Amazonia would be 2,800,000 hectares/year. As farmers using more advanced methods in Amazonia easily manage to produce 5,000 kg of grains per hectare, the per capita area needed for farming is only 400 m², a mere cultivated 840,000 hectares, using technology to ensure high productivity, could feed the entire population of Amazonia, an insignificant amount compared to the 71 million hectares already deforested by 2006. This provides a clear indication that zero deforestation could be attained in Amazonia by using technology, concentrating on the already deforested frontier, instead of incorporating new areas, far from population centers.

Despite this huge potential, in 2002 it was estimated that 46 million Brazilians were affected by poverty throughout the country, between 7 to 9 million alone in Amazonia. Hunger in Brazil is not an endemic problem, it is political and economic in nature, that is, it is not due to lack of production capability or calamities or to a regime of scarcity. Studies unanimously point out that the problem of hunger in Brazil has been due to lack of income for people to properly feed themselves, a reflection of the inequality of income in Brazil. This is aggravated by high levels of unemployment, feeble rates of economic growth and poorly effective public policies regarding food security.

Household budget studies performed throughout the country demonstrated that poor households spent from 70% to 80% of their earnings on food purchases. Enhanced public policies on supporting family-based agriculture began to be implemented in 2003, in addition to social policies as well as others to increase basic food production, causing a drop in food prices and thus promoting a real increase in wages and income distribution. Except for the poorest segments of urban populations in Amazonia, food security has only been a serious issue in the rural areas during periods of natural catastrophes, such as flooding and the major drought along the Amazon River in 2005. Government-supplied food is often provided in settlement projects and land squattages, due to lack of productive alternatives and job opportunities (BECKER, 2004). By 2002, there were some 1,354 settlement projects throughout Amazonia, occupying over 231,000 km², involving 231,815 families (BRANDÃO JÚNIOR & SOUZA JÚNIOR, 2006).

Beginning in 2003, the Brazilian government set up the *Bolsa Família* ("Family Grant") Program, which, in October 2006, provided assistance to a total of 11,118,929 families, out of which 1,873,045 families were living in Amazonia, corresponding to 17% of the national total, providing a monthly stipend of a minimum amount of US$ 23.30 and a maximum amount of US$ 44.27, depending on the number of children aged 15 years or under and the state of poverty. This policy has reduced the poorest segment of the country's population by nearly 20%.

Nearly 51% of Brazilian poor are concentrated in non-metropolitan urban areas, while 23% live in metropolitan areas and 26% in rural areas. Regionally speaking, 26% of the country's poor are concentrated in the Southeast, 10% in the South, 5% in the Midwest, 9% in the North and 50% is concentrated in the Northeast.

A number of government assistance and welfare programs have been implemented over the last 50 years in Brazil to address nutritional deficiencies of the poorest segments of the population. Among the longest running, we highlight the School Lunch Program, established in 1940, and which currently serves some 37 million children in public schools, almost 1/5 of the country's population. Generally speaking, the focus of these programs has been investment in human resources and welfare, along with poverty alleviation programs, especially welfare programs to provide assistance to rural smallholders, land reform and rural development.

Studies on family agriculture in Amazonia and Northeastern Brazil have shown that produce sold represents 34% of total income earned by agriculture per se, household consumption valued at market prices represents 19%, selling their labor corresponds to 23% and retirement and community-based public service benefits (lunch providers, teachers, health agents etc.) represents 17%, while community joint efforts and aid from children and relatives who live outside the communities accounts for 7%. These results indicate that greater public investments must be made in hinterland communities, opening more schools, health clinics and, perhaps, involving communities in recovery of side roads, environmental surveillance etc. Insofar as 17% of family agriculture income is from public transfers, the role of the government is important in generating new jobs and enhancing the well-being of communities (MENEZES, 2002). Institutionalized payment for environmental services might also be considered in specific cases.

25.2 Amazonia: Physical, human and political environment

The continental biome of Amazonia covers nine countries and includes an area estimated at 6.4 million square kilometers in size, 63% or 4 million square kilometers of which is located in Brazil (Figure 25.1). The remaining 37% (2.4 million square

Plate 25.1 Location of Amazonia in Latin America. (modified from www.images.google.com.br) (See *Colour plate 25.1*).

kilometers) are distributed among Peru (10%), Colombia (7%), Bolivia (6%), Venezuela (6%), Guyana (3%), Suriname (2%), Ecuador (1.5%) and French Guyana (1.5%). The continental Amazon River Basin corresponds to 44% of the surface area of South America and 5% of the Earth's land mass. It is the largest tropical forest on the planet, equivalent to 1/3 of tropical rainforest reserves and the world's largest gene bank (FENZL & MATHIS, 2004; LENTINI *et al.*, 2005). Despite the fact that 63% of continental Amazonia is located in Brazil, it is noteworthy that the headwaters of the Amazon River and its tributaries are located in neighboring countries, which means that there is a need for Amazonian countries to form a group to ensure its preservation (KINOSHITA, 1999) (Figure 25.1).

For policy planning purposes, nine states, representing 60% of Brazilian territory, were defined in law in 1953 as Brazilian Legal Amazonia: Acre, Amapá, Amazonas, Mato Grosso, Pará, Rondônia, Roraima and Tocantins and part of Maranhão. Despite the stereotyped image of "peoples of the forest", Amazonia is practically urban. The urbanization of Brazilian society occurred in Amazonia as in the rest of the country, where 68% of the population is now urban. This percentage reached 90% in Amapá, 79% in Mato Grosso, 76% in Roraima, 75% in Amazonas, 74% in Tocantins, 67% in Pará and Acre and 64% in Rondônia. The State of Roraima presents the lowest demographic density with 1.45 inhab./km^2 and the State of Maranhão the highest,

Table 25.1 Area; total, rural and indigenous population (2006); demographic density in the states of Amazonia and major regions (2000); HDI and families serviced by the *Bolsa Familia* program.

State	Area (km²)	Total population	Rural population	Indigenous population	Demographic density Inhab/km²	HDI	Families serviced (Oct. 2006)
Pará	1,247,702.7	6,188,685	2,072,911	20,185	4.96	0.723	499,797
Acre	152,522.0	557,337	187,541	9,868	3.66	0.697	54,721
Amazonas	1,570,946.8	2,840,889	732,411	83,966	1.79	0.713	203,066
Roraima	224,118.0	324,152	77,420	30,715	1.45	0.746	31,525
Amapá	142,815.8	475,843	52,262	4,950	3.34	0.753	20,936
Tocantins	277,297.8	1,155,251	296,863	7,193	4.17	0.710	109,159
Rondônia	237,564.5	1,377,792	494,744	6,314	5.81	0.735	94,506
Maranhão	331,983.3	5,651,475	2,287,405	18,371	17.02	0.636	716,604
Mato Grosso	903,357.9	2,504,353	516,627	25,123	2.77	0.773	142,731
Amazonia	5,088,308.8	21,075,777	6,718,184	206,685	4.14	0.721	1,873,045
North	3,852,967.6	12,900,704	3,914,152	163,191	3.35	0.725	1,013,710
Northeast	1,561,177.8	47,741,711	14,759,714	77,585	30.58	0.676	5,530,124
Southeast	927,286.2	72,412,411	6,851,646	12,084	78.09	0.782	2,919,591
South	577,214.0	25,107,616	4,780,924	29,474	43.50	0.811	1,085,104
Midwest	1,612,077.2	11,636,728	1,540,568	57,988	7.22	0.793	570,400
Brazil	8,514,876.6	169,799,170	31,847,004	340,322	19.94	0.766	11,118,929

Source: Basic data IBGE, www.ibge.gov.br, www.undp.org.br

at 17.02 inhab./km². Demographic density in Amazonia contrasts sharply to that in Southeastern Brazil, where it is 78.09 inhab./km², followed by Southern Brazil with 43.50 inhab./km² and the Northeast with 30.58 inhab./km². The high density of these regions is always a push factor for displacement of contingents of the population towards Amazonia in search of public assets and new opportunities not found in their places of origin (Table 25.1). One can safely say that cities such as Manaus and Belém, with 2 million inhabitants, represent large population centers located along the planet's equator.

This region is home to the tropical rainforest that Alexander von Humboldt (1769–1859) called '*Hylea*', being characterized by its singular biodiversity (HOMMA, 2003). It is estimated that every 250 hectare area of Amazon forest contains roughly 750 different tree species, 120 species of mammals, 400 types of birds, 100 varieties of reptiles, 60 amphibians, 43 types of ants and others. This number may be increased further to 950 bird species, 300 mammals, 100 amphibians, 2,500 fish species and 30 million invertebrates, depending on new discoveries.

History has witnessed a succession of cycles based on extraction of its natural resources. There was the cocoa cycle (*Theobroma cacao*) that began at the time City of Belém was founded (1616) and lasted until the period of Brazil's independence (1822). The opportunity presented by cocoa biodiversity was lost when, in 1746, it was taken to Bahia and from there to the African continent and to Asia, which became the new sites for large-scale production. This was the first case of biopiracy in Amazonia of an active element of the economy.

After the biodiversity of cocoa came the rubber tree (*Hevea brasiliensis*), which lasted until the rational plantations in southeast Asia were operating, from seed taken

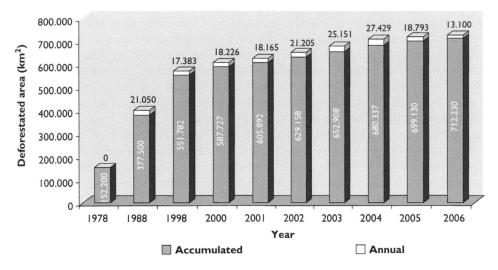

Figure 25.1 Estimated deforested land in the Amazon region in Brazil (www.obt.inpe.br/deter)

by Henry Wickham (1846–1928) in 1876 (HOMMA, 2003). This was the second case of biopiracy and reminders of this period of opulence include the opera houses known as Theatro da Paz and Theatro do Amazonas, the building of railways, the floating port of Manaus, among others. The rubber tree was subsequently planted around the world, totaling more than 8.3 million hectares of plantations, resulting in the fact that, presently, Brazil imports 65% of its rubber consumption. The same thing occurred with neighboring countries in the case of tomatoes (*Lycopersicum esculentum*), potato (*Solanum tuberosum*) and tobacco (*Nicotiana tabacum*), which became products of worldwide consumption. Corn (*Zea mays*) is another example of a plant known to the Incas, Mayas and Aztecs, and cassava (*Manihot esculenta*) used by indigenous peoples, which were spread by the Portuguese to Africa and Asia.

Later came the biodiversity cycles of rosewood (*Aniba rosaeodora*) and Brazil nut (*Bertholletia excelsa*), which expanded and reached a peak, became vulnerable and listed as endangered species. Currently, timber logging and açaí (*Euterpe oleracea*) are the most biodiversity-based products, along with cupuaçu (*Theobroma grandiflorum*), peach palm (*Bactris gasipaes*), guaraná (*Paullinia cupana*), ornamental and frozen fish, shrimp and others. Throughout this history, a number of exotic biodiversity elements have been introduced, as well, such as cattle, water buffalo, jute (*Corchorus capsularis*), black pepper (*Piper nigrum*), papaya (*Carica papaya*), jambo (*Syzygium malaccensis*), mangosteen (*Garcinia mangostana*), durian (*Durio zibethinus*), rambutan (*Nephelium lappaceum*), melon (*Cucumis melo*), and others. In the case of jute and black pepper, introduced by Japanese immigrants, coming from former British colonies, represent the other side of biopiracy of species from Amazonia, and had strong impact on regional economy, even though they have recently become relatively less important.

Chronic deforestation of the Amazon Rainforest (Figure 25.2) has become a national and international concern. Concrete measures must be taken to reach *zero deforestation* in order to avert repetition of what took place with the Atlantic Rainforest in

Brazil, reduced to less than 7% of its original forest cover (www.obt.inpe.br). In 1975, when preliminary deforestation assessments for Amazonia were released, based on the Landsat satellite launched on 07/23/1972, the deforested area of Amazonia went from 15 million hectares to over 71 million hectares (2006), equivalent to the surface area of France, the Netherlands, Belgium and Israel or 16% of Amazonia. This does not mean that 95% of Amazonia will be completely deforested by 2020, as several scientific journals projected in early 2001 (LAURANCE et al., 2001; SCHNEIDER et al., 2000). Data from the 2000 Demographic Census showed that 81.22% of the country's overall population and 69.70% of the population of Northern Brazil already live in urban centers. This means that, as a result of the urbanization process, there is insufficient labor available to effect such a sizeable deforestation. What is quite possible is that, if current deforestation rates are maintained, the area deforested may double, reaching 1/4 of Amazonia by 2030. It does not, however, justify this chronic annual deforestation rate that varies between 1.1 and 2.9 million hectares, indicating the need to adopt stricter enforcement measures and to establish new alternatives.

In addition to the biophysical consequences caused by suppressing the forests, such as erosion, contamination of rivers, extinction of species and loss of environmental services that have impacts worldwide, deforestation has negative social impacts such as land conflicts, poverty, social inequality and innumerable problems in the field of public health.

25.3 Agriculture in Amazonia – macroeconomic view

Grain harvests in Brazil have quadrupled in the last 30 years, while the area used for agriculture, which in 1973 was roughly 24 million hectares, has not even doubled in size, reaching 41 million hectares in 2003. Supply of three million tonnes of beef, pork and fowl multiplied almost six-fold during the same period, totaling 17.8 million tonnes. The most expressive increase was in chicken, which rose sharply from 217,000 to 7.6 million tonnes. Increases were also seen in fresh produce, fruit, flowers, fibers and forest essences. In 2003, Brazil was already the world's major exporter of tobacco, orange juice (*Citrus sinensis*), sugar, alcohol, beef, tanned leather and garments, in addition to coffee (*Coffea* spp). In 2004, Brazil surpassed the United States as the world's largest chicken exporter. Agribusiness currently corresponds to 33% of the national GDP, contributes to 42% of export receipts and employs 37% of the economically active population.

The states that comprise the Amazon Region of Brazil are characterized, macro economically, by their low participation in the country's GDP (Gross Domestic Product). Considering that Amazonia, in 2004 alone, was responsible for 7.8%, percentages that individual states such as Rio Grande do Sul and Paraná easily surpass, just to cite two examples (Table 25.2), the agricultural GDP share of the states from Amazonia is quite small. With the exception of the States of Pará and Mato Grosso, the remaining states of Amazonia contribute only negligible amounts. This leads to questioning the high environmental and social costs of agricultural activities in Amazonia, when related to levels of deforestation and rural violence. The State of Paraná has a GDP that is triple those of states such as Pará and Mato Grosso or half of that of Santa Catarina (PRODUTO, 2006; ELECTRONORTE, 2006).

Table 25.2 Share in national GDP of states in Legal Amazonia, total amount, per capita, state and national share in agriculture, rural population, and population engaged in agricultural activities (2004).

State/Region	GDP share %	GDP US$ 1,000,000	Per capita GDP US$	Agriculture share in GDP %	Agriculture share in GDP Brazil %	Rural population* %	Active agriculture population* %
Rondônia	0.6	3,566	2,283	15.3	0.9	35.91	33.14
Acre	0.2	1,187	1,882	5.9	0.1	33.65	25.63
Amazonas	2.0	13,135	4,185	3.6	0.8	25.78	24.98
Roraima	0.1	682	1,786	3.8	0.0	23.88	17.56
Pará	1.9	12,515	1,827	22.8	4.6	33.50	26.93
Amapá	0.2	1,361	2,487	4.6	0.1	10.98	8.86
Tocantins	0.3	1,745	1,382	12.9	0.4	25.70	27.27
Mato Grosso	1.6	10,224	3,719	40.8	6.6	20.62	20.87
Maranhão	0.9	6,056	1,006	20.1	2.0	40.49	43.15
Amazonia	7.8	50,472	2,284		15.5	31.80	
North	5.3	34,192	2,379		6.9	30.30	26.45
Northeast	14.1	90,929	1,803		14.3	30.96	30.32
Southeast	54.9	355,102	4,590		31.7	9.48	9.11
South	18.2	117,769	4,422		31.7	19.07	19.10
Midwest	7.5	48,577	3,804		15.4	13.27	13.68
Brazil	100.0	646,569	3,561	9.5	100.0	18.78	17.56

Source: Basic data IBGE, www.ibge.gov.br, www.undp.org.br
Average exchange rate 1US$ = R$2.7323;
*Rural population and active population refers to year 2000.
Population active in agriculture as a percentage of persons 10 years of age or older occupied in the week of reference, by gender, groups of hours normally worked per week, in agriculture, ranching, forestry, hunting and fishing activities in 2000.

The transfer of regional wealth in the per capita GDP shows that the Northern Region only outranks the Northeast, which, despite having a GDP almost three times greater than that of the North, it is diluted by a larger population. The *per capita* GDP of the State of Amazonas, due to the industrial park at the Free Trade Zone of Manaus, ranks highly in national terms, below Brasília, Rio de Janeiro, São Paulo, Rio Grande do Sul and Santa Catarina (Table 25.2).

The performance of the states that comprise Amazonia are specific and characterize agricultural and industrial predominance, as well as important third sector activities, particularly public services (Table 25.2). Agriculture is negligible in the makeup of state GDP of the State of Amazonas, where the transformation industry is responsible for over half of its GDP. One can say that the States of Mato Grosso and Pará are predominantly agricultural, with over 1/4 of the wealth produced therein coming from primary activities. States with lower share in production of national wealth, third sector jobs, especially in the States of Roraima (58.2%), Amapá (44.4%) and Acre (42.7%), are important.

The present decline in population growth rate in Brazil, which should grow at an annual rate of 1% for the period from 2005–2014, added to rural migration to the city in the last few years, should reduce pressure on deforestation in Amazonia. There is already a significant decline in deforestation in Amazonia (Plate 25.2), even more than the decline in the international price of soybean, a commodity that has been

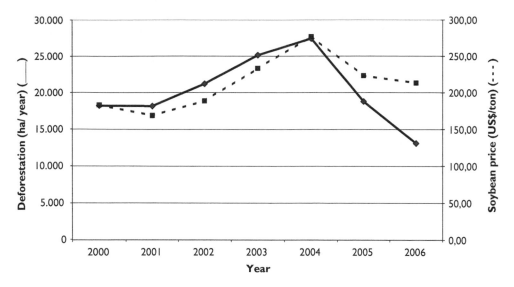

Plate 25.2 Deforestation in Amazonia versus international soybean prices since year 2000 (www.obt.inpe.br/deter and www.ibge.gov.br). (See *Colour plate 25.2*)

attributed the responsability for deforestation. The Northern and Northeastern regions are the ones that still present the highest percentages of rural population in Brazil. Of all of the states in Legal Amazonia, Maranhão has the largest relative percentage of its population living in the rural areas and the State of Amapá has the highest rate of urbanization (Table 25.1). The decrease in percentages of the rural population and the share of agriculture in state GDP shows the low profitability of primary sector activities, except for in the States of Pará and Mato Grosso. The predominance of an extractivist economy and insufficient downstream integration are the main reasons for the low value attached to agricultural GDP.

25.4 Productive macro systems in Amazonia

When one divides worldwide arable land by the population that existed in 1900, one finds that there was a little over 2 hectares of land per capita for food production. In 1960, per capita arable land had dropped to 1.2 hectares and in 2000 had fallen to 0.5 hectare. Estimates are that this will drop to 0.3 hectare by 2025. One third of land in China is desert, another third is mountainous and only one third is available for agricultural activities (DIMÁRZIO, 2004). The United States is reaching the limit of its farmable land, from 270 to 280 million hectares, and Europe is no different. In Brazil, without including Amazonia, 380 million hectares could be used for agriculture, if one includes pasture lands (220 million), annual crops (59 million), perennial crops (8 million), planted forests (5 million) and 106 million in unused lands fit for agriculture. This shows the huge potential of Brazilian agriculture for the world food security in the future. In the case of Amazonia, there are over 51 million hectares of planted and native pasturelands, showing the predominance of this type of activity, whereas annual crops occupy over 8.7 million hectares and permanent crops 648,000 hectares.

Table 25.3 Area planted (x 1,000 hectares) and relative share of areas with annual, perennial crops and pastures, in selected countries and Brazilian Amazonia, 2002.

Country or region	Annual Crop		Permanent Crop		Pastures	
	Area	Relative share	Area	Relative share	Area	Relative share
Brazil	58,980	7.76	7,600	1	197,000	25.92
Amazonia	8,722	13.46	648	1	51,149*	78.93*
Canada	45,744	338.84	135	1	29,000	214.81
China	142,621	12.58	11,335	1	400,001	35.29
Costa Rica	225	0.75	300	1	2,340	7.80
Indonesia	20,500	1.55	13,200	1	11,177	0.85
India	161,715	19.25	8,400	1	11,062	1.32
Malaysia	1,800	0.31	5,785	1	285	0.05
Australia	48,300	161.00	300	1	398,400	1.328.00
United States	176,018	85.86	2,050	1	233,795	114.05
Argentina	33,700	25.92	1,300	1	142,000	109.23
Japan	4,418	12.84	344	1	428	1.24
World	1,404,052	10.78	130,257	1	3,485,339	26.76

Source: Basic data FAO (www.fao.org), IBGE (www.ibge.gov.br)
* Related to the year of 1996 (planted and native pasture).

As a long-term public policy for deforested areas, it would be interesting to expand areas with permanent crops and reduce the amount of pastureland (Table 25.3).

Using permanent crop areas as a reference indicator, one sees that certain countries specialize in annual crops, others in pastures and still others in perennial crops. Land availability in association with soil quality and favorable climatic conditions, and overcoming manpower constraints, is what made the United States, Canada and Australia specialize in grain production and livestock raising (Table 25.3 and Table 25.4). Indonesia, Malaysia and Costa Rica, on the other hand, are characterized by development of permanent tropical crops, such as rubber and cocoa (which were originally from Amazonia), oil palm (*Elaeis guineensis*), black pepper, coconut (*Cocos nucifera*), and others. Land availability in Brazil is characterized by relative expansion of ranching and grain production, despite the large absolute area for permanent crops such as coffee, orange, cocoa, reforestation, banana, oil palm, black pepper, cashew (*Anacardium occidentale*) etc.

Table 25.4 presents land use in states of Amazonia and other selected states, according to destinations of use by annual and perennial crops and pastures for the years of 1980, 1985 and 1995–1996, where one may observe the predominance of ranching and annual crops, except in the case of Mato Grosso (REBELLO, 2004).

Based on the Agriculture/Livestock Raising Census of 1995–1996, pasture land in the State of Pará was 28.41 times larger than areas dedicated to perennial crops. Likewise, the area used for annual crops is 2.08 times greater than perennial crops, a reflection of the advance of ranching in the state. Other states that have large portions of their lands dedicated to pastures are the States of Acre (37.18), Maranhão (65.90), Mato Grosso (133.02) and Tocantins (491.75), in relation to their respective areas dedicated to perennial crops. Due to their characteristics, these

Table 25.4 Relative index of area planted with temporary and permanent crops and pastures in states within Amazonia and other selected states, 1980, 1985 and 1995–1996.

State	1980			1985			1995–1996		
	Annual crop	Permanent crop	Pasture	Annual crop	Permanent crop	Pasture	Annual crop	Permanent crop	Pasture
Pará	2.95	1.00	19.54	2.44	1.00	21.06	2.08	1.00	28.41
Acre	2.37	1.00	11.61	3.03	1.00	19.12	3.60	1.00	37.18
Amapá	1.81	1.00	27.46	1.05	1.00	29.47	1.04	1.00	25.15
Amazonas	2.28	1.00	3.92	1.45	1.00	4.07	1.40	1.00	5.40
Rondônia	1.19	1.00	4.42	1.46	1.00	5.11	0.70	1.00	11.49
Roraima	6.15	1.00	389.73	4.98	1.00	267.81	1.07	1.00	23.95
Tocantins (1)	0.00	0.00	0.00	0.00	0.00	0.00	10.86	1.00	491.75
Maranhão	21.46	1.00	77.51	14.11	1.00	63.09	9.20	1.00	65.90
Mato Grosso	10.97	1.00	113.87	14.59	1.00	120.09	17.25	1.00	133.02
São Paulo	2.36	1.00	5.84	3.04	1.00	6.15	2.84	1.00	6.62
Paraná	5.39	1.00	5.80	8.65	1.00	9.55	15.38	1.00	21.44

Note: (1) founded in 1988.
Source: Rebello (2004).

areas may, in a certain manner, become future areas for the expansion of soybean (*Glycine max*) in these states. Heavy investments are needed to recover these pasture areas, due to the need to expand pastures for the vegetative growth of herds in these states.

25.5 Food security program for Amazonia

Contrary to the image that only destruction reigns in Amazonia, academic literature rarely mentions the region's productive park, not only agricultural and industrial, but of services, as well. With regards to the agricultural segment, Amazonia represents an important center for agricultural production, not only locally but nationally and internationally. Table 25.5 lists the main crops, livestock, fish and important activities developed in the Amazon Region.

Agricultural activities not listed include leafy vegetables (native and exotic), small livestock (goats, sheep etc.), native fruit trees (cupuaçu, açaí, peach palm etc.) and exotic fruit trees [lemon (*Citrus* spp., tangerine (*Citrus nobilis*), mango (*Mangifera indica*), avocado (*Persea americana*), guava (*Psidium guayava*), watermelon (*Citrullus vulgaris*), melon, mangosteen, rambutan (*Nephelium lappaceum*), durian (*Durio zibethinus*), jambo etc.], and diverse plants such as noni (*Morinda citrifolia*), neem (*Azadirachta indica*). Native timber species such as mahogany (*Swietenia macrophylla*), parika (*Schizolobium amazonicum*), andiroba (*Carapa guianensis*), freijó (*Cordia goeldiana*), etc., and exotic tree species such as eucalyptus (*Eucalyptus* spp.), acacia (*Acacia mangium*), teak (*Tectona grandis*), gmelina (*Gmelina arborea*), and African mahogany (*Khaya ivorensis*), are being planted in monocultures or as components of agroforestry systems.

These activities undoubtedly contribute towards the 71 million hectares of deforestation (2006). Nevertheless, already existing knowledge and technologies show that

Table 25.5 Primary sector production in Amazonia and in Brazil (Average 2000–2004).

Crops/livestock	Brazil	Amazonia	% in relation to Brazil
Animal food products			
Eggs (1,000 dozen)	2,594,472	118,275	4.56
Fowl (individual)	901,002,842	58,573,057	6.50
Milk (1,000 liters)	21,529,699	2,081,587	9.67
Pork (head)	32,295,235	3,383,670	10.48
Fish farming (ton)	269,698	35,358	13.11
Beef (head)	186,335,480	58,461,053	31.37
Buffalo (head)	1,123,440	793,230	70.61
Plant food products			
Orange (ton)*	17,686,323	267,311	1.51
Papaya (ton)	1,621,548	47,453	2.93
Sugarcane (ton)	369,204,268	14,433,578	3.91
Beans (ton)	2,968,648	193,452	6.52
Coffee (ton)	2,901,914	214,988	7.41
Corn (ton)	40,067,837	3,692,390	9.21
Passion fruit (ton)*	480,769	51,772	10.77
Coconut (ton)	1,742,816	254,682	14.61
Banana (ton)*	6,562,754	1,350,746	20.58
Pineapple (ton)	1,423,271	366,326	25.74
Cocoa bean (ton)	184,651	50,289	27.23
Soybean (ton)	42,861,016	12,566,304	29.32
Hulled rice (ton)	11,075,274	3,353,913	30.28
Urucum (ton)	12,168	3,707	30.47
Cassava root (ton)	22,931,378	7,555,172	32.95
Guaraná (ton)	3,965	1,378	34.75
Cotton (ton)	2,562,877	1,337,945	52.20
Oil Palm fruit (ton)	794,859	623,602	78.45
Black pepper (ton)	56,082	49,006	87.38
Planted timber			
Planted firewood (m³)	36,950,608	281,603	0.76
Planted charcoal (ton)	2,158,025	29,426	1.36
Planted timber in logs (m³)	80,750,747	3,279,744	4.06
Cellulose paper wood (ton)	45,235,359	2,791,428	6.17
Planted rubber (m³)	149,267	37,378	25.04

Source: Basic data IBGE (www.ibge.gov.br).
Note: *Average refer to period from 2001–2004.

production can be both maintained and increased, without the need to conduct further deforestation, by enhancing current technological standards.

As Amazonia is home to 11% of Brazil's population, and many agricultural products are produced at levels below this percentage and are imported from other parts of the country, the major ones being corn, beans (*Phaseolus vulgaris*), coffee, pork, milk, eggs, sugar, fuel ethanol (*Saccharum officinarum*), oranges, potato, tomato, onion (*Allium cepa*), garlic (*Allium sativum*), and edible oil. Some products depend heavily on foreign imports, especially wheat (*Triticum aestivum*) and powdered milk, in the case of the Free Trade Zone of Manaus. Naturally, those foods that do not adapt ecologically to the region, such as wheat, potato, onion, garlic and apple (*Malus sp.*), will continue to be imported.

Amazonia is also a major exporter of foods and raw materials, many of them typically regional, to other regions of Brazil and abroad, as occurs with soybean, beef, black pepper, urucum (*Bixa orellana*), cocoa, oil palm, Brazil nut, açaí, cupuaçu, banana, pineapple (*Ananas comusus*), cassava, rice (*Oryza sativa*), cowpea (*Vigna unguiculata*), cotton (*Gossypium* spp.), guaraná, passion fruit (*Passiflora edulis*), palm heart, without mentioning the timber complex and others. For the oil palm Brazil imports 2/3 of what it consumes and the largest plantations in the country are located in Amazonia. The region could expand production in already deforested areas with an immediate need of 320,000 hectares, thereby contributing towards the recovery of these areas to comply with current legislation stating that diesel fuel be mixed with 2% vegetable oil.

Contrary to perennial crops, except in the case of palm oil for biodiesel, where small fractions of the land are enough to saturate the local, regional, national and international markets, dimensions of land used for annual crops are much larger. Theoretically, in terms of land size, if we were to place tropical perennial crops of the major producing countries in Amazonia, only 3% of the region would be enough to double world production of these crops. Since 16% of Amazonia has already been deforested, one sees that Amazonian agriculture can increase its share of national and international production without the need to expand deforestation.

There is a brisk trade in food crops between the states of Legal Amazonia. Dependency is high in certain states such as Amapá, Amazonas and Acre, which import large amounts of food from Pará, Roraima and other parts of the country. The low deforestation rates in the States of Amapá and Amazonas are the result of importing food from deforested areas in other states. As 21 million people live in Amazonia, there must be sufficient food production to supply both rural and urban populations.

There are a number of products that are not listed in official statistics, which are "invisible products", important in survival and income generation and job strategies, especially of those in family-scale agriculture in Amazonia. These include native fruits, fish, game, firewood etc. Even a part of the products tabulated in official statistics is withheld for domestic consumption, indicating that agricultural and livestock production in Amazonia is underestimated in relation to more advanced locations due to improved data collection systems.

The most important products taken from the forest are timber, Brazil nut, rubber, açaí fruit, rosewood and babassu (*Orbygnia* spp.) (Table 25.6). Non-industrial fishing in Amazonia should be mentioned, due to its great importance in feeding riverine populations and in the volume of fish caught at the national level. A large part of the fish and shrimp caught in Amazonia is exported to other parts of Brazil and abroad.

Amazonian production is primarily extractivist: almost 80% of the timber is from native forests in Brazil, 64% of which is sold on the domestic market and 36% exported (Table 25.6). Timber exports from Brazilian Amazonia represent from 2% to 3% of world timber product exports (LENTINI *et al.*, 2005). There is a large amount of charcoal production using native forests to supply 15 pig-iron plants, located in the States of Pará and Maranhão, which require timber from deforestation of roughly 105,000 hectares/year, mainly performed by small-scale farmers and using waste material from sawmills. It is interesting to note that firewood consumption in Amazonia is low, despite being highly available, since even the poorest families prefer bottled gas and there is a comprehensive distribution of cooking gas even in remote locations.

Table 25.6 Forest and fishing products from Amazonia (average 2001–2004).

Products	Brazil	Amazonia	% in relation to Brazil
Timber			
Extractivist firewood (m³)	48,659,979	1,327,031	2.73
Extractivist charcoal (ton)	1,905,406	864,229	45.36
Extractivist timber (m³)	20,625,692	16,299,949	79.03
Non-timber			
Brazil nuts (ton)	28,428	28,428	100.00
Açaí (ton)	124,493	124,490	99.99
Extractivist rubber (ton)	4,356	4,248	97.52
Babassu (ton)	115,351	106,602	92.42
Heart of Palm (ton)	14,621	14,159	96.84
Rosewood (ton)	29	29	100.00
Buriti (ton)	414	374	90.34
Jaborandi (ton)	710	686	96.62
Piaçava (ton)	95,458	8,582	8.99
Copaiba (ton)	458	458	100.00
Fishing			
Non-industrial extractivist fishing (ton)	505,255	279,201	55.26
Industrial extractivist fishing (ton)	240,962	19,647	8.15

Source: Basic data IBGE (www.ibge.gov.br)

There are dozens of non-timber forest products that are consumed as food and raw materials, the surplus of which is exported to other parts of the country and abroad, many never appearing in official statistics. Among these are native fruits such as açaí, cupuaçu, peach palm, bacuri (*Platonia insignis*), uxi (*Endopleura uchi)*, tucuman (*Astrocarium vulgare*), buriti (*Mauritia flexuosa)*, taperebá (*Spondias mombin*), muruci (*Byrsonima crassifolia)* and piquiá (*Caryocar glabrum*). Several leafy vegetables such as jambu (*Spilanthes oleracea)*, vinagreira (*Hibiscus sabdariffa)*, cariru (*Amaranthus viridis)*, maxixe (*Cucumis anguria)* and others, part of regional cuisine, are beginning to become known nationally and internationally. Medicinal plants including copaiba (*Copaifera* spp.), Andiroba (*Carapa guianensis*) and jaborandi (*Pilocarpus microphyllus*), and fibers such as piaçava (*Leopoldinia piassaba*) and others are exported.

Collection of non-timber forest products such as cocoa, rubber, Brazil nuts and rosewood was important in bringing civilization to Amazonia, in settlement processes and providing economic support to the region and the country. Suffice it to say that extractivist rubber from Amazonia was once the third ranking export product in Brazil, for a period of thirty years (1887–1917), surpassed only by coffee and sugar. Brazil currently imports 1/3 of the cocoa it consumes and 3/4 of its natural rubber, indicating opportunities for family-based agriculture, through a policy to attain self-sufficiency in these products. The extensive waterway network of over 20,000 km of navigable rivers for deep keel vessels enabled access to natural resources and availability of fish and turtles, guaranteeing protein supply, which began to run into conflict with population and market growth, resulting in the exhaustion of many of these resources.

There are an estimated 200,000 extractivists who work collecting timber and non-timber forest products in Amazonia, particularly babassu, timber, açai fruit, Brazil nuts, and others. Except for açai, income share from extractivist collection is less than 25% and all gatherers depend on other activities to ensure survival.

These results indicate that, under current extractivist conditions, this activity by itself is unable to avert deforestation and burnings in Amazonia. The extractivist movement gained momentum worldwide with the murder of union leader, Chico Mendes (1944–1988), but it must be stressed that extractivism of plant material is of limited supply capacity due to low density of plants in the forest. Market growth has led to the development of rational planting, as has taken place with several plants from Amazonia such as cinchona (*Chinchona calisaya* and *C. ledgeriana*), cocoa, and rubber, and is currently occurring with cupuaçu, açai, peach palm, Brazil nut, jaborandi, pepper (*Piper hispidinervum*), curauá (*Ananas erectifolius*), making economic sustainability of communities that depend on these activities difficult. Mankind has, over the last ten thousand years, managed to domesticate some 3 thousand plant species. Some of the other variables that affect extractivism is the discovery of synthetic replacements as happened with rosewood (synthetic linalool), timbó (*Derris nicou* and *D. urucu*) and their replacement by synthetic inseticides. Extractivist economy was important in the past, it is important in the present, but its future options must be seen in making use of the benefits of domestication, in order to buy time, until other technological alternatives appear or as complementary income.

Therefore fauna and flora management techniques, based on solid scientific research and flexibility to change strategies according to markets, is of great importance. Concrete results of açai management performed by riverine populations at the mouth of the Amazon River and pirarucu (*Arapaima gigas*) management at the Mamirauá Sustainable Development Reserve (RDSM) demonstrate the possibilities for implementing programs that add value to and enhance living conditions of local populations, and of establishing strategic partnerships with governmental and nongovernmental organizations to develop proposals for sustainable use of natural resources. The concern lies in the fact that, with market growth, as is already taking place with açai, the management programs used may end up increasing ecological risk to flora and fauna, by homogenizing ecosystems.

25.6 New plants and animals

Several plants from the New World, including potatoes, corn, tobacco, tomato, avocado, rubber, cinchona and others, became universal and are planted and consumed worldwide. Even today, several Amazonian plants are becoming universal – this is happening with guaraná, cupuaçu, açai, peach palm, bacuri, jambu, curauá etc. Among aquatic species, we would mention pirarucu, peacock bass (*Cichla* spp.), tambaqui (*Colossoma macropomum*) and turtles (*Podocnemis expansa*, *Podocnemis unifilis*) etc. In the future, new plants, fish and animals from Amazonia may be incorporated into the productive chain due to their nutritional and functional values, leaving behind extractivism for rational forms of production and management. The high price of many conventional fruits and vegetables, including bell pepper (*Capsicum annuum*), papaya, passion fruit and tomato, in the city of Manaus, a major research challenge will be to make these crops productive in the region, similar to what was done for soybean to grow in the Brazilian tropics.

In addition to the already mentioned plants and animals, new genetic resources from Amazonia may be used as promising foods in the future. Some plants that have been quite important in the past, such as babassu, lost their relevance as an annual oleaginous crop, like cotton, peanut (*Arachis hypogaea*), soybean, corn and sunflower (*Helianthus annuus*), may become important in recovering areas that should not have been cleared, such as river banks, slopes etc. The patauá palm (*Oenocarpus bataua*), produces an oil similar to olive oil (*Olea europaea*). Expanded planting of the Brazil nut tree, replacing extractivist supplies, is already attracting the interest of farmers, as a tree for reforestation in recovering cleared areas and for Brazil nut production.

A major possibility for income and job creation reserved for Amazonia is related to vegetable oil production from perennial plants such as oil palm, andiroba, copaiba, tucuman and others, to mix with diesel oil (biodiesel) or as a compound in the production process of diesel, resulting in less polluting and higher quality oils (H-bio), a field in which Brazil is pioneer. Such programs could act to recover degraded areas and provide socioeconomic inclusion for family farmers.

Within this context, agricultural research should receive incentives to expand on these possibilities, developing agriculture with plants native to the region and taking advantage of the enormous water resources the region has for fish farming. This agriculture and fish farming could provide new meaning to the underused areas already deforested and promote recovery of areas that should never have been cleared.

25.7 The future of food security in Amazonia

Amazonian society faces three major challenges: (1) to protect the largest amount of its area to ensure biodiversity, water resources, and global climate balance; (2) to ensure the survival of the population that lives in the region and enable its sustainable development over time, and (3) to maintain sovereignty over an area that represents 60% of the nation's territory.

The current deforested area of Amazonia (Figure 25.2) estimated at 71 million hectares (2006), would be more than enough to feed the population that lives in the region, now and in the future, using only part of this area and recovering areas that should not have been cleared according to environmental law and conservation and preservation principles. *Zero deforestation* will depend on technological improvements of agricultural practices adopted and resolving issues of poverty and formal education, which are problems that afflict Brazilian society. Reducing the costs of recovering degraded areas, which currently is quite high (US$ 500.00/hectare) and pushes farmers to use cheaper and unsustainable practices (US$ 200.00/hectare), could have positive effects on conservation and preservation in Amazonia.

The Brazilian government has expanded the Conservation Units and Indigenous Lands in Amazonia as a means to preserve the Amazon Rainforest. There are 405 Indigenous Lands in Amazonia, with 103,483,167 hectares and a population of roughly 206,685 indigenous people, representing 20.67% of land area in the Amazon or 98.61% of all Indigenous Land in the country in terms of size (Table 25.7). The availability of 500 hectares per capita has raised voices in protest, some saying that there is "too much land for too few indigenous people" and others pointing to the hunger and malnutrition in many indigenous settlements when linked to the market

Table 25.7 Conservation Units and Indigenous Lands in Amazonia.

Type of use	Area (ha)	% Area Amazonia	% Area Brazil
Conservation Units	**51,733,218**	**10.27**	**6.05**
Full Protection	**24,327,223**	**4.83**	**2.85**
Ecological Station	6,654,919	1.32	0.78
National Park	14,076,048	2.79	1.65
Biological Reserve	3,596,256	0.71	0.42
Sustainable Use	**27,405,995**	**5.44**	**3.21**
Environmental Protection Area	365,006	0.07	0.04
Area of Relevant Ecological Interest	19,012	0.00	0.00
National Forest	19,111,549	3.79	2.24
Extractive Reserve	7,910,428	1.57	0.93
Indigenous Lands	**103,483,167**	**20.67**	**12.11**
Total	**155,216,385**	**30.81**	**18.16**

Source: Basic data IBAMA (www.ibama.gov.br), www.mma.gov.br and www.funai.gov.br).

economy. Amazonia is home to 61% of the country's indigenous population, estimated at over 340,000 inhabitants.

Conservation Units in Amazonia are divided into full protection – which must remain untouched, and include Ecological Stations, National Parks and Biological Reserves, representing over 24 million hectares – and those that can allow economic activities, such as National Forests, Environmental Protection Areas, Areas of Relevant Ecological Interest, Sustainable Development Reserves and Extractive Reserves, which represent over 27 million hectares (Table 25.7). Thus, protected areas in Amazonia (Indigenous Lands and Conservation Units) represent over 1/3 of Brazilian Amazonia. The establishment of these areas continues, following a complex process under pressure from private interests, specific groups, and tracked by national and international institutions, as well as movements of local and foreign public opinion, which interact with national public policies at different levels (MIRANDA *et al.*, 2006; BARRETO *et al.*, 2005).

Conservation Units can become efficient instruments for promoting and conserving biodiversity, and their current importance has expanded due to sustainable use conservation units. Chief among the models of Conservation Units is the category called Sustainable Development Reserves (RDS) (QUEIROZ, 2006).

The first RDS established in Brazil was the Mamirauá Sustainable Development Reserve (RDSM). The traditional population has participated in activities designed to conserve biodiversity, protect endangered species, use local natural resources in a sustainable manner and provide sustainable development to riverine communities. These activities are conducted through a participatory process, with involvement of the local population in different stages of land and resource management (IDSM, 2006). The major characteristics of this type of conservation unit are the following: maintenance of local population, which participates in natural resource management activities and surveillance of the reserve; the possibility of fauna and flora management based on solid scientific research; flexibility to change strategies according to market; maintenance of private property; implementation of programs to enhance living conditions of local population and establishment of strategic partnerships with governmental and nongovernmental organizations for developing proposals for sustainable use of natural resources. The results of ten years of investments in this area enable an assessment of the advantages of this category of conservation unit, and indicate that results are indeed significant, both from the standpoint of biodiversity conservation as well as in terms of improved quality of life for local inhabitants.

It should be mentioned that establishing Conservation Units and Indigenous Lands may be a precautionary instrument in areas where there is no pressure of occupation, but has proven to be ineffective in already occupied areas. Environmental destruction is also occurring endogenously within Conservation Units and Indigenous Lands, showing them to be as ineffective as the Maginot Line (1931–1936), which was built by the French to counter the advance of German troops during World War II (MIRANDA, 2006; BARRETO *et al.*, 2005).

Despite this, it is possible to conduct agricultural activities in Amazonia with a minimum of deforestation. Some states in Amazonia, such as Amazonas and Amapá, *zero deforestation* could easily be reached, but they would continue importing products from deforested areas in Pará, Mato Grosso and Roraima. The buffering effect of the Free Trade Zones of Manaus, Macapá and Santana, have both promoted urbanization and greatly decreased deforestation. This indicates that the problem of deforestation and burnings in Amazonia are not independent of, but rather linked to poverty in Northeastern Brazil, which leads to migration to the Amazon Region. Another factor is the importing of timber from Amazonia by other parts of Brazil and the world. This is why Amazonia must be considered within the context of a national policy.

The general understanding is that ensuring conservation and preservation of Amazonia will depend on seeking out new technological alternatives that focus on partial use of deforested areas and recover areas that should never have been cleared. Below is a list of five categories of alternatives considered as priorities to ensure food security and food preservation of Amazonia in the future:

1. **Reduction of deforestation and burnings**
   Reduction of chronic deforestation and burnings can take place by partial use of the internal and already deforested frontier, which spans over 71 million hectares in 2006. Food and raw material production can be rendered compatible with preservation of Amazonia, with income generation and job creation.

2. **Increase sustainability of natural resource use**
   Several renewable natural resources in Amazonia are being used faster than their regeneration capacity. Appropriate management techniques need to be developed for logging and use of other natural resources, where biological sustainability does not always ensure economic sustainability and vice-versa.

3. **Increase sustainability of agricultural activities**
   There are two concurrent types of agriculture in Amazonia, one using advanced farming techniques and, at the other extreme, traditional or swidden agriculture, based on slash-and-burn. Constraints must be overcome so that agriculture can remain in the same spatial area and avert constant incorporation of new areas. Considering that a typical family farmer in Amazonia cuts down 2 hectares of dense forest and farms it for 2 years, then leaving it to fallow to cut down a new area, he would need 12 hectares and 12 years to return to the original site to again

cut down the vegetation. If technological enhancements were to enable farming the area for 3 years, by increasing only one year, he would then need 10 hectares and 15 years to return to the original site, thus reducing deforestation by 17%.

4. **Creation of new technological and economic alternatives**

   There is a need for ongoing discoveries and domestication, to make use of the wealth of Amazonian biodiversity, instead of the random and negligent manner it has been addressed heretofore. More intensive activities of land use and labor, such as ranching and reforestation, as well as those where mechanization is not possible at some phase of the productive process (harvesting of oil palm, cocoa, black pepper, açai, cupuaçu, tapping rubber trees, etc) are major opportunities for Amazonia and family-based agriculture.

5. **Expand knowledge of ecosystems and their interrelations**

   The efforts invested in many unsustainable agricultural activities in Amazonia are the result of a lack of knowledge of the ecosystem, in addition to a lack of economic alternatives and appropriate technological practices especially exploring the large water resources existing in the region.

References

BARRETO, P., SOUZA JÚNIOR, C., NOGUERÓN, R.; ANDERSON, A., & SALOMÃO, R. (2005) *Pressão humana na floresta amazônica brasileira.* Belém: WRI; Imazon, 84p.

BECKER, B.K. (2004) *Amazônia: geopolítica na virada do III milênio.* Rio de Janeiro: Garamond, 172p.

BRANDÃO JÚNIOR, A. & SOUZA JÚNIOR, C. (2006) *Desmatamento nos assentamentos de reforma agrária na Amazônia. Belém, Imazon,* (O Estado da Amazônia, 7).

DIMÁRZIO, J.A. (2004) O agronegócio brasileiro é muito competitivo. In: *CONGRESSO DE AGRIBUSINESS,* 6, 2004. Rio de Janeiro, Anais...., Rio de Janeiro: Sociedade Nacional de Agricultura, 2004. p.13–18.

ELECTRONORTE (2006) *Cenários macroeconômicos para a Amazônia 2005–2025.* Brasília, 2006. 170p.

FENZL, N. & MATHIS, A. (2004) Pollution of natural water resources in Amazonia: sources, risks and consequences. ARAGÓN, L.E. & CLÜSENER-GODT, M. (Ed.). *Issues of local and global use of water from the Amazon.* Montevidéo: UNESCO, p.57–75.

HOMMA, A.K.O. (2003) *História da agricultura na Amazônia: da era pré-colombiana ao terceiro milênio.* Brasília: Embrapa Informação Tecnológica, 2003. 274p.

IDSM (Instituto de Desenvolvimento Sustentável Mamirauá),Tefé, AM. (2006) Disponível em: http://www.mamiraua.org.br. Acesso em 10 de novembro de 2006.

KINOSHITA, D.L. (1999) *Uma estratégia para inserção soberana da América Latina na economia globalizada: a questão amazônica.* São Paulo: IFUSP, 4f. (mimeografado).

LAURANCE, W.F., COCHRANE, M.A., BERGEN, S., FEARNSIDE, P.M., DELAMÔNICA, P., BARBER, C., D'ANGELO, S., & FERNANDES, T. (2001) The future of the Brazilian Amazon. *Science,* **291** (5503), 438–439, 19 Jan. 2001.

LENTINI, M.; PEREIRA, D.; CELENTANO, D.; PEREIRA, R. (2005) *Fatos florestais da Amazônia 2005.* Belém: Imazon, 2005. 140p.

MENEZES, A.J.E.A. (2002) de. *Análise econômica da "produção invisível" nos estabelecimentos agrícolas familiares no Projeto de Assentamento Agroextrativista Praialta e Piranheira, município de Nova Ipixuna, Pará.* 130 f. Dissertação (Mestrado em Agriculturas Familiares e Desenvolvimento Sustentável) – Universidade Federal do Pará.

MIRANDA, E.E. (2006) Situação da região amazônica pelo monitoramento com satélites. In: *CONGRESSO BRASILEIRO DE SOJA*, 4, Londrina, 2006. Anais..., Londrina: Embrapa Soja, 2006. p.86–91.

MIRANDA, E.E.; MORAES, A.V.C.; OSHIRO, O.T. (2006) *Queimadas em áreas protegidas da Amazônia em 2005*. Campinas: Embrapa Monitoramento por Satélite, 13p. (Embrapa Monitoramento por Satélite. Comunicado Técnico, 19).

PRODUTO (2006) interno bruto dos municípios 2002–2004, www.ibge.gov.br. Acesso em: 22 nov. 2006.

QUEIROZ, H.L. (2005) A Reserva de Desenvolvimento Sustentável Mamirauá: um modelo de alternativa viável para a proteção e conservação da biodiversidade na Amazônia. *Revista de Estudos Avançados, São Paulo*, **19** (54), 183–203, mai./ago. 2005.

REBELLO, F. K. (2004) *Fronteira agrícola, uso da terra, tecnologia e margem intensiva: o caso do Estado do Pará*. Belém: UFPA; Centro Agropecuário, Embrapa Amazônia Oriental, 2004. 223f. Dissertação (Mestrado em Agriculturas Familiares e Desenvolvimento Sustentável) – Universidade Federal do Pará.

SCHNEIDER, R.R.; ARIMA, E.; VERÍSSIMO, A.; BARRETO, P.; SOUZA JÚNIOR, C. (2000) *Amazônia sustentável: limitantes e oportunidades para o desenvolvimento rural*. Brasília: Banco Mundial; Belém: Imazon, 2000. 58p.

Chapter 26

Governance of food security in the Philippines through community based watershed management approach

G.P. Antolin Jr., F.M. Serrano & E.F. Bonzuela
Bureau of Soils and Water Management, Quezon City, Philippines

26.1 Introduction

Seventy percent of the rural communities in the Philippines largely depend on agriculture sector. Thus, food security and poverty alleviation programs are related to livelihood opportunities that focus on agri-related economic activities. The dilemma facing the development workers and practitioners nowadays is the growing competition of use of resources between extensive and intensive agricultural activities (including animal production, use of chemicals and modern inputs) and its resulting negative externalities that pose threat to the environment and people's health.

This situation is apparent in areas where marginalized rural sector dwells in the uplands, especially in the watershed areas. The watershed serves multifunction, of serving as agricultural production area and ecological region, which most of the time are competing and conflicting. Most upland people living in the watershed areas are estimated to be at a range of 9 to 17 million. They are subsistence upland farmers who till their lands to produce their daily food requirements and to generate income for buying their other needs. Their activities are found to contribute to the diminishing forest cover. Thus, watershed management is a crucial task for community developer and research practitioners.

26.2 Food security through watershed management

Food security is ensuring that household incomes especially those of the poor are enough to purchase adequate food at reasonable prices regardless of whether these are imported or locally produced. Aside from this, the entitlement of households on consumer basket largely relies on the supply of agricultural production, e.g. increasing food grain, to suffice the growing demand of the nation, and by reducing the unit cost of production through advanced machineries and technologies.

Watersheds are important sources of organic matter, nitrogen, and other nutrients which enrich the soils in the watersheds, the floodplains and farms irrigated by streamflows and the coastal areas where most watersheds empty its rivers. The valuable contribution of the watershed is pronounced in agricultural productivity and, conclusively food security of the country.

Activities related to sustainable management of watershed resources executed by human, agriculture, and environment and natural resources (including ecological and bio-diversity component) sectors of the watershed influences food security: as food source, as income source, and as provider of ecological services vital for the production and maintenance of the overall productivity of terrestrial ecosystems.

26.3   Negative externalities in the watershed areas

As mentioned above, most upland communities abuse the valuable contribution of the natural ecosystem to their productivity. Worsening scenario of deforestation and deterioration of the soil and water resources are evident. When forests are diminished, the watershed fails to sustain food production and support other income generating activities evident in its upland, lowlands and coastal areas. The landless and small-scale farmers who heavily rely on upland farming and other forest-based livelihood activities are most affected. Loss of biodiversity, soil erosion, and impaired hydrology, are the most direct and severe damages inflicted on the watersheds through deforestation. It also drains the genetic pool which is essential in improving the resistance and productivity of economically important food crop species.

In line with this, the advent of the project that promotes environmental conservation while increasing agricultural production for watershed dwellers is necessary. The Bureau of Soils and Water Management (BSWM) of the Department of Agriculture (DA) with the support of the Bureau of Agricultural Research (BAR) and technical assistance of the International Crop Research Institute for Semi-Arid Tropics (ICRISAT) launched the project.

One strategy to address the key issues of food sustainability and poverty reduction is through watershed management. This approach focuses on important concerns of both ecological and environmental issues while improving the economic status of the marginalized upland rural areas.

The majority of the inhabitants in the uplands are subsistence farmers who depend on rainfed agriculture as the primary source of their income. But rainfed agriculture is considered complex, diverse and risk prone, and characterized by low levels of productivity and input usage. Soil erosion is widespread in these areas with devastating impacts on the people who depend on upland agriculture. As such, poverty, unemployment, poor health and sanitation are widespread. It is recognized that community-based watershed management approach is a promising strategy to prevent further ecological deterioration of the uplands. The promotion of relevant farm, non-farm and off-farm livelihood opportunities is essential in conserving and improving the agricultural production base.

The project: *Community-based Watershed Management Approach in Improving Livelihood Opportunities in Selected Areas in the Philippines* basically aims at improving the livelihood opportunities through community-based watershed management through the achievement of the following project objectives:

- To promote sound soil and water conservation and management technologies in minimizing land degradation through community participation
- To promote technologies and conduct of training to minimize land degradation through community-based approach
- To provide employment opportunities for local community through various natural resource-based livelihood activities
- To empower rural community with technical know-how on sustainable watershed management.

Four sites were identified for this project. These are: (1) Sitio Parungao, Barangay Sapang Bulak, Doña Remedios Trinidad, Bulacan, (2) Sta. Maria, Ilocos Sur, (3) Casipo, San Clemente, Tarlac and (4) Talibon, Bohol.

The biophysical characterization of each project site was undertaken to assess their current status as well as their potential natural and human resources and how these resources could be developed in a way that is consistent with the objectives of watershed management and that would consequently provide livelihood opportunities to the community. Part of this activity involved undertaking various methodologies characterizing the site: topographic survey, soil and land use survey, as well as agro-socio-economic survey. The agro-socio-economic survey provided information about the farmers and the community relative to their social and economic conditions as well as their agricultural practices. A community assessment was also executed for each project site. The project acknowledges that each project site is unique not only in terms of physical characteristics but more so in community composition, tradition and cultural practices.

26.4 Community empowerment through capacity building

Human resource is one of the vital factors in the development and conservation of the watershed. For the rural communities to manage the watershed and improve their quality of life, relevant capacity building activities are imperative. In all project sites, capacity building activities were undertaken such as trainings. This endeavor necessitates an appraisal of stakeholder's capacity building needs done through consultations, individual assessments and group assessment. The BSWM has addressed their needs through the conduct of the following trainings:

- Training on **soil and water conservation technologies** – this tackled the establishment of contour farming technology which is most applicable in sloping landscape. It consists of hedgerows which are planted with permanent trees (fruit trees are preferred for their economic value) and alleys which are planted with cash crops to provide farmers with cash while waiting for the benefits from the permanent crops. Part of the training is to provide farmers with knowledge in developing their own site development plan.
- Other relevant trainings were conducted with the aim of **conserving the natural resources,** providing the farmers with additional sources of farm income, or decreasing agricultural production cost. These are the following:

1. Training on organic farming technologies – farm composting makes use of available agricultural wastes like crop residues and animal manure. The compost is a substitute for commercial inorganic fertilizer and reduces capital for farm inputs.
2. Training on the use of Soil Test Kit in diagnosing deficiency of soil in Nitrogen-phosphorus-potassium (NPK).
3. Mushroom culture was introduced as a means of producing organic fertilizer while providing another opportunity to generate income.
4. Biogas production as a way of producing organic fertilizer and biogas for cooking – production of biogas is a measure to reduce the cutting of trees as a source of energy for cooking.

5. Goat and duck production trainings were conducted to diversify farm income sources.
6. Inland fish culture, especially for tilapia, was introduced to give additional income to farmers and supplement their dietary requirements.
7. Use of Trichoderma for enhancing rice straw composting.
8. Use of N fixing bacteria (Bio-N)
9. Use of the "Tipid-Abono*" strategy to maximize profit in rice production.
10. Training on the production of nata de coco, vinegar making and vegetable pickles, macaroons baking, cassava flour and honey production.
11. Candle making and laundry detergent making were introduced as source of income during dry season when vegetable gardening is not possible.

* This is a cost reduction strategy through judicious and optimum fertilization for using the right mixtures of fertilizers and the timely application of the most appropriate type and amounts of fertilizers that will ensure optimum crop yields and better income without causing decline in soil fertility.

26.5 Capacity building of the community through training

Factors considered in the trainings selection are the market potential as livelihood, available local resources and the available expertise from the BSWM. The participants in the training included the farmer-cooperators, other interested farmers in the adjacent communities, farmer-leaders, baranggay officials, members of the farmers' organizations. Women were encouraged to attend and participate from the above-mentioned trainings. The major project proponents (i.e. academe, other government agencies in the site, local government units, non-government organizations) acted as facilitators and resource persons during the training in addition to BSWM experts.

The success of any capability building activity can be gauged in two ways. First, through the number of trained farmers adopting soil and water conservation technologies and the initial area covered by the technologies. The farms of the pilot adopters will serve as a technology demonstration to showcase soil and water conservation techniques for the untrained farmers in nearby areas. Another more important measure of success is in relation to the sustainable promotion of the technologies, through the number of farmers who were not formally trained but replicate the technologies in their own lots.

26.6 Soil and water conservation techniques as part of watershed management

The project recognizes the concept of community-based approach. The participants or the involved farmers are not merely consulted; they make the decisions for themselves. The BSWM acted only as facilitator that provided feasible options wherein the farmers in the communities could be assisted in making the right decisions.

Implementation of soil and water conservation and management technologies were agreed upon by the farmer-cooperators and were implemented in the three sites namely: Bulacan, Tarlac, and Bohol to address both the production problem at the farm level and conservation objectives at micro watershed level.

These sites represent the upland and hilly landscapes of the watershed where erosion should be prevented. In the three sites mentioned above, the contour farming technology was implemented as this was identified to be the most appropriate considering the topography of the area (Table 26.1).

26.7 Project sustainability

One of the factors that would dictate the sustainability of the project is the capacity of the community/farmers' organization to handle the complex social and technical responsibilities in operating the project. That is why, it is very important that the project beneficiaries or cooperators are empowered and capacitated.

Sustainable poverty reduction would require the cooperation between government, organizations of poor people, non-government organizations, and the rest of civil society. Hence, the governance component of poverty reduction is crucial in projects such as this that focuses on the environment and food sustainability. The quality of governance is critical to poverty reduction . Good governance is reflected by participatory and pro-poor policies as well as sound macroeconomic management. Participatory governance, as understood here, goes beyond the framework of accountability, as it also refers to the transformation of relationships between rulers and the ruled in ways that recognize and respect autonomy, and that include more accountable, participatory-based, and egalitarian political practices.

The sense of responsibility realized by all stakeholders is a necessary component in attaining the aim of achieving adequate natural and productive resources available throughout time. Poverty reduction is understood here not as an individual problem but as a community problem, and a function of building social capital and participatory governance mechanisms. Sustainable poverty reduction therefore refers to the systematic or programmatic attempts to increase the number of people who are able to get out of poverty, and sustain that success over long periods of time for the eventual goal of poverty eradication or elimination, the ultimate target of human development.

In all project sites, farmers' association had been responsible in all the activities of the project. All of them were also empowered through various trainings, seminars and consultations. Likewise, the project provided physical infrastructures that would facilitate crop protection and other livelihood activities in their respective areas. The association's sustainability as a partner in development can be assured. The association provides free labor for the development of the irrigation system and other efforts needing collective action and effort.

Table 26.1 Planting of shrubs and fruit trees along the contours.

Picture/Figure	Activity	Description	Remarks
	1. Trimming of grass on the contours	After the land preparation and planting of crops on the alleys, grasses along the contours were trimmed for easy movement for planting of fruit trees.	Bulacan, Tarlac, Bohol
	2. Preparation of fertilizer to be applied	3 kgs of organic fertilizer mixed with 0.7 kgs of complete fertilizer (14-14-14) were applied as basal for each hole of fruit trees along the hedge rows. Amount of fertilizer were computed and prepared based on the analyses of the soil in the area.	This was done in Tarlac for a specific fertilizer recommendation for the fruit trees to be planted (i.e. calamansi, jackfruit and mango)
	3. Digging of holes	Holes to be planted with jackfruit and mango were 2 ft depth, while holes for calamansi, papaya and banana were dugout of about 1 ft.	Undertaken in Tarlac dugout of about where calamansi, jackfruit and mango are chosen by the farmer-cooperators for hedgerows

Table 26.1 Continued

Picture/Figure	Activity	Description	Remarks
	4. Planting of fruit trees	Planting distance of jackfruit × jackfruit, and mango × mango was 10 m and 15 m, respectively. Each planted fruit trees were installed with bamboo pegs to protect from damaged.	Done in Tarlac
	5. Planting of shrubs	Papaya and banana were planted on the embankment of SFR. Planting distance for papaya and banana Is 4 m.	Done in Tarlac

Chapter 27

Overview and integration

U. Aswathanarayana (Editor)

Section 3: Governance of food security in different agroclimatic and socioeconomic settings

In Chap. 22, Bhargava and Chakrabarti hold that knowledge can play a vital in achieving food security in India. About 200 million people in India (about one-fifth of the population) are under-nourished. Food security in India can be achieved only through making the farmers knowledgeable. Three kinds of knowledge are involved : (i) General knowledge that would make them reasonably informed citizens of the country and the world; (ii) Specialized knowledge, e.g., on seeds; optimal use of fertilizers, water and power; management of live-stock; output channels; market information; water management such as rain water harvesting; technologies for value addition; time and labour saving devices; and energy plantations, and (iii) Knowledge for immediate use – that is, updated information, for example, on hydroclimatic calendar, which can be obtained using satellite remote sensing, and GIS, with farmer-specific data assimilation. The knowledge packages are to be customized depending upon the agroclimatic setting, water resources, soil health cards, socioeconomic environment, food habits, etc. of a given village. Young, motivated farmers may be trained in ways of implementing the knowledge packages, through the *panchayats* (village councils).

In Chap. 23, Zhengbin & Ping give a detailed account of how China is addressing the food and water security issues. The population of China which is expected to reach 1.5 billion by 2020 would need about 600 million tonnes of food grains, i.e., 100 million tonnes more than the present food production. Any strategy to achieve this goal of food sufficiency has to be strongly science and technology oriented. As 81% of the water resources are in southern China, while 64% of arable land is in northern China, the country has embarked upon a massive South – North transfer of water involving Yangtze and Huang He rivers. China has 53.3 million ha of irrigated farmland. By planting water-saving, high-yielding, and drought-resistant varieties, and by reducing the frequency and scale of irrigation, it is possible to save 80–100 billion m^3 of irrigation water annually, which water can be used to grow 80–100 million tonnes of extra food grains. By 2030, it is planned to improve the efficiency of utilization rate of irrigation water from its present 30–40% to 60%–70%, and to raise the water productivity from 0.87 kg of food grains/m^3 to 1.5 kg/m^3.

The World Bank estimates that the urban water demand will increase by 60%, from the current 50 billion m^3 to 80 billion m^3. Meanwhile industrial water consumption is expected to increase by 62%, from 127 billion m^3 to 206 billion m^3. Thus, the challenge facing China is how to grow more food with lesser amount of *land* and *water*.

The strategy to enhance agricultural productivity is customized to the agroclimatic and biophysical settings in various zones. Of all the arable land in China, the extent of fields with low-yields, middle yields and high yields amount to 41.6%, 29.6% and 28.8% respectively. Through increasing inputs such as water conservation, fertilizer use and other factors, China proposes to raise the productivity of 2 million ha of low-middle-yield fields every year, and thereby enhance the food output by 40 million tonnes annually. The cultivation of "super rice" can increase the yield by 30 million tonnes; and that of the super maize can also increase the output by 40 million tonnes. Dryland farming in areas of low rainfall, grassland development and stock breeding in Qinhai-Tibet Plateau areas, Loess Plateau areas, and karst areas in Southwest China, etc. is proposed. About 70% of the food-insecure people in China live in arid and semi-arid areas of Northwest China which is characterized by low soil fertility and water scarcity. This situation is sought to be ameliorated through measures such as, reducing and exempting agricultural taxes, increasing financial inputs, building water conservancy and other construction projects, including the tapping of glacial water resources.

In Chap. 24, Lunkad and Sharma deal with the adverse impact of Green Revolution in Haryana, India, on groundwater, land and soil. The Green Revolution enabled the small state of Haryana to become the "wheat granary" of India – the state, though occupying 1.3% geographical area and containing 2% of the population of India, produces 13% wheat and about 3% quality rice of India besides other cereals, oil seeds, sugarcane and cotton. Haryana paid a heavy price for this impressive agricultural development – there has been serious degradation of basic geo-resources, i.e., water-soil-land. The number of hand pumps and tube-wells in Haryana has increased from 25,000 in 1966 to 564,000 in 1998, i.e. a rise of more than 22 times in 32 years. The utilization of shallow groundwater is 100–200 times more than that of deep groundwater. Intensification of agriculture has led to a water table decline and soil salinization in both the semiarid and arid regions, but more so in the latter. Thus, one third of the irrigated land in entire Haryana is salinity affected. Some areas (dark zone) recorded a decline of 3 to 12 m. in water level during the period from 1974 to 1998. Seepage through unlined canals and excessive irrigation in areas of shallow groundwater table are the main causes of waterlogging. The trend of increasing N-fertilizer consumption and groundwater use for irrigation in Haryana vis-à-vis rising nitrate levels (>50 mg/l on an average) in shallow groundwater are indicative of high nitrate pollution levels. About 3% of agricultural land is lost for food production as a consequence of urbanization, industrialization and the creation of Special Economic Zones. Apart from measures to improve water use efficiency in agriculture, obtaining surplus surface water of Satluj River by early completion of Satluj–Yamuna Link Canal (SYL), would enhance the possibility of artificial recharge of groundwater in the dark zone of Eastern Haryana.

In Chap. 25, Homma *et al.* give a fascinating account of the efforts that are being made by Brazil to achieve food security to the indigenous people of Amazonia. Amazon is the largest tropical forest on earth, and the world's largest gene bank. Many of the agricultural and forest products which are used all over the world (such as, cocoa, rubber, tomato, potato, tobacco, corn and cassava) originally came from Brazil and have been spread by the Portuguese in Africa and Asia. The current deforested area of Amazonia, estimated at 71 million hectares, would be more than enough to feed the population that lives in the region, now and in the future, using only part of this area. Amazonia is home to 61% of the country's indigenous population, estimated at over 340,000 inhabitants. Food availability in Brazil is high (340 kg/per capita/year), and represents almost one third more than the minimum nutritional needs of the country's population. Still 46 million Brazilians are food-insecure, including 7 to 9 million in Amazonia. The problem of hunger in Brazil has been due to lack of income for people to properly feed themselves. Brazil seeks to protect the Amazon Rain Forest, while at the same time alleviating the hunger of the indigenous people, through the adoption of the following strategy: Food and raw material production can be rendered compatible with preservation of Amazonia, by reducing the necessity for slash-and-burn agriculture, with income generation and job creation. To achieve this goal, the Government of Brazil established Sustainable Development Reserves (RDS), such as the one in Mamirauá. The traditional population has participated in activities designed to conserve biodiversity, protect endangered species, use local natural resources in a sustainable manner and provide sustainable development to riverine communities. These activities are conducted through a participatory process, with involvement of the local population in different stages of land and resource management.

In Chap. 26, Antolin *et al.* describe how the Philippines are trying to achieve food security through community-based watershed management. Activities related to sustainable management of watersheds involving human, agriculture, and environment and natural resources (including ecological and bio-diversity component) sectors of the watershed influences food security: as food source, as income source, and as provider of ecological services vital for the production and maintenance of the overall productivity of terrestrial ecosystems. Empowerment of the stakeholders through knowledge is a critically important part of the project. Hence, the project includes the training of the stake-holders in soil and water conservation technologies but also in getting additional income through provision of activities/services such as organic farming, use of soil test kit, technical advice about fertilizer use, biogas production, rearing of goats and ducks, inland fisheries, etc.

Author index

Subject Index

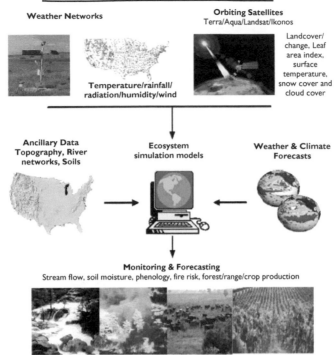

Terrestrial Observation and Prediction Systems

Weather Networks

Temperature/rainfall/
radiation/humidity/wind

Orbiting Satellites
Terra/Aqua/Landsat/Ikonos

Landcover/
change, Leaf
area index,
surface
temperature,
snow cover and
cloud cover

Ancillary Data
Topography, River
networks, Soils

Ecosystem
simulation models

Weather & Climate
Forecasts

Monitoring & Forecasting
Stream flow, soil moisture, phenology, fire risk, forest/range/crop production

Plate 1.1 The Terrestrial Observation and Prediction System (TOPS) integrates a wide variety of data sources at various space and time resolutions to produce spatially and temporally consistent input data fields, upon which ecosystem models operate to produce ecological nowcasts and forecasts needed by natural resource managers.

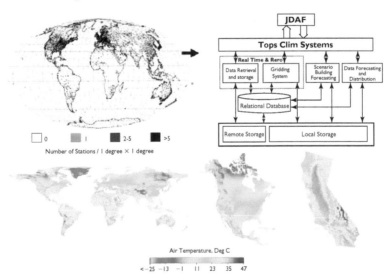

Plate 1.2 Flow diagram of the SOGS. Three main components that comprise the system are: data retrieval and storage, interpolation and output handling. Data retrieval is configured to automatically retrieve the most recent data available and insert those data into the SQL database. Interpolation methods are modular and allow maximum flexibility in implementing new routines as they become available. Outputs are generated on the prediction grid that is determined by the latitude, longitude, elevation and mask layers. Another key feature of the SOGS implementation is scenario generation where long-term normal station data can be perturbed according to climate model forecasts. Weather data from over 6000 stations distributed globally is ingested into TOPS database where it is gridded to a variety of resolutions, globally at 0.5 degree, continental U.S at 8 km and at 1 km over California.

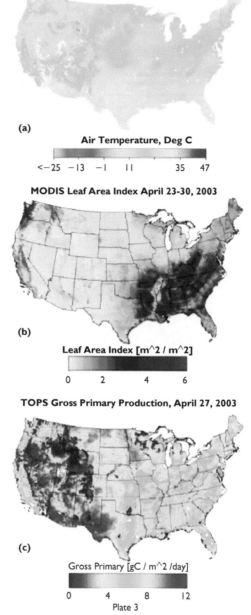

Gridded Air Temperature, April 27, 2003

(a)

Air Temperature, Deg C

<−25   −13   −1   11   35   47

MODIS Leaf Area Index April 23-30, 2003

(b)

Leaf Area Index [m^2 / m^2]

0   2   4   6

TOPS Gross Primary Production, April 27, 2003

(c)

Gross Primary [gC / m^2 /day]

0   4   8   12

Plate 3

Plate 1.3 Examples of TOPS nowcasts. (a) Patterns of maximum air temperature over the conterminous U.S., produced using over 1400 stations on April, 27, 2004. (b) MODIS derived leaf area index after pre-processing through TOPS, and (c) Model estimated gross primary production, the amount of photosynthate accumulated on April 27, 2004 over the conterminous U.S.

Forecasted Irrigation, September 7th, 2005

(a)

mm water

0   15   30

Near Realtime Biospheric Monitoring, May 2005

(b)   Changes in Vegetation Prodcution and Sea Surface Temperatures

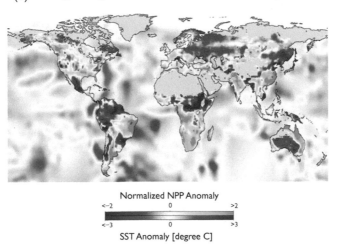

Normalized NPP Anomaly

<−2   0   >2

<−3   0   >3

SST Anomaly [degree C]

Plate 1.4 (a) Application of TOPS over Napa valley vineyards showing the recommended irrigation amounts to keep the vines at a stress level of −12bars for the week of September 7, 2004. (b) A global application of TOPS for monitoring and mapping net primary production anomalies over land and sea surface temperature anomalies over global oceans. NPP and SST anomalies for May 2005 are based on monthly means from 1981–2000.

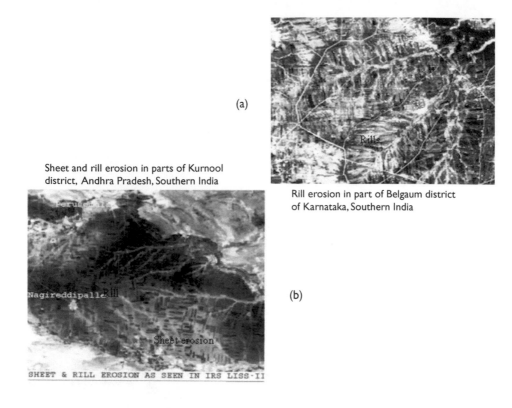

(a)

Sheet and rill erosion in parts of Kurnool district, Andhra Pradesh, Southern India

Rill erosion in part of Belgaum district of Karnataka, Southern India

(b)

Plate 2.1 Sheet and rill erosion as seen in satellite images.

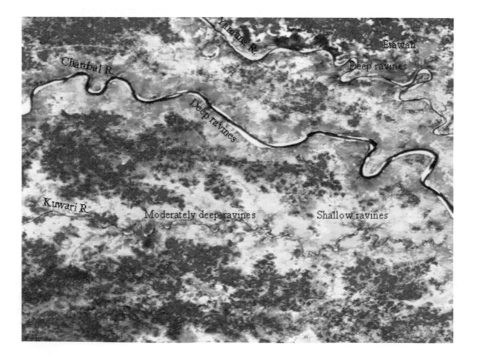

Plate 2.2 Ravines as seen in IRS-1D LISS-III image acquired in February, 2006.

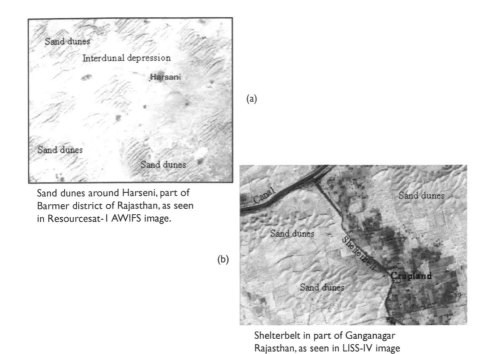

(a)

Sand dunes around Harseni, part of Barmer district of Rajasthan, as seen in Resourcesat-1 AWIFS image.

(b)

Shelterbelt in part of Ganganagar Rajasthan, as seen in LISS-IV image

Plate 2.3 Wind erosion features as seen in Resourcesat-1 AWIFS and LISS-IV images.

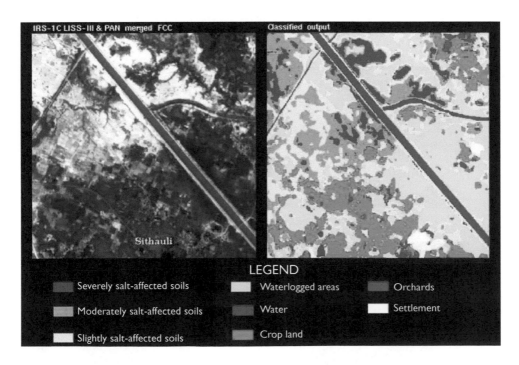

LEGEND

Severely salt-affected soils	Waterlogged areas	Orchards
Moderately salt-affected soils	Water	Settlement
Slightly salt-affected soils	Crop land	

Plate 2.4 Map showing salt affected soil in part of Rai Bareli dt., U.P.

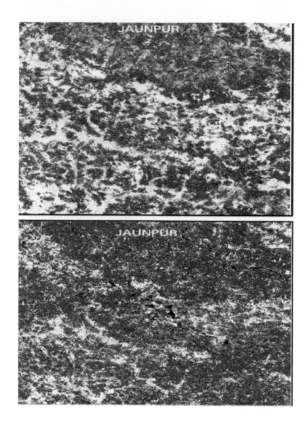

Plate 2.5 Salt-affected soils in part of sharda Sahayak command area (Indo-Gangetic plains), Jaunpur (Uttar Pradesh).

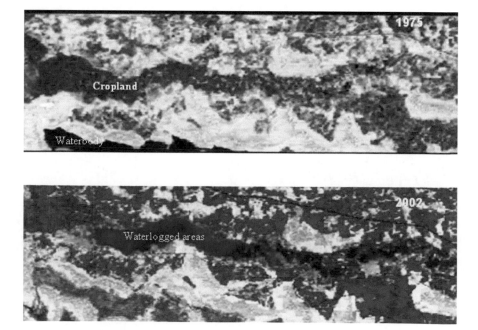

Plate 2.6 Waterlogging around Suratgarh, Ganganagar district, Rajasthan due to seepage from canal.

Kharif rice area	Rabi Fallow area	Fallow land as % of Kharif rice area
3.88 Mha	1.22 Mha	31.4 %

Plate 3.1 Spatial distribution of *kharif* rice and fallow lands in Orissa state.

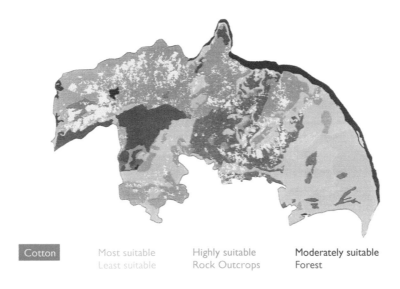

Cotton   Most suitable   Highly suitable   Moderately suitable
Least suitable   Rock Outcrops   Forest

Plate 3.2 Suitability regimes for cotton cultivation in Guntur district, AP.

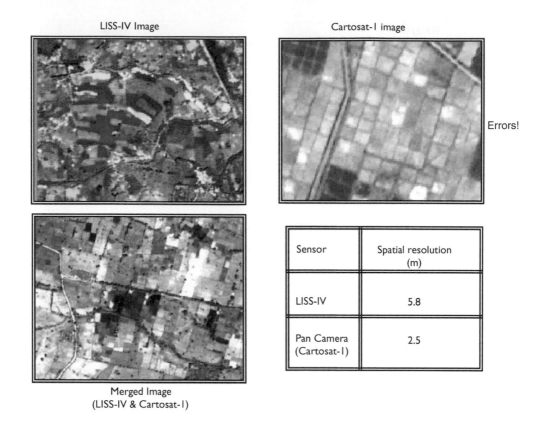

LISS-IV Image

Cartosat-1 image

Errors!

Sensor	Spatial resolution (m)
LISS-IV	5.8
Pan Camera (Cartosat-1)	2.5

Merged Image
(LISS-IV & Cartosat-1)

Plate 3.3 Indian Earth Observation Systems of high spatial resolution.

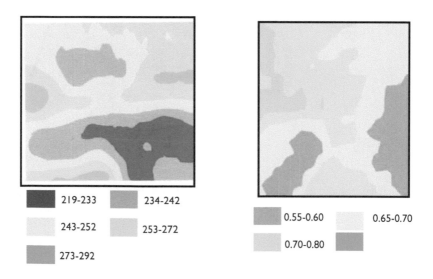

219-233   234-242

243-252   253-272

273-292

0.55-0.60   0.65-0.70

0.70-0.80

Plate 3.4 Intrafield variablity of soil-N (kg/ha) and yield of soybean (g/m^2).

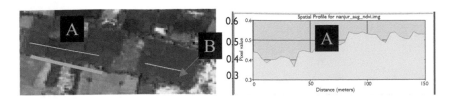

Plate 3.5 Intrafield variability of NDVI in sugarcane.

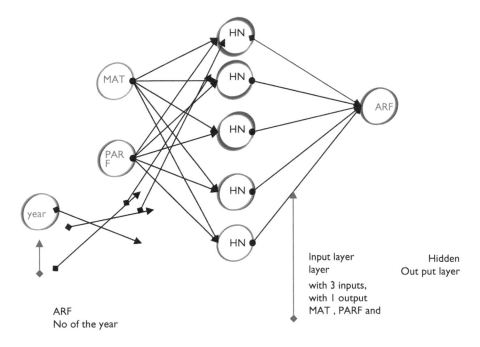

Input layer
layer

with 3 inputs,
with 1 output
MAT , PARF and

Hidden
Out put layer

ARF
No of the year

Plate 4.1 Artificial neural network.

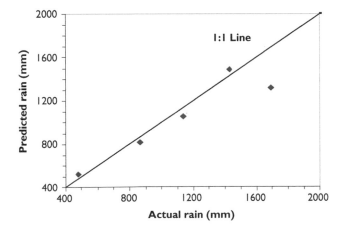

Plate 4.2 Actual vs. predicted rain for different years by ANN.

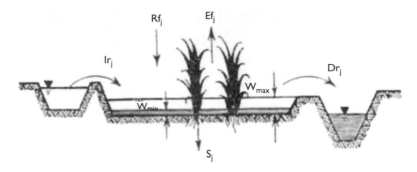

Plate 8.1 Components of the water balance in an individual rice field. (Ir = Irrigation; Rf = Rainfall; Et = Evapotranspiration; S = Seepage and Percolation; Dr = Drainage; Wmax = Maximum depth of water; Wmin = Minimum depth of water required; j = Time period considered).

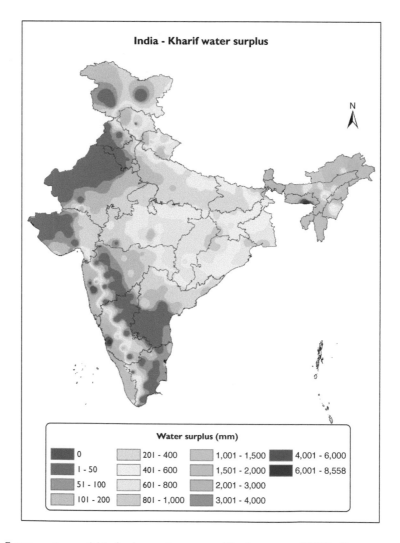

Plate 9.1 Excess water available for harvesting as runoff in the states of SAT India.

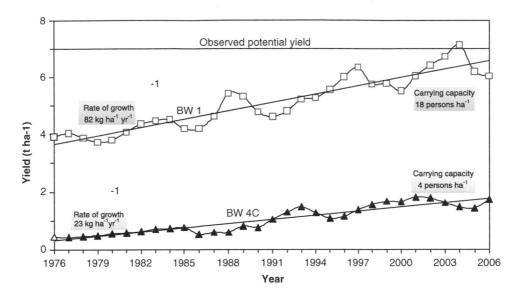

Plate 9.2 Three-year moving average of sorghum and pigeon pea grain yield under improved management and on farmers' fields in a deep Vertisol catchment, Patancheru, India.

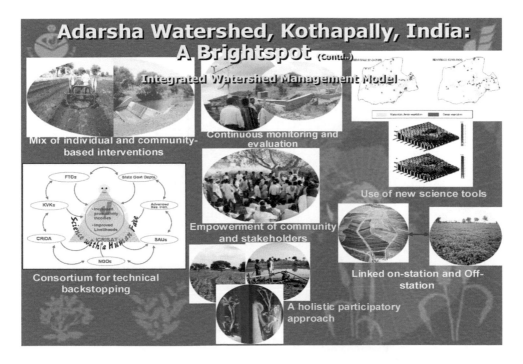

Plate 9.3 An innovative consortium model for integrated watershed management.

Plate 11.1 Indonesian farmer holding two rice plants of same variety and maturity, the one on left grown with SRI practices and the one on right grown with conventional practices. Picture courtesy of Shuichi Sato, Nippon Koei.

Plate 11.2 Farmer in Dông Trù village, Hanoi province, Vietnam, holding up two rice plants, SRI (on left) and non-SRI (on right). Behind her are the fields from which the plants were taken, shown after a typhoon passed over the village. Note lodging of the non-SRI rice crop on right and resistance to lodging of the SRI crop on left. Picture courtesy of Elske van de Fliert, FAO IPM Program, Hanoi.

Plate 11.3 Comparison of typical finger millet (*Eleusine coracana*) plants grown with different crop management methods. On left is modern variety (A404) grown with adaptation of SRI methods; in center, same variety with conventional methods; on right, local variety grown with conventional methods. Picture courtesy of Ashish Anand, PRADAN Khunti team, Jharkhand state, India.

Sri Ragi (finger millet), rabi 2004-05
60days after sowing-varieties 762 and 708

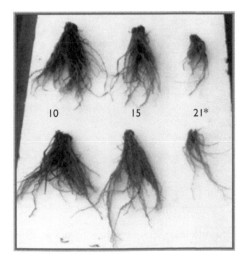

Results of trials being
being done by ANGRAU

VR 762

VR 708

* Age at which seedlings were
transplanted from nursery

Plate 11.4 Roots of two varieties of finger millet (*Eleusine coracana*) at 60 days of age, having been transplanted at 10, 15 or 21 days after emergence. As with SRI, seedlings transplanted beyond 15 days demonstrate less growth potential. Picture courtesy of Dr. A. Satyanarayana, director of extension, Acharya N. G. Ranga Agricultural University, Hyderabad.

Plate 11.5 Two rice plants of same age (52 days) and variety (VN 2084), grown by Luis Romero, San Antonio de los Baños, Cuba, as explained in text. Picture courtesy of Dr. Rena Perez.

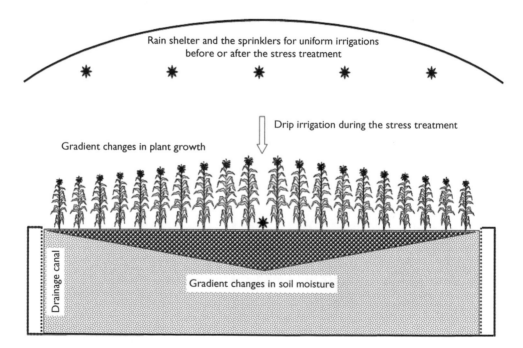

Plate 13.1 Illustration of the facility for drought tolerance screening showing the gradient changes in both soil moisture and the growth of the plants.

Plate 24.1 A satellite view of Haryana showing degraded land in the west.
Source: HARSAC (2006).

Plate 25.1 Location of Amazonia in Latin America. (modified from www.images.google.com.br)

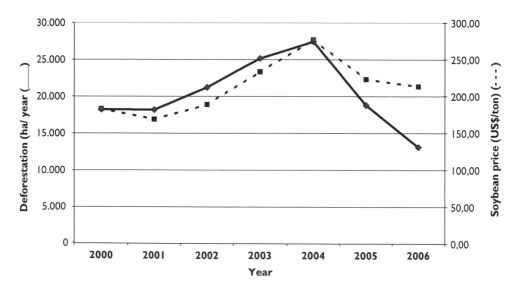

Plate 25.2 Deforestation in Amazonia versus international soybean prices since year 2000 (www.obt.inpe.br/deter and www.ibge.gov.br).

T - #0491 - 071024 - C16 - 246/174/16 - PB - 9780367388447 - Gloss Lamination